VOL.1 MODELING SURFACE AND SUB-SURFACE FLOWS

ELSEVIER

COMPUTATIONAL MECHANICS PUBLICATIONS

D
628.1610'515353
COM

OTHER TITLES IN THIS SERIES

4 J.J. FRIED
GROUNDWATER POLLUTION

5 N. RAJARATNAM
TURBULENT JETS

7 V. HÁLEK AND J. ŠVEC
GROUNDWATER HYDRAULICS

8 J. BALEK
HYDROLOGY AND WATER RESOURCES IN TROPICAL AFRICA

10 G. KOVÁCS
SEEPAGE HYDRAULICS

11 W.H. GRAF AND C.H. MORTIMER (EDITORS)
HYDRODYNAMICS OF LAKES: PROCEEDINGS OF A SYMPOSIUM 12-13 OCTOBER 1978, LAUSANNE, SWITZERLAND

13 M.A. MARIÑO AND J.N. LUTHIN
SEEPAGE AND GROUNDWATER

14 D. STEPHENSON
STORMWATER HYDROLOGY AND DRAINAGE

15 D. STEPHENSON
PIPELINE DESIGN FOR WATER ENGINEERS
(completely revised edition of Vol.6 in the series)

17 A.H. EL-SHAARAWI AND S.R. ESTERBY (EDITORS)
TIME SERIES METHODS IN HYDROSCIENCES

18 J. BALEK
HYDROLOGY AND WATER RESOURCES IN TROPICAL REGIONS

19 D. STEPHENSON
PIPEFLOW ANALYSIS

20 I. ZAVOIANU
MORPHOMETRY OF DRAINAGE BASINS

21 M.M.A. SHAHIN
HYDROLOGY OF THE NILE BASIN

22 H.C. RIGGS
STREAMFLOW CHARACTERISTICS

23 M. NEGULESCU
MUNICIPAL WASTEWATER TREATMENT

24 L.G. EVERETT
GROUNDWATER MONITORING HANDBOOK FOR COAL AND OIL SHALE DEVELOPMENT

25 W. KINZELBACH
GROUNDWATER MODELLING

26 D. STEPHENSON AND M.E. MEADOWS
KINEMATIC HYDROLOGY AND MODELLING

27 A.H. EL-SHAARAWI AND R.E. KWIATKOWSKI (EDITORS)
STATISTICAL ASPECTS OF WATER QUALITY MONITORING -
PROCEEDINGS OF THE WORKSHOP HELD AT THE CANADIAN CENTRE FOR INLAND WATERS, OCTOBER 1985

28 M.K. JERMAR
WATER RESOURCES AND WATER MANAGEMENT

29 G.W. ANNANDALE
RESERVOIR SEDIMENTATION

30 D. CLARKE
MICROCOMPUTER PROGRAMS FOR GROUNDWATER STUDIES

31 R.H. FRENCH
HYDRAULIC PROCESSES ON ALLUVIAL FANS

32 L. VOTRUBA
ANALYSIS OF WATER RESOURCE SYSTEMS

33 L. VOTRUBA AND V. BROŽA
WATER MANAGEMENT IN RESERVOIRS

34 D. STEPHENSON
WATER AND WASTEWATER SYSTEMS ANALYSIS

36 M.A. CELIA, L.A. FERRAND, C.A. BREBBIA, W.G. GRAY AND G.F. PINDER (EDITORS)
VOL.2 NUMERICAL METHODS FOR TRANSPORT AND HYDROLOGIC PROCESSES - PROCEEDINGS OF THE VIITH INTERNATIONAL CONFERENCE ON COMPUTATIONAL METHODS IN WATER RESOURCES, MIT, USA, JUNE 1988

COMPUTATIONAL METHODS IN WATER RESOURCES

VOL.1 MODELING SURFACE AND SUB-SURFACE FLOWS

Proceedings of the VII International Conference, MIT, USA, June 1988

Edited by

M.A. Celia
Massachusetts Institute of Technology, Cambridge, MA, USA
L.A. Ferrand
Massachusetts Institute of Technology, Cambridge, MA, USA
C.A. Brebbia
Computational Mechanics Institute and University of Southampton, UK
W.G. Gray
University of Notre Dame, Notre Dame, IN, USA
G.F. Pinder
Princeton University, Princeton, NJ, USA

ELSEVIER
Amsterdam - Oxford - New York - Tokyo 1988

Co-published with

COMPUTATIONAL MECHANICS PUBLICATIONS
Southampton - Boston

Distribution of this book is being handled by:

ELSEVIER SCIENCE PUBLISHERS B.V.
Sara Burgerhartstraat 25, P.O. Box 211
1000 AE Amsterdam, The Netherlands

Distributors for the United States and Canada:

ELSEVIER SCIENCE PUBLISHING COMPANY INC.
52 Vanderbilt Avenue
New York, N.Y. 10017, U.S.A.

British Library Cataloguing in Publication Data

International Conference on Computational
　Methods in Water Resources (7th : 1988 :
　Cambridge, Mass.)
　Computational methods in water resources.
　Vol.1 : Modeling surface and sub-surface flows
　1. Natural resources. Water. Analysis
　I. Title II. Celia, M.A. III. Series
　628.1'61'01515353
　ISBN 1-85312-006-5

Library of Congress Catalog Card number 88-70628

ISBN 0-444-98912-9(Vol.35) Elsevier Science Publishers B.V.
ISBN 0-444-41669-2(Series)
ISBN 1-85312-006-5　　　　Computational Mechanics Publications UK
ISBN 0-931215-73-0　　　　Computational Mechanics Publications USA

Published by:

COMPUTATIONAL MECHANICS PUBLICATIONS
Ashurst Lodge, Ashurst
Southampton, SO4 2AA, U.K.

This work is subject to copyright. All rights are reserved, whether the whole or part of the material is concerned, specifically those of translation, reprinting, re-use of illustrations, broadcasting, reproduction by photocopying machine or similar means, and storage in data banks.

© Computational Mechanics Publications 1988
© Elsevier Science Publishers B.V. 1988

Printed in Great Britain by The Eastern Press, Reading

The use of registered names, trademarks, etc., in this publication does not imply, even in the absence of a specific statement, that such names are exempt from the relevant protective laws and regulations and therefore free for general use.

PREFACE

This book forms part of the edited proceedings of the Seventh International Conference on Computational Methods in Water Resources (formerly Finite Elements in Water Resources), held at the Massachusetts Institute of Technology, USA in June 1988. The conference series originated at Princeton University, USA in 1976 as a forum for researchers in the emerging field of finite element methods for water resources problems. Subsequent meetings were held at Imperial College, UK (1978), University of Mississippi, USA (1980), University of Hannover, FRD (1982), University of Vermont, USA (1984) and the Laboratorio Nacional de Engenharia Civil, Portugal (1986). The name of the ongoing series was modified after the 1986 conference to reflect the increasing diversity of computational techniques presented by participants.

The 1988 proceedings include papers written by authors from more than twenty countries. As in previous years, advances in both computational theory and applications are reported. A wide variety of problems in surface and sub-surface hydrology have been addressed.

The organizers of the MIT meeting wish to express special appreciation to featured lecturers J.A. Cunge, A. Peters, J.F. Sykes and M.F. Wheeler. We also thank those researchers who accepted our invitation to present papers in technical sessions: R.E. Ewing, G. Gambolati, I. Herrera, D.R. Lynch, A.R. Mitchell, S.P. Neuman, H.O. Schiegg, and M. Tanaka. Important contributions to the conference were made by the organizers of the Tidal Flow Forum (W.G. Gray and G.K. Verboom) and the Convection-Diffusion Forum (E.E. Adams and A.M. Baptista) and by K. O'Neill who organized the Special Session on Remote Sensing. The conference series would not be possible without the continuing efforts of C.A. Brebbia, W.G. Gray and G.F. Pinder, who form the permanent organizing committee.

The committee gratefully acknowledges the sponsorship of the National Science Foundation and the U.S. Army Research Office and the endorsements of the American Geophysical Union (AGU) the International Association of Hydraulic Research (IAHR), the National Water Well Association

(NWNA), the American Institute of Chemical Engineers (AIChE), the International Society for Computational Methods in Engineering (ISCME), the Society for Computational Simulation (SCS) and the Water Information Center (WIC).

Papers in this volume have been reproduced directly from the material submitted by the authors, who are wholly responsible for them.

M.A. Celia
L.A. Ferrand
Cambridge (USA) 1988

CONTENTS

SECTION 1 - FEATURED LECTURES

Some Examples of Interaction of Numerical and Physical   3
Aspects of Free Surface Flow Modelling
J.A. Cunge

Vectorized Programming Issues for FE Models   13
A. Peters

Parameter Identification and Uncertainty Analysis for   23
Variably Saturated Flow
J.F. Sykes and N.R. Thomson

Modeling of Highly Advective Flow Problems   35
M.F. Wheeler

SECTION 2 - MODELING FLOW IN POROUS MEDIA

2A - Saturated Flow

Cross-Borehole Packer Tests as an Aid in Modelling   47
Ground-Water Recharge
J.F. Botha and J.P. Verwey

The Boundary Element Method (Green Function Solution) for   53
Unsteady Flow to a Well System in a Confined Aquifer
Xie Chunhong and Zhu Xueyu

Finite Element Solution of Groundwater Flow Problems by   59
Lanczos Algorithm
A.L.G.A. Coutinho, L.C. Wrobel and L. Landau

Finite Element Model of Fracture Flow   65
R. Deuell, I.P.E. Kinnmark and S. Silliman

Finite Element Modeling of the Rurscholle Multi-Aquifer   71
Groundwater System
H-W. Dorgarten

On the Computation of Flow Through a Composite Porous Domain   77
J.P. du Plessis

Two Perturbation Boundary Element Codes for Steady   83
Groundwater Flow in Heterogeneous Aquifers
O.E. Lafe, O. Owoputi and A.H-D. Cheng

A Three-Dimensional Finite Element - Finite Difference Model   89
for Simulating Confined and Unconfined Groundwater Flow
A.S. Mayer and C.T. Miller

Galerkin Finite Element Model to Simulate the Response of   95
Multilayer Aquifers when Subjected to Pumping Stresses
A. Pandit and J. Abi-Aoun

Finite Element Based Multi Layer Model of the "Heide Trough"   101
Groundwater Basin
B. Pelka

Three-Dimensional Finite Element Groundwater Model for the   107
River Rhine Reservoir Kehl/Strasbourg
W. Pelka, H. Arlt and R. Horst

2B - Unsaturated Flow

An Alternating Direction Galerkin Method Combined with   115
Characteristic Technique for Modelling of Saturated-
Unsaturated Solute Transport
Kang-Le Huang

Finite-Element Analysis of the Transport of Water, Heat and   121
Solutes in Frozen Saturated-Unsaturated Soils with Self-
Imposed Boundary Conditions
F. Padilla, J.P. Villeneuve and M. Leclerc

A Variably Saturated Finite-Element Model for Hillslope   127
Investigations
S.T. Potter and W.J. Gburek

A Subregion Block Iteration to 3-D Finite Element Modeling   133
of Subsurface Flow
G.T. Yeh

2C - Multiphase Flow

Numerical Simulation of Diffusion Rate of Crude Oil Particles   141
into Wave Passes Water Regime
M.F.N. Abowei

A Decoupled Approach to the Simulation of Flow and Transport   147
of Non-Aqueous Organic Phase Contaminants Through Porous
Media
H.W. Reeves and L.M. Abriola

INVITED PAPER

The Transition Potentials Defining the Moving Boundaries in Multiphase Porous Media Flow
H.O. Schiegg — 153

An Enhanced Percolation Model for the Capillary Pressure-Saturation Relation
W.E. Soll, L.A. Ferrand and M.A. Celia — 165

2D - Stochastic Models

A High-Resolution Finite Difference Simulator for 3D Unsaturated Flow in Heterogeneous Media
R. Ababou and L.W. Gelhar — 173

Solving Stochastic Groundwater Problems using Sensitivity Theory and Hermite Interpolating Polynomials
D.P. Ahlfeld, G.F. Pinder — 179

Supercomputer Simulations of Heterogeneous Hillslopes
A. Binley, K. Beven and J. Elgy — 185

A Comparison of Numerical Solution Techniques for the Stochastic Analysis of Nonstationary, Transient, Subsurface Mass Transport
W. Graham and D. McLaughlin — 191

Modelling Flow in Heterogeneous Aquifers: Identification of the Important Scales of Variability
L.R. Townley — 197

2E - Saltwater Intrusion

Modelling of Sea Water Intrusion of Layered Coastal Aquifer
A. Das Gupta and N. Sivanathan — 205

A Comparison of Coupled Freshwater-Saltwater Sharp-Interface and Convective-Dispersive Models of Saltwater Intrusion in a Layered Aquifer System
M.C. Hill — 211

Can the Sharp Interface Salt-Water Model Capture Transient Behavior?
G. Pinder and S. Stothoff — 217

SECTION 3 - MODELING SURFACE WATER FLOWS

3A - Tidal Models

A Consistency Analysis of the FEM: Application to Primitive and Wave Equations 225
J. Drolet and W.G. Gray

A Comparison of Tidal Models for the Southwest Coast of Vancouver Island 231
M.G.G. Foreman

Computation of Currents due to Wind and Tide in a Lagoon with Depth-Averaged Navier-Stokes Equations (Ulysse Code) 237
J.M. Hervouet

The Shallow Water Wave Equations on a Vector Processor 243
I.P.E. Kinnmark and W.G. Gray

Testing of Finite Element Schemes for Linear Shallow Water Equations 249
S.P. Kjaran, S.L. Hólm and S. Sigurdsson

INVITED PAPER
Long Term Simulation and Harmonic Analysis of North Sea/English Channel Tides 257
D.R. Lynch and F.E. Werner

Tidal Motion in the English Channel and Southern North Sea: Comparison of Various Observational and Model Results 267
J. Ozer and B.M. Jamart

Experiments on the Generation of Tidal Harmonics 275
R.A. Walters and F.E. Werner

A 2D Model for Tidal Flow Computations 281
C.S. Yu, M. Fettweis and J. Berlamont

3B - Lakes and Estuary Models

A Coupled Finite Difference - Fluid Element Tracking Method for Modelling Horizontal Mass Transport in Shallow Lakes 289
P. Bakonyi and J. Józsa

Hydrodynamics and Water Quality Modeling of a Wet Detention Pond 295
D.E. Benelmouffok and S.L. Yu

Solving the Transport Equation using Taylor Series Expansion and Finite Element Method
C.L. Chen
301

Cooling-Induced Natural Convection in a Triangular Enclosure as a Model for Littoral Circulation
G.M. Horsch and H.G. Stefan
307

System Identification and Simulation of Chesapeake Bay and Delaware Bay Canal Hydraulic Behavior
B.B. Hsieh
313

A Layered Wave Equation Model for Thermally Stratified Flow
J.P. Laible
319

A Simple Staggered Finite Element Scheme for Simulation of Shallow Water Free Surface Flows
S. Sigurdsson, S.P. Kjaran and G.G. Tómasson
329

Improved Stability of the "CAFE" Circulation Model
E.A. Zeris and G.C. Christodoulou
337

3C - Open Channel Flow and Sedimentation

An Implicit Factored Scheme for the Simulation of One-Dimensional Free Surface Flow
A.A. Aldama, J. Aparicio and C. Espinosa
345

Practical Aspects for the Application of the Diffusion-Convection Theory for Sediment Transport in Turbulent Flows
W. Bechteler and W. Schrimpf
351

Computing 2-D Unsteady Open-Channel Flow by Finite-Volume Method
C.V. Bellos, J.V. Soulis and J.G. Sakkas
357

Eulerian-Lagrangian Linked Algorithm for Simulating Discontinuous Open Channel Flows
S.M.A. Moin, D.C.L. Lam and A.A. Smith
363

SECTION 4 - SPECIAL SESSION ON REMOTE SENSING AND SIGNAL PROCESSING FOR HYDROLOGICAL MODELING

On Thin Ice: Radar Identification of thin and not so thin Layers in Hydrological Media
K. O'Neill
371

Satellite Observations of Oceans and Ice 379
K.C. Jezek and W.D. Hibler

Applications of Remote Sensing in Hydrology 383
T. Schmugge and R.J. Gurney

SECTION 1 - FEATURED LECTURES

Some Examples of Interaction of Numerical and Physical Aspects of Free Surface Flow Modelling

J.A. Cunge
CEFRHYG, BP172X - 38042 Grenoble, France

INTRODUCTION

Mathematical models of free surface flows are commonly built for engineering purposes. As in other fields of Computational Hydraulics users are making increasingly greater demands concerning simulation codes. In the near future most models will be built and run in the same way as CAD software is used today for structures. This trend will impose quality constraints on software developers who must supply users with safe codes including not only numerical solution of equations but also analysis of physical and computational features and operating aids. While moving towards Intelligent Knowledge Based Systems (or 5th-generation codes), the developers must bear in mind that the role of physical considerations in developing efficient, industrial software of practical use for engineering is essential.

SIMPLIFIED 1-D EQUATIONS OR 'MUSKINGUM' REVISITED

Unsteady open channel flow equations (de Saint-Venant or dynamic wave equations) can be written as:

$$\frac{\partial y}{\partial t} + \frac{1}{b}\frac{\partial Q}{\partial x} = 0 \tag{1}$$

$$\frac{1}{g}\left(\frac{\partial u}{\partial t} + u\frac{\partial u}{\partial x}\right) + \frac{\partial h}{\partial x} - S_0 + S_f = 0 \tag{2}$$

where $y(x,t)$ = free surface elevation; $b(h)$ = free surface width; $Q(x,t)$ = discharge; $u(x,t)$ = mean flow velocity; $h(x,t)$ = depth; S_0 = river bed slope; $S_f(h, Q)$ = friction slope.

If the inertia terms (enclosed in parentheses) of Eq. (2) are small compared with the friction slope, they can be neglected. The system of two equations, freed of its inertia terms, becomes:

$$\frac{\partial y}{\partial t} + \frac{1}{b}\frac{\partial Q}{\partial x} = 0 \quad ; \quad \frac{\partial h}{\partial x} - S_0 + S_f = 0 \quad (3)$$

Friction slope S_f can be expressed as a function of the discharge by the conveyance factor K. Using the Manning-Strickler formula the following equations can be written:

$$S_f = \frac{Q|Q|}{K^2} \quad ; \quad K = k \, A \, h^{2/3} \quad (4)$$

where $A(h)$ = cross sectional area; k = Strickler coefficient.

Writing $\partial A/\partial h = b$ and assuming all functions derivable it is possible to eliminate y and h from Eq. (3) thus obtaining the following relationship (for derivation see, e.g., Todini 1987):

$$\frac{\partial Q}{\partial t} + C \frac{\partial Q}{\partial x} - D \frac{\partial^2 Q}{\partial x^2} = 0 \quad (5)$$

$$C = \frac{Q}{bK}\frac{dK}{dh} + \frac{D}{b}\frac{\partial b}{\partial x} = C_k + \frac{D}{b}\frac{\partial b}{\partial x} \quad ; \quad D = \frac{Q}{2b\,S_f} \quad (6)$$

If the change in depth $\partial h/\partial x$ is small compared to the river slope S_0 or friction slope S_f, it can be neglected in Eq. (2) which then expresses a single-valued relationship $Q = Q(h)$ (or $Q = Q(A)$). The original system of Eqs. (1) - (2) can be replaced by the kinematic wave equation:

$$\frac{\partial Q}{\partial t} + C_k \frac{\partial Q}{\partial x} = 0 \quad ; \quad C_k = \frac{dQ}{dA} = \frac{d}{dA}(K\sqrt{S_0}) = \frac{Q}{A}\left(1 + \frac{2}{3}\frac{A}{bh}\right) \quad (7)$$

The following properties of the equations are to be noted:
- System (1) - (2) is hyperbolic with two families of characteristics and requires two boundary conditions for subcritical flow, one upstream, one downstream.
- System (3) is parabolic and needs two boundary conditions, one upstream, one downstream.
- Eq. (7) is a first order p.d.e. of advective transport without damping and needs only one boundary condition, upstream.

In the past engineers have always used simplified methods which are not necessarily consistent with mathematical formulation. For a hydrologist, flood waves propagate 'from upstream to downstream' and discharge hydrographs are 'routed' along a river course.

This is true in terms of geographical distances and categories but things are different when free surface elevations are to be computed for engineering structure design or for definition of water management policies. One of the most popular approaches in traditional engineering is the

MUSKINGUM method, the blind application of which can lead to serious errors.

Considering a space (x,t) with two cross-sections (j, j+1) distant Δx one from the other (see fig. 1), then MUSKINGUM method enables (cf. Eq. 8) computation of the discharge Q_{j+1}^{n+1} at the lower section j+1 at the time level n+1 if the discharges Q_j^n, Q_j^{n+1} and Q_{j+1}^n are known. Thus, if the whole hydrograph $Q(j,t)$ is known, the hydrograph $Q(j+1, t)$ can be computed provided that the celerity C_k and damping coefficient X are known:

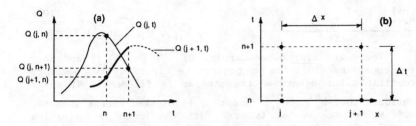

Figure 1. MUSKINGUM method. (a) Hydrograph space (Q,t); (b) time-distance space (x,t).

$$Q_{j+1}^{n+1} = \frac{X+0.5Cr}{(1-X)+0.5Cr} Q_j^n + \frac{0.5Cr-X}{(1-X)+0.5Cr} Q_j^{n+1} + \frac{(1-X)-0.5Cr}{(1-X)+0.5Cr} Q_{j+1}^n \quad (8)$$

where $Cr = C_k \frac{\Delta t}{\Delta x}$ is the advection Courant Number.

It has been shown (Cunge[1]) that for X = 0.5, Eq. (8) is a consistent finite difference approximation of kinematic wave Eq. (7). When $0 \leq X \leq 0.5$, Eq. (8) is consistent with diffusion Eq. (5).

This property explains how generations of hydrologists were able to obtain damping of the discharge peaks between j and j+1 while using an approximation of pure translation Eq. (7). It is possible (Cunge[1]) to choose the coefficient X in Eq. (8) in such a way, that, <u>for a given moment in time</u>, this equation can approximate diffusive wave equations including coefficients C_k and D. If the system of Eq. (3) is linearised in the neighbourhood of a certain situation at which $Q = \overline{Q}$, $S_f = \overline{S}_f$, $K = \overline{K}$, then the choice is:

$$X = \frac{1}{2} - \frac{\overline{Q}}{2\overline{b}\ \overline{C}_k\ \overline{S}_f\ \Delta x} \quad (9)$$

This approach has been widely adopted but in most cases with a serious flaw. Indeed, as was implied in the original paper, the choice of $X \neq 0.5$, according to Eq. (9), means that this coefficient is variable in time, for a given river reach.

To define it properly, it is necessary to solve the system (3) i.e. to find, at every time $n.\Delta t$, the values of y and h as well as the discharge Q. In other terms, the linearisation is to be made in the neighbourhood of a gradual variable flow. Common application of Eq. (9) is based on the linearisation near a <u>uniform</u> flow, i.e. $S_f = S_0$. This enables computation from upstream to downstream but must be wrong when the flow is gradually variable. Consider the simple four-point implicit scheme of finite differences applied to the system of Eq. (3); putting $S_f = f(Q,h)$:

$$\frac{h_j^{n+1} - h_j^n + h_{j+1}^{n+1} - h_{j+1}^n}{2\Delta t} + \frac{1}{b} \frac{Q_{j+1}^{n+1} - Q_j^{n+1} + Q_{j+1}^n - Q_j^n}{2\Delta x} = 0$$

$$\frac{h_{j+1}^{n+1} - h_j^{n+1} + h_{j+1}^n - h_j^n}{2\Delta x} + S_f(h_j^{n+1}, h_{j+1}^{n+1}, Q_j^{n+1}, Q_{j+1}^{n+1}) - S_0 = 0 \quad (10)$$

There are <u>four</u> unknowns: $(h, Q)_j^{n+1}$, $(h, Q)_{j+1}^{n+1}$ and only <u>two</u> equations (10). The system can be closed by two boundary conditions but otherwise the problem is ill-posed. It seems obvious, that h_j^{n+1}, h_{j+1}^{n+1} cannot be eliminated except when the MUSKINGUM-Cunge method is used with the rough approximation $S_f = S_0$. It is interesting to note the presence of Δx in the denominator of Eq. (9), which thus directly influences the damping coefficient. If the distance between two cross-sections increases, the damping effect decreases. Physically this seems absurd, but it is reasonable considering that the approximation is consistent with Eq. (7) when Δx is small. Another warning signal can be found in the consistency analysis between Eq. (5) and the MUSKINGUM formula (8). Developing all terms of Eq. (8) in a Taylor series and neglecting higher order terms the second derivative can be solved as follows:

$$\frac{\partial^2 Q}{\partial x^2} \approx - \frac{1}{C_k \Delta t \Delta x} \left[(Q_{j+1}^{n+1} - Q_{j+1}^n) - (Q_j^{n+1} - Q_j^n) \right] \quad (11a)$$

$$\frac{\partial^2 Q}{\partial x^2} \approx - \frac{1}{C_k \Delta x} \left[\left(\frac{\partial Q}{\partial t}\right)_{j+1} - \left(\frac{\partial Q}{\partial t}\right)_j \right] \approx - \frac{1}{C_k} \frac{\partial^2 Q}{\partial x \partial t} \quad (11b)$$

It is to be noted that if the kinematic wave Eq. (7) is true, then indeed $\frac{\partial^2 Q}{\partial x \partial t} = -C_k \frac{\partial^2 Q}{\partial x^2}$; this explains the possibility of approximating a second space derivative with two points j, j+1. Also, if $C_k = \Delta x/\Delta t$, then, assuming Eq. (7) is true:

$Q_j^{n+1} = Q_{j-1}^n$, $Q_{j+1}^{n+1} = Q_j^n$, and (11a) becomes:

$$- \frac{1}{\Delta x^2} (-Q_{j-1}^n + 2 Q_j^n - Q_{j+1}^n) \approx \left(\frac{\partial^2 Q}{\partial x^2}\right)_j^n$$

However the consistency should be sought with Eq. (5) and not Eq. (7), but in this case, a third derivative of discharge would be introduced.

In conclusion, 'you cannot have it both ways'. Either the kinematic wave Eq_1 (7) without damping is used, or as indicated by Cunge[1], the coefficient X must be a variable based on h(x,t) and downstream influence. Recently an original method for solving Eq. (3) was published which is consistent with the concept of a well-posed problem and encompasses, in an elegant way, the original idea of the MUSKINGUM-Cunge method (Todini and Bossi[2]). Obviously it requires a downstream boundary condition.

2-D FINITE DIFFERENCE SCHEME

Considering the 2-D de Saint-Venant equations of free surface unsteady flow (neglecting Coriolis and momentum diffusion terms):

$$\frac{\partial \vec{V}}{\partial t} + \vec{V} \text{ grad } \vec{V} + g \text{ grad } z = \frac{\vec{\tau_b}}{\rho} \quad (12a)$$

$$\frac{\partial z}{\partial t} + \text{div}(h\vec{V}) = 0 \quad (12b)$$

where $\vec{V} = \{u(x,y,t), v(x,y,t)\}$ = velocity vector in (x,y) plane, $z(x,y,t)$ = free surface elevation, $h(x,y,t)$ = water depth, $\tau_b = g S_f$, $S_f = \{S_{fx}, S_{fy}\}$ = friction slopes:

$$S_{fx} = \frac{u\sqrt{u^2+v^2}}{k_x^2 h^{4/3}} \quad ; \quad S_{fy} = \frac{v\sqrt{u^2+v^2}}{k_y^2 h^{4/3}} \quad (13)$$

where (k_x, k_y) = Strickler coefficients. Eqs. (12) are of the hyperbolic type and the mixed problem is well posed over a domain in (x,y) space when the following boundary conditions are supplied for subcritical flow (Daubert and Graffe[3]):
. one condition along the boundary where the water flows out of the domain, or if the boundary is closed;
. two conditions when there is inflow.

The following paragraphs comment on three points of practical interest when using discrete approximations to Eqs. (12): finite difference, well posedness, boundary conditions when finite difference process splitting methods are used, and the 'small depth' problem.

Historically Eqs. (12) were initially solved using finite difference methods and neglecting advective terms $\vec{V}$ grad $\vec{V}$. Equations were written in terms of discharges $QX = uh$, $QY = vh$ and elevations z, and a staggered mesh (Hansen[4]) was applied (see Fig. 2).

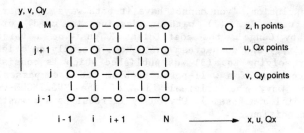

Figure. 2

To neglect advective terms simplifies the mixed problem: only one condition is needed, regardless of the type of boundary. The depths can be evaluated at points distinct from the point of the discharges, except for the friction slope terms. If $z(t)$ is imposed on the boundary, the limit passes through $x = i-1$ line (cf. Fig. 2), i.e. through z-points. If the inflow or outflow is imposed, it is enough to impose a normal discharge. If QX is imposed at $i-1/2$, there is no need to evaluate z at $i-1$ line. For N y-lines and M x-lines in a basin with closed boundaries (QX = 0, QY = 0) there are (N-2)(M-2) unknown levels z and (N-1)(M-2) + (N-2)(M-1) unknown discharges, i.e. $3NM - 5(N+M) + 8$ unknowns. On the other hand, there are (M-2)(N-2) continuity equations plus (N-3)(M-2) dynamic equations for QX and (N-2)(M-3) equations for QY. Two (M-2) + (N-2) imposed boundary discharges QX, QY close the system which is then well-posed in finite difference terms. A different situation is faced when advective terms are retained in the equations. Both velocities (u, v) are needed at points with elevation z and it is not easy to define their consistent finite difference analogs. However real trouble arises at the boundaries. The deficiencies of certain approaches are described by Gerritsen[5]. He points out that, even for closed boundaries, the difficulties force the modeller to adopt an artificial approach. For example, the need to compute the tangential velocity along a closed boundary leads to the introduction of additional equations on the boundary, obtained through differentiation of basic ones.

For open boundaries (inflow or outflow) things are even more complex. Many existing finite difference models show the situation described by Gerritsen[5] for the finite element method: a numerical solution is always found, even when no (or all) variables are prescribed at the open boundaries. In most practical cases only water levels $z(t)$ or normal discharges $Q_n(t)$ are known to the user, who does not realise that his solution is mortgaged by the algorithm built into his software.

Consider an implicit, finite difference 2-D scheme analogs to the 4-point 1-D conservative discretization. At every computational point three variables $z(x,y,t)$, $u(x,y,t)$ and $v(x,y,t)$ are to be computed. Consider a computational grid as shown in Fig. 3, corresponding to a closed domain (i.e. the boundary condition is $u_n = 0$ on the boundary).

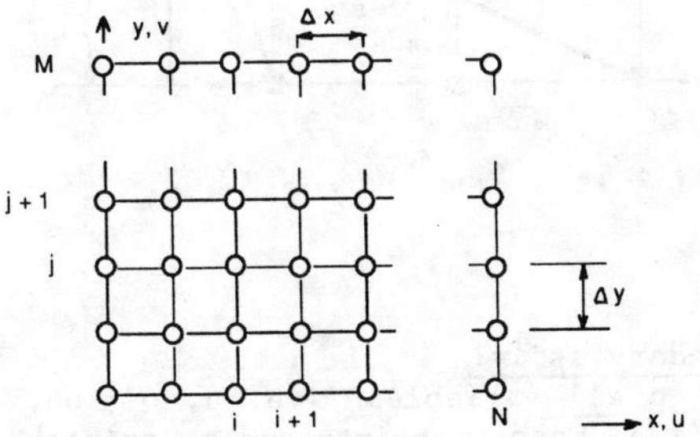

[Figure 3: staggered grid with axes y,v and x,u, spacings Δx, Δy, indices up to M, N]

Figure. 3

The number of unknowns $(u,v,z)_{i,j}^{n+1}$, (i=1, ... N; j=1, ... M) is 3 MN. There are three finite difference equations for every control volume ($\Delta x \Delta y \Delta t$) defined by (i,j), (i+1, j), (i+1, j+1), (i, j+1) space points and two time levels (n, n+1). This amounts to $3(N-1)(M-1) = 3NM - 3M - 3N + 3$ equations. There are also 2N + 2M boundary conditions $u_n = 0$. This leaves a deficit of M+N-3 equations. Thus some of the information contained in Eqs. (12) and their appropriate boundary conditions.

The splitting-process operator method applied to Eqs. (12) is described by Benqué et al.[6]. It enables accurate treatment of advective terms. First the intermediate values of u, v are found by solving the advective operator, by the method of characteristics. Then the continuity-propagation operator is solved using modified (u, v) velocities. We shall illustrate certain difficulties concerning the existence of solution(Chenin-Mordojovich).

The following 1-D splitted equations:

$$\frac{\partial u}{\partial t} + u \frac{\partial u}{\partial x} = 0 \qquad \text{1st step} \quad (14)$$

$$\frac{\partial h}{\partial t} + \frac{\partial (uh)}{\partial x} = 0 \; ; \; \frac{\partial u}{\partial t} + g \frac{\partial h}{\partial x} = f(x,t,h,u) \qquad \text{2nd step} \quad (15)$$

and a staggered (as in the 2-D case) computational grid as shown in Fig. 4 are now considered.

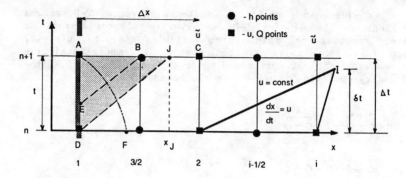

Figure. 4

Advection step - inflow boundary at i=1

Assuming that at time level n <u>all</u> variables (i.e. u, Q = uh, h) are known at <u>all</u> points (i.e. both, u-points and h- points), it is first to be noted that there is a limit for the time step resulting from the possibility of non-existence of the solution of Eq. (14). The characteristics of Eq. (15) are

straight lines $\frac{dx}{dt} = u$, on which $u(x,t)$ is constant. A single-valued solution is not guaranteed beyond the intersection of two characteristics (cf. point I in Fig. 5). Hence splitting is allowed only for:

$$\Delta t < \delta t = \left(\frac{\partial u}{\partial x}\right)^{n-1} \tag{16}$$

Assuming that this condition is satisfied, and also assuming that $Q(x=0, t)$ is the known boundary condition to be imposed at point i = 1, it is clearly possible to compute the velocities u for $x \geq x_J$, where DJ is a characteristic drawn from the point $D(1,n)$. Indeed, $(u, h, Q)_D^n$ are known as previously mentioned. Thus $u_2, u_3, \ldots, u_i, \ldots u_n$ can be computed, but not u_1 because at the boundary the discharge

Q_1^{n+1} is known (imposed), h_1^{n+1} is unknown and the approximation $u_1 = Q_1^{n+1}/h_1^n$ is not satisfactory. It is also to be noted that $u_{3/2}$ can only be computed if the Courant Number for advection at the boundary satisfies $Cr = u_1^n \Delta t/\Delta x \leq 0.5$, for the same reason. If $Cr \geq 0.5$, u_E is not known and it cannot be interpolated because u_A is not known either.

Propagation step
The splitting discretization for this step is:

$$\frac{h^{n+1} - h^n}{\Delta t} + \alpha \frac{\partial \left[(u + \Delta u_p)h^{n+1}\right]}{\partial x} + (1-\alpha) \frac{\partial (u^n h^n)}{\partial x} = 0 \quad (17a)$$

$$\frac{\Delta u_p}{\Delta t} + \alpha g \frac{\partial h^{n+1}}{\partial x} + (1-\alpha)g \frac{\partial h^n}{\partial x} = \alpha f^{n+1} + (1-\alpha)f^n \quad (17b)$$

$0 \leq \alpha \leq 1$

Substitution of (17b) into (17a) and discretization over the h-point grid leads to a system of equations in $h^{n+1}_{i+1/2}$, i = 1,2,..., n-1. Its solution does not enable computation of the depth h^{n+1}_1 (at the boundary). Shokin and Kompaniets[8] enumerate some 30 schemes in which extrapolations or 'extra' conditions are derived by different authors to solve this difficulty! Actually there is only one 'pure' approach, which is seldom accepted because it complicates the algorithm of the solution: to use the characteristic equations derived from the full hyperbolic system. In the 1-D case of Fig. 4 this involves tracing the backward characteristic AF, and given Q_A, solving for x_F and h_A the system of two equations:

$$\frac{dx}{dt} = u - \sqrt{gh} \quad ; \quad \frac{D(u + 2\sqrt{gh})}{Dt} = \varphi(u, h) \quad (18)$$

where (u, h) are functions of both $(u, h)_A$ and $(u, h)_F$. This is the only way of respecting the equations within the shadowed triangle DAJ and especially at the boundary point A. It is to be noted that in the 2-D problem, the characteristics of the full 2-D system of Eqs. (12) are to be used. This seems to be complex but it solves another problem: second boundary conditions to be imposed can be properly expressed and tangential velocity correctly computed.

The 'small depth' problem concerns the simulation of drying-up or flooding of large areas such as tidal flats and marshes. It is also encountered in dam-break problems, emptying of reservoirs, drying-up of some river or irrigation canal reaches, etc. The simulation difficulties arise when the bottom slope is small and computational grid points spacing is large. In most published approaches, by finite elements or finite differences, computational points are 'removed' or 'put back' into the domain when the water depth is, respectively, smaller or greater than predefined value. Such an approach can lead to very serious errors in volume and even become meaningless when large areas are concerned. The simulation of the drying up period is essential. When the depth if flow is small, inertia and advective terms are small compared to friction slope. Special algorithms should be used (see Usseglio-Polatera[9]) while Eqs. (12) progressively lose their hyperbolic character to become parabolic. On the other hand, the finite difference analog of Eqs. (15), which is perfectly symmetric when the right-hand terms f (x,t,u,h) are nil, loses this quality when f is great. This may well result in ill-conditioning of matrices.

CONCLUSIONS

Although they have been used now for nearly 30 years, mathematical models of open channel flow still present grey areas. Progress in user-friendly interfaces (especially graphical input-output processors) must not hide the need for combined hydraulic and numerical competence that is necessary to build such systems and interpret their results.

REFERENCES

1. Cunge J.A. (1969), On the subject of a flood propagation computation method (MUSKINGUM method), Journ. of Hydr. Res., Vol. 7, No 2, pp. 205-230.

2. Todini E. and Bossi A. (1986), PAB (Parabolic and Backwater): An unconditionally stable flood routing scheme particularly suited for real-time forecasting and control. Journ. of Hydr. Res., Vol. 24, No 5, pp. 405-424.

3. Daubert A. and Graffe O. (1967), Quelques aspects des écoulements presque horizontaux à deux dimensions en plan et non permanents. La Houille Blanche, Vol. 22, No 8, pp. 847-860.

4. Hansen W. (1956), Theorie zur Errechnung des Wasserstandes und der Strömungen in Randmeeren nebst Anwendungen, Tellus, Vol. 8, No 3, pp. 287-300.

5. Gerritsen H. (1982) Accurate boundary treatment in shallow water flow computations. Waterloopkundig Lab., Delft Hydraulics Laboratory.

6. Benqué J.P., Cunge J.A., Feuillet J., Hauguel A. and Holly F.M. Jr (1982), New method for tidal currents computation, ASCE, Journ. of Waterway Div., Vol. 108, No WW3.

7. Chenin-Mordojovich M.I. (1987), Private communication.

8. Shokin Yu.I. and Kompaniets L.A. (1987), A catalogue of the extra boundary conditions for the difference schemes approximating the hyperbolic equations, Computers and Fluids, Vol. 15, No 2, pp. 119-136.

9. Usseglio-Polatera J.M. (1988), Dry beds and small depths in 2-D codes for coastal and river engineering, First Inter. Conf. in Africa on Computer Methods and Water Resources, Rabat, Morocco.

Vectorized Programming Issues for FE Models
A. Peters
Institut fuer Wasserbau und Wasserwirtschaft, Aachen University of Technology, Federal Republic of Germany

ABSTRACT

In order to obtain top performances on the novel computer architectures it is worth reconsidering the conventional approaches. This paper presents several programming issues involved in the optimization of FE programs on supercomputers. The discussion focuses on vector processing because most of the commercially available high-speed computers are of this variety. The presented programming constructs are applied for exemplification to the FE solution of the irrotational ideal fluid flow equation.

INTRODUCTION

The recent advances in hardware technology, which have enabled the production of supercomputers, show no sign of abatement. All the indicators are that changes during the next years will be even greater, particularly due to novel forms of hardware architectures involving vector and parallel capabilities[11,13].

In the case of software for these computers the progress has not been so radical. Most of the programs tailored to suit the hardware of conventional computers are not likely to exceed very modest speed-up factors on vector and parallel processors.

To make effective use of the novel architectures the gap between hardware and software must be constantly bridged. This is not an easy task since algorithms are machine dependent and the required changes are not trivial. However, software designers contend that a reasonable fraction of the performance of a wide variety of architectures can be achieved through the use of certain programming constructs and the isolation of machine-dependent code within some modules[5,17].

This paper is neither a detailed presentation of the supercomputers in use today nor an attempt to survey parallel algorithms. Each of these tasks would require a volume to itself.

Instead, Section 2 describes some general approaches used to increase the computational speed of many of today's supercomputers. The discussion focuses on vector processing, because most of the commercially available high-speed computers are of this variety. Section 3 presents several programming issues involved in the optimisation of FE codes on vector computers. The programming constructs used for exemplification have been selected from a FE system[21]. The system is implemented on a CRAY X-MP and on a CDC-CYBER 205 and has been used to solve large-size problems of groundwater flow and contaminant transport through porous media[22].

2 ADVANCED COMPUTER ARCHITECTURES

2.1 Algorithmic Structures

A wide variety of approaches is used to increase the speed of computations at all levels of hardware design. At the lowest level, impressive technological improvements have been achieved. At the highest, new algorithmic structures which allow the concurrent execution of several instructions have been introduced.

The organisation of almost all commercially available high-speed computers bases on the following approaches[7] :

- Sequential; each computation is performed sequentially and one at a time. The organisation which follows this structure is usually referred to as serial or scalar processor.

- Group-Sequential; many computations are arranged in groups, which can be executed concurrently, but the groups are executed sequentially. The organisation of array processors and vector processors is based on this approach.

- Loosely-Coupled; in this approach the algorithm consists of several sequential streams with a small number of dependencies among them. Each of the streams is executed in a separate processor and the processors are connected to control the dependencies. This type of organisation is called multi-processor.

Of course, individual designs may combine all these approaches. For example CRAY X-MP[24] is available with one, two or four CPUs, which each contain one scalar and one vector processor, so it combines sequential processing, vector processing and multi-processing.

An excellent discussion about different possible algorithm structures, their application, speed gain and degradation factors is given in[7]. Readers interested in more detail on the development and organisation of parallel computers are directed to the books[11,14]. An introduction to a wide variety of parallel architectures is given in[12].

2.2 Vector Processing

Figure 1 shows the basic organisation of a vector computer. The instruction processor fetches the instructions from memory, decodes and issues them to the scalar and vector processor. Sequential computations are executed in the scalar processor. Since the organisation of the scalar and instruction processors follows similar concepts to that developed for conventional computers, they are not discussed here. A detailed presentation can be found in[18].

The organisation of a vector processor follows the group-sequential approach : several computations are initiated with one vector instruction. While in an array computer several processors are used, in a vector computer the computations inside a group are executed in a pipelined fashion. The idea behind pipelining is that of an assembly line : if the same process is going to be repeated many times, throughput can be greatly increased by dividing the process into a sequence of sub-tasks (segments) and maintaining the flow of operands in various states of completion. Figure 2 illustrates the possible organisation of an execution pipeline[11]. The throughput is determined by the delay of the slowest segment.

Usually the execution steps are significantly faster than the data fetch from memory. In order to output operands at a sufficient rate, the memory is organized in several modules which can be accessed simultaneously. Vector instructions need simplified address computations and make possible the concurrent access of whole groups of operands stored in consecutive positions. Such a group of operands is called vector. In some organisations[24] a local memory can store partial results in the processor to avoid their unnecessary transfer to and from the main memory.

2.3 Degradation Factors and Remedies

Several factors can make the actual throughput of a program executed in a vector computer significantly lower than the maximum throughput. A detailed discusion of the degradation factors can be found in[8]. The most important of them are summarized here :

- Start-up of Pipelines; the maximum throughput of a pipeline is obtained after the pipeline has been loaded, reconfigurated and filled with operands. The amount of time required to output the first result is called start-up time. The throughput as a function of vector length for a typical state-of-the-art supercomputer is represented in Figure 3. It becomes obvious that to make the start effort negligible large amounts of computations must be performed.

- Data Dependencies; the key to utilizing a vector processor with several arithmetic units is to keep all the arithmetic

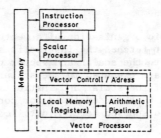

Figure 1 : Basic organisation of a vector computer

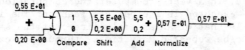

Figure 2 : Possible organisation of an execution pipeline

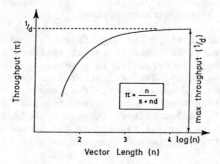

Figure 3 : Throughput of a pipeline as a function of vector length (d = the segment delay, s = start-up time)

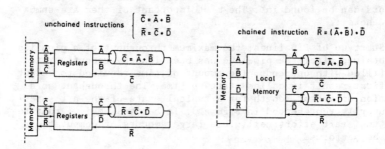

Figure 4 : Execution of a chained instruction

units busy and to avoid unnecessary memory references[5]. If there are data dependencies between instructions, one arithmetic unit must wait until another has finished the work. In some cases[24] this degradation of performance can be reduced by chaining. Chaining is a technique whereby the results from an arithmetic unit are forwarded without delay to the next unit. Chaining not only reduces the memory traffic but makes possible the concurrent work of several pipelines as well. An example of two chained vector instructions is illustrated in Figure 4.

- Conditional Processing; simple vector instructions cannot perform on vectors whose elements depend on a condition. In these cases the execution can be controlled with bit vectors. A bit vector has as many components as the vector it controls. The instruction is executed only for those elements of the vector for which the bit vector has the value 1.

- Limitations in Addressing Capabilities; the memory organisation makes possible the concurrent addressing of several vector elements stored in consecutive addresses. Random access and indirect indexing, as they occur in sparse matrix computations, are not adequate for vector processing. Many computer manufacturers have developed hardware facilities which make possible the handling of certain irregular data structures with vector functions :

By means of the GATHER function[2,4] all elements m of a vector B are gathered from the n elements of a vector A according to the pointer vector K :

$B(i)=A(K(i))$ $\qquad$ i=1,m $\quad$ K(i)$\leq$n

SCATTER[2,4] performs the inverse operation :

$B(K(i))=A(i)$ $\qquad$ i=1,m $\quad$ K(i)$\leq$n

Unfortunately the SPAXPY function[4] :

$B(K(i))=B(K(i))+s*A(i)$ $\qquad$ i=1,m $\quad$ K(i)$\leq$n $\quad$ s=scalar

is not available in all vector libraries and for given data structures (e.g., if K(i) = K(i+1)) generates errors.

3. VECTORISATION OF FE PROGRAMS

The vectorisation of FE programs can be very challenging particularly due to the large amount of sparse matrix computations required by the assemblage and the solution of the equilibrum equations. Nevertheless, it is possible to develop effectively vectorized codes through the use of certain data structures and vector functions. For exemplification, several programming issues applied to the FE solution of the irrotational ideal fluid flow equation :

$$\frac{\partial^2 \phi}{\partial x^2} + \frac{\partial^2 \phi}{\partial y^2} \equiv \nabla^2 \phi = 0$$

have been chosen. The approaches presented could have easily been extended to three dimensions and unsteady flow to deal with more general problems[19,21].

3.1 Data Structures and Pre-Processing

It is assumed that the flow domain is sub-divided into m triangular elements with linear approximation functions.

The input data structure is easy to generate and check (Figure 5). In order to permit the vector processor to develop its full potential is necessary to impose a secondary data structure. The secondary data structure is obtained through pre-processing from the primary one and contains the coordinates rearranged in vectors with m elements and a complete topologic description of the mesh.

The coordinates are rearranged using the GATHER function and the vectors of node-element connections:

$$X_i(e) = X(K_i(e)) \quad Y_i(e) = Y(K_i(e)) \quad e = 1,m \quad i = 1,3$$

The topologic description gives all the index relations associated with a given mesh. These are :

- node-element connections K_i i=1,3
- element-global matrix connections L_{ij} i,j=1,3
- index map of the global matrix T

The elements of the vectors L_{ij} and T depend on the chosen storage scheme of the global matrix, which in many FE applications has a random sparse structure. Because of the large number of address computations the generation of the vectors L_{ij} and T is usually performed in the scalar processor. For problems which involve a large number of simulations, e.g. sensitivity studies, it is of advantage to generate the topologic description by a separate pre-processing program[20].

3.2 Element Matrices and Assemblage of Global Matrix

The components of the derivative matrices $B_i(e)$, $C_i(e)$ and the areas of elements $F_i(e)$ are computed using chained vector instructions of length m :

$$F(e) = .5 * \sum_{ijk} (X_i(e) * (Y_j(e) - Y_k(k)) \quad i,j,k=1,3 \quad e=1,m$$

$$B_i(e)=.5*(Y_j(e)-Y_k(e))/F(e)$$
$$C_i(e)=.5*(X_j(e)-X_j(e))/F(e) \qquad i,j,k=1,3 \quad e=1,m$$

Figure 6 shows the generation of the element matrices. Instead of performing short sequences of element-by-element computations, as in most conventional programs, chained vector instructions have been introduced :

$$D_{ij}(e)=(B_i(e)*B_j(e)+C_i(e)*C_j(e))*F(e) \qquad e=1,m \quad i,j=1,3$$

The finer the mesh, the higher is the performance of the vector instructions. If there are several arithmetic units which work concurrently, suplementary speed gain can be achieved avoiding data dependencies between consecutive vector instructions.

Each vector D_{ij} is matched by an index vector L_{ij}, which defines the positions of the elements of D_{ij} in the matrix of the equations system A. Since A is sparse, only a compacted form of it A' is kept in memory. The c+1 colons of A' are stored consecutively so that A' can be considered as a vector with n*(c+1) elements. Figure 7 illustrates for a small example (n=7,c=3), how the compacted matrix A' and the corresponding index map T look like. A' can be assembled using the vector function SPAXPY and the vectors of element-global matrix connections L_{ij}:

$$A'(L_{ij}(i))=A'(L_{ij}(i))+D_{ij}(i) \qquad i=1,n \quad i,j=1,3$$

3.3 Solution of the Equations System

For the solution of the equations system the MCG (modified conjugated gradient) has been chosen. MCG is "decisively superior to any other technique"[9] for large-size diagonally dominant matrices. A review of the MCG method centered around the Cholesky technique for preconditioning can be found in[10]. The incomplete factorization technique is very powerful in scalar computers but vectorizes well only if the matrix has a regular sparsity. A fully vectorized algorithm restricted to matrices that come from FD approximations has been proposed in[25].

The application of polynomial preconditioners[1,6,15,23] seems to be very attractive on supercomputers. However, much work is needed to find good polynomials[16].

A central problem of the MCG implementation is the optimization of the matrix-vector product :

B = A * X

For the case that matrix A is stored in A' (s. Figure 7 for

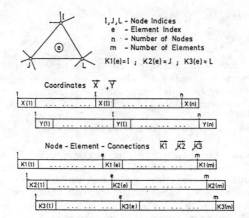

Figure 5 : Primary data structure

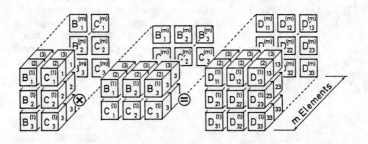

Figure 6 : Generation of element matrices

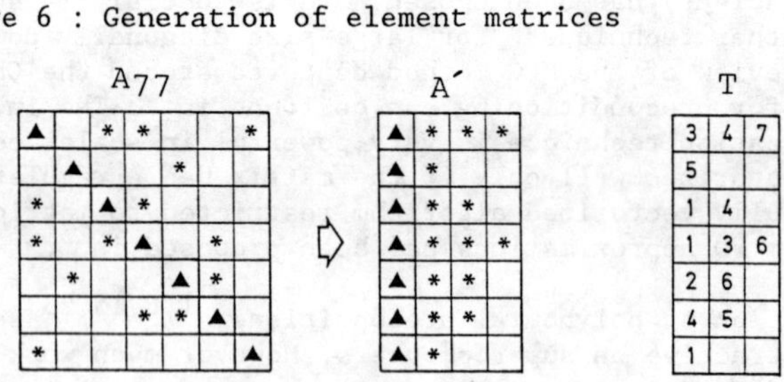

Figure 7 Storage scheme of the global matrix A
(A´ = the compacted form of A and
T = the index map)

exemplification) the following vectorized scheme is proposed :

First, the elements of the input vector X are rearranged using the GATHER function and the index map T :

XX(i)=X(T(i))   i=1,n*c

Then, the multiplication is performed by a single chained instruction of length n :

B(i)=A'(i)*B(i)+ $\sum_k$ A'(n*k+i)*XX(n*k+i) k=1,c   i=1,n

4. CONCLUSION

In order to obtain top performances on the novel computer architectures it is worth reconsidering the conventional approaches. This is not an easy task since through years of experience users are trained to think in a scalar way. "So what needs actually to be vectorized is not an existing code, but our own mind"[26].

REFERENCES

1. Adams L. (1983), An M-Step Preconditioned Conjugate Gradient Method for Parallel Computation. Proc.Int. Conf. on Parallel Processing (Ed. H.I Siegel) pp.36-43) IEEE.
2. CDC (1984), Fortran 200 Version 1,Reference Manual 60480200 Rev. C, Rechenzentrum Ruhr Universitaet Bochum.
3. Concus P. G.H. Golub and G. Meurant (1985), Block Preconditioning for the Conjugate Gradient Method. SIAM J.Sci.Stat. Comput. Vol.6, No1.,pp.220-252.
4. CRAY Research Inc. (1984), CRAY X-MP and CRAY 1 Computer Systems, Fortran (CFT) Reference Manual SR 0014.
5. Dongarra J.J and D.C. Sorensen (1986), Linear Algebra on High Performance Computers. Parallel Computing 85 (Ed. M.Feimeier et. al.), pp.3-32. North-Holland.
6. Dubois P.F A. Greenbaum and G.H. Rodrigue (1979), Approximating the Inverse of a Matrix for Use on Iterative Algorithms on Vector Processors. Computing 22, pp.257-268.
7. Ercegovac M. and T.Lang (1986), General Approaches for Achieving High Speed Computations. Supercomputers (Ed. S. Fernbach), pp.1-28. North-Holland.
8. Ercegovac M. and T. Lang (1986), Vector Processing. Supercomputers (Ed. S. Fernbach), pp.29-57. North-Holland.
9. Gambolati G. (1980), Perspective on a Modified Conjugate Gra-

dient Method for the Finite Element Solution of Linear Subsurface Equations. FE in Water Resources (Ed. S.Y. Wang) pp.2.15-2.31. Rose Printing Co., Inc.
10. Gresho P.M (1986), Time Integration and Conjugate Gradient Methods for the Incompressible Navier-Stokes Equations. FE in Water Resources (Ed. A. Sa da Costa) pp.3-27. Springer Verlag.
11. Hockney R.W. and J.R. Jesshope (1981) Parallel Computers, Adam Hilger Ltd.
12. Hockney R.W. (1986), Parallel Computers: Architecture and Performance. Parallel Computing 85 (Ed. M.Feilmeier) pp.33-70
13. Hossfeld F. (1987), Vector-Supercomputers, presented at the 11th Bochum Conference on Supercomputers and Applications.
14. Hwang H. and F.A Briggs (1983), Computer Architecture and Parallel Processing, Mc Graw Hill.
15. Johnson O.G. C.A. Micchelli and G. Paul (1983), Polynomial Preconditions for Conjugate Gradient Calculations, SIAM J. Num. Anal. 20, pp.362-376.
16. Jordan T.L. (1983), Conjugate Gradient Preconditioners for Vector and Parallel Processors. Elliptic Problem Solvers II (ed. G.Birkhoff), pp.127-139. Academic Press.
17. Kascic M.J. (1986), Vectorisation as Intelligent Processing. Supercomputers (Ed. S. Fernbach),pp.59-67. North-Holland.
18. Kogge P.M. (1981), The Architecture of Pipelined Computers. Mc Graw Hill, 1981.
19. Pelka W. and A. Peters (1986), FE Groundwater Models on Vector Computers. Int.J.Num.Meth.Fluids, Vol.6 pp.913-25.
20. Peters A. (1987) A Vectorized FE System for the Solution of Large Groundwater Problems. Laminar and Turbulent Flow (Ed. C. Taylor), pp.182-193. Pineridge Press.
21. Peters A. (1988) Die Entwicklung eines vektorisierten FE Systems mit Anwendung in der Grundwasserhydraulik. PhD Thesis. Institut f. Wasserbau, RWTH Aachen.
22. Rouve et. al. (1986-1988), several unpublished reports. Institut f. Wasserbau, RWTH Aachen.
23. Saad Y. (1985) Practical Use of Polynomial Preconditionings for the Conjugate Gradient Method. SIAM J.Sci.Stat.Comput., Vol.6, Nr.4, pp. 865-881.
24. Thomson J.R.(1986), The CRAY 1, the CRAY X-MP, the CRAY 2 and Beyond: The Supercomputers of Cray Research. Supercomputers (Ed. S.Fernbach), pp.69-81. North-Holland.
25. Van der Vorst H.A. (1982), A Vectorisable Version of Some ICCG Methods SIAM J.Sci.Stat.Comput., Nr.3, pp.350-56.
26. Zabolitzky J. (1982), Vector Programming of Monte-Carlo and Numerical Problems, Proc. 1982 Conf. on CYBER 200 in Bochum (ed. U. Bernutat-Buchmann), pp.165-174. Bochumer Schriften zur Parallelen Datenverarbeitung.

Parameter Identification and Uncertainty Analysis for Variably Saturated Flow

J.F. Sykes and N.R. Thomson
Department of Civil Engineering, University of Waterloo, Waterloo, Ontario, N2L 3G1 Canada

ABSTRACT

This study investigates variably saturated groundwater flow in the vicinity of a sanitary landfill. A Conjugate gradient method with an objective function that includes both pressure terms and travel time terms is used for parameter identification. The uncertainty in calculated travel time is estimated using both a moment method and a Latin hypercube direct parameter sampling method. The adjoint operator technique is an important component of both the parameter identification procedure and the moment method uncertainty analysis.

INTRODUCTION

Parameter optimization and uncertainty analysis are becoming important aspects of groundwater flow and solute migration studies. The objective of the analyst is the determination of the hydrologic and contaminant transport parameters that minimize the uncertainty in estimates of the response of the conceptualized system of interest. Estimated responses can include piezometric heads, interstitial velocities, travel paths, times of travel and solute concentrations. Observations of these quantities are generally used in conceptual model development, and for parameter identification or optimization. The uncertainties in estimated system responses are a function of errors or approximations used in the development of the site conceptual model, errors in system observations, uncertainties in estimated parameters, and errors in the choice of the mathematical representation.

This study presents an integration of conceptual model development, parameter optimization and uncertainty analysis in the assessment of groundwater flow at a sanitary landfill southwest of Midland Ontario. At the site, the regional topography slopes westward towards Nottawasaga Bay, at the south east end of Lake Huron. Limestone bedrock occurs at a depth of approximately 90m, while the overburden consists of three major units: a lower till of approximately 40m thickness, a sand layer of approximately 36m to 40m thickness and a localized deposit of silt, clay and till (see Figure 1). Site data include piezometric heads at observation wells and chemical measurements giving contaminant plume definition. Waste disposal began at the site in 1971, the observed plume currently extends westward to approximately point A on Figure 1. An objective of the site conceptualization is the maximization of the sensitivity of the system responses to changes in the system parameters. This results in increased efficiency of parameter optimization algorithms and a decrease in parameter uncertainty, as small changes of parameter values will significantly degrade goodness of fit measures.

There is considerable literature on parameter optimization techniques. An excellent review on the subject was conducted by Yeh[1], while the work of Cooley[2,3] and Carrera and Neuman[4] have provided some valuable insight. In this study, a Conjugate gradient method is used for parameter identification. The goodness of fit measure includes two components: a contribution from the difference between calculated and observed heads at the site, and a contribution from calculated and observed travel time as estimated from the plume migration. The adjoint operator technique is used as part of the parameter updating procedure.

Various techniques are available to handle the effects of parameter uncertainty in groundwater flow. These techniques can be divided into two broad categories: full distribution analyses, and first and second moment analyses (Dettinger and Wilson[5]). The most common full distribution techniques are the method of derived distributions, the more widely known Monte Carlo method and the Latin hypercube method (Iman and Conover[6]).

First and second moment analyses are based on the underlying assumption that all important information about the random variable or parameter is contained in only two values, the mean and the variance-covariance. The two most common approaches are the perturbation method and the Taylor series expansion. Perturbation methods involve the formulation and subsequent analysis of an equation that describes the perturbation. The method can lead to complicated results when a number of parameters are uncertain. The Taylor series method is based on the expected values of a truncated Taylor series expansion. Generally, only the first order approximation is utilized for the analysis of the first and second moments; however, second order approximations have been used to determine better estimates (Dettinger and Wilson[5]).

For remedial investigation and landfill studies, the more practical (i.e., computationally feasible) choice for the uncertainty analysis is the first and second moment method based on Taylor series expansion. The ob-

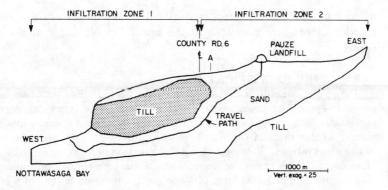

Figure 1: Site conceptualization and travel path.

jective of this paper is to demonstrate a version of this technique for the determination of travel time uncertainty in variably saturated groundwater flow. The uncertainty calculation provided by the developed moment method is based on the adjoint operator technique and is validated using the Latin hypercube method. Only Type II uncertainty which is the result of employing the correct model with uncertain parameters is considered. The groundwater flow and solute migration model GWPGM3 (Sykes[7]) is used for all optimization and uncertainty calculations.

GOVERNING EQUATIONS

The steady state variably saturated groundwater flow equation is

$$\frac{\partial}{\partial x_i}\left[K_{ij}K_r\left(\frac{\partial P}{\partial x_j}+\frac{\partial z}{\partial x_j}\right)\right]=0 \;\; i,j=1,2 \tag{1}$$

where K_{ij} is the saturated hydraulic conductivity tensor; K_r is the relative hydraulic conductivity which is a function of the pressure P and restricted to the interval $(0,1]$; z is the elevation above a reference datum; and x_i represents the two-dimensional Cartesian coordinate system x_1 and x_2. The boundary conditions associated with (1) are

$$P(\Gamma_1)=\hat{P} \text{ on } \Gamma_1 \tag{2}$$

and

$$-K_{ij}K_r\left(\frac{\partial P}{\partial x_j}+\frac{\partial z}{\partial x_j}\right)n_i=\hat{q}_n \text{ on } \Gamma_2 \tag{3}$$

where $\hat{P}$ is a prescribed pressure on Γ_1, $\hat{q}_n$ is a prescribed normal flux on Γ_2, n_i represents the components of the unit outward normal, and $\Gamma=\Gamma_1+\Gamma_2$ represents the boundary of the domain of consideration. In most situations, $\hat{q}_n$ represents net infiltration at the air-soil interface. The components of the interstitial velocity are

$$v_i=-\frac{K_{ij}K_r(P)}{\theta(P)}\left(\frac{\partial P}{\partial x_j}+\frac{\partial z}{\partial x_j}\right) \tag{4}$$

where $\theta = \theta(P)$ is the moisture content defined as a unique non-hysteretic function of pressure restricted to the interval $[0, \phi]$, and ϕ is the soil porosity.

PARAMETER IDENTIFICATION

Parameter identification techniques have been divided into two categories (Neuman[8]): direct and indirect methods. The latter category is particularly amenable to problems where the number of observations are scarce. In this method the results of the numerical model are compared to the observations. This results in the formation of an error or objective function for which the set of optimal model parameters will minimize its magnitude. The classical form of the objective function is that of a least square error function. Since the observed information for the particular site under investigation in this study consists of two physically different quantities (i.e., pressure and travel time) a modified least square error function is utilized. The objective function based on these distinct pieces of information may be written in a continuous form as

$$\min R(\alpha, P) = \int_A \lambda_P (P' - P)^2 \delta(\mathbf{x}' - \mathbf{x})\, dA$$
$$+ \lambda_T (T' - \int_S |v|^{-1}\, dS)^2 \qquad (5)$$

where P' is the observed pressure at location $\mathbf{x}'$, P is the calculated pressure from the model, $|v| = (v_x^2 + v_y^2)^{1/2}$ represents speed, S is the travel path of the particle from the point of entry into the domain to the point of concern, T' is the observed travel time along the travel path, and λ_p and λ_T are the weights associated with the observed pressure and the observed travel time data, respectively.

The algorithm for the minimization of (5) may be based on a number of techniques of which the Gauss-Newton and the Conjugate gradient methods are the most popular. Regardless of which of these methods are employed, the general sequence of iterates remains unchanged. Starting with an initial set of parameters, α^0, the initial solution of pressures is obtained from a discretized form of (1). Using these results the discrete form of (5) is determined. Following this, an updating direction is obtained which reflects the direction that the parameters should change in order to minimize the objective function as expressed by

$$\alpha^{k+1} = \alpha^k + \gamma^k \mathbf{d}^k \quad \text{for } k = 1, 2, \ldots \qquad (6)$$

where $\mathbf{d}^k$ is a vector representing the updating direction, γ^k is a scalar which represents the step size, and k is the iteration number. The magnitude of the step size may be determined by any one of a number of line search procedures (e.g., trial and error, quadratic interpolation, and Newtons method), but in the end the step size must be chosen such that $R(\alpha^{k+1}) < R(\alpha^k)$ (i.e., the magnitude of the objective function is reduced). From this point, the iteration procedure is repeated until some user defined convergence tolerance has been satisfied. The variations in all of the parameter identification methods, generally, lies in the approach used to determine the updating direction, $\mathbf{d}^k$. For example, the updating direction obtained from the Gauss-Newton method is based on the approximation of the Hessian matrix determined from estimates of the Jacobian matrix (i.e., the derivatives of the pressure at all the observation points with respect to the parameters). On the other hand, the Conjugate gradient methods only requires the generation of the objective function gradient vector (with respect to the parameters). A computationally efficient way to determine the entries in this gradient vector is through the use of the adjoint state technique. This is the parameter identification approach adopted in this study. A line search method based on a modified version of Newtons method as outlined by Neuman[9] is employed to determine the step size, γ^k, at each iteration.

UNCERTAINTY ANALYSIS

Consider the first order Taylor series expansion of the function represented by (5), about the vector of system parameters, α_o. The mathematical expectation of the resulting expression produces the mean of $R(\alpha_o)$. The variance or second moment of the first order Taylor series expansion produces

$$Var(R) \simeq \left[\frac{dR}{d\alpha^T}\right] cov(\alpha) \left[\frac{dR}{d\alpha}\right] \qquad (7)$$

where $cov(\alpha)$ is the covariance matrix of the vector of system parameters α, and the vector $dR/d\alpha$ is evaluated at α_o and are thus local derivatives. The variance of R as expressed by (7) is a function of the uncertainty in the parameters α and the influence or sensitivity of the functional value (5) to α in the neighbourhood of α_o.

SENSITIVITY COEFFICIENTS

Sensitivity coefficients are required for both the parameter identification procedure and for the uncertainty analysis. Sensitivity coefficients represent the change in a specified performance measure, the value of the objective function in this study, to changes in parameter values. There exist a number of ways to generate this information; however, in this study the adjoint state technique is employed (Sykes et al.[10]). The adjoint state governing equation corresponding to the variably saturated groundwater flow equation (1) may be expressed as

$$\frac{\partial}{\partial x_j}\left[K_{ji}K_r\frac{\partial \psi^*}{\partial x_i}\right] - K_{ij}\frac{\partial K_r}{\partial P}\left[\frac{\partial P}{\partial x_j} + \frac{\partial z}{\partial x_j}\right]\frac{\partial \psi^*}{\partial x_i} = \frac{\partial R(\alpha, P)}{\partial P} \qquad (8)$$

where ψ^* is the adjoint state variable which corresponds to pressure. Associated with (8) are the boundary conditions

$$\psi^* = 0 \quad \text{on } \Gamma_1 \qquad (9)$$

and
$$\frac{\partial \psi^*}{\partial x_j} n_j = 0 \quad \text{on } \Gamma_2. \tag{10}$$

The right hand side of (8) represents the partial derivative of the objective function (5) with respect to the dependent variable, P. Expanding this term yields

$$\frac{\partial R(\alpha, P)}{\partial P} = -2\lambda_P (P' - P)\delta(\mathbf{x}' - \mathbf{x})$$
$$+ 2\lambda_T (T' - \int_S |v|^{-1} dS) \cdot \int_S |v|^{-3} v_i \frac{\partial v_i}{\partial P} dS \tag{11}$$

In the numerical framework, (11) represents the load vector, and in terms of the discrete finite element formulation the derivative of the velocity components with respect to the nodal pressure P_k is

$$\frac{\partial v_i}{\partial P_k} = \left(\frac{-K_{ij}}{\theta^{gp}} \frac{\partial K_r^{gp}}{\partial P^{gp}} + \frac{K_{ij} K_r^{gp}}{(\theta^{gp})^2} \frac{\partial \theta^{gp}}{\partial P^{gp}} \right) N_k^{gp} \frac{\partial N_l^{gp}}{\partial x_j} P_l - \frac{K_{ij} K_r^{gp}}{\theta^{gp}} \frac{\partial N_k^{gp}}{\partial x_j} \tag{12}$$

where the superscript gp designates gauss point values.

Once the adjoint state variable, ψ^*, has been determined, the sensitivity coefficient may be calculated from

$$\frac{dR(\alpha, P)}{d\alpha_k} = \int_A \left\{ \frac{(\partial R(\alpha, P)}{\partial \alpha_k} - \psi^* \left[\frac{\partial}{\partial x_i} \left(\frac{\partial K_{ij}}{\partial \alpha_k} K_r (\frac{\partial P}{\partial x_j} + \frac{\partial z}{\partial x_j}) \right) \right] \right\} dA$$
$$+ \int_{\Gamma_2} \psi^* \hat{q}_n \, d\Gamma_2 \tag{13}$$

where the first term in the direct effect and is expressed as

$$\frac{\partial R(\alpha, P)}{\partial \alpha_k} = -\int_A 2\lambda_T (T' - \int_S |v|^{-1} dS) (\int_S |v|^{-3} v_i \frac{\partial v_i}{\partial \alpha_k} dS) dA \tag{14}$$

RESULTS

The sandy aquifer shown in the regional cross-section of Figure 1 is approximately 6 kilometers in length, extending from Nottawasaga Bay in the west to the recharge zone in the east. A prescribed hydrostatic boundary condition was used to model groundwater outflow at the western boundary. The lower boundary of the spatial domain was assumed to be a flow line. Two infiltration zones were conceptualized for the upper surface: the first represents the zone overlying the till west of County Rd. 6, the second zone represents recharge to the thicker sandy aquifer east of the road (see Figure 1). This interpretation is consistent with the occurrence of wetter lands and many small streams west of County Rd. 6. Comparatively, the eastern region has less surface water and thus less surface runoff. The observed contaminant plume at the site has migrated westerly, with the furthest extent reaching approximately point A of Figure 1 in a period of approximately 17 years.

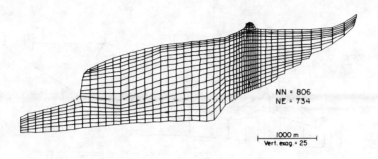

Figure 2: Spatial discretization.

The conceptualization of Figure 1 is based on spatial data; piezometric head data from observation wells were not used to generate prescribed boundary conditions. This decreases the importance of boundaries and increases the importance of both applied fluxes and assigned hydraulic conductivity values.

The spatial discretization of the cross-section of Figure 1 is shown in Figure 2. The nonlinear flow equation was solved using the Galerkin finite element method with a pseudo Newton Raphson matrix solution scheme and direct solver. The optimum parameters derived using the Conjugate gradient method are outlined in Table 1. The hydraulic conductivity anisotropy ratio was assumed to be 10 for all zones. The parameters minimize both the difference in calculated and observed head and the difference between the calculated travel time to point A in Figure 1 and the observed time. The weights for the head component of (5) are equal and sum to 0.5. The travel time weight was assigned a value of 0.5. For the sand zone, a change of the hydraulic conductivity by 1% results in a significant degradation of the objective function, (5). A change of 10% in the infiltration value for the eastern zone also results in a degradation of the results. The estimated head distribution corresponding to the parameters of Table 1 is depicted in Figure 3. The travel path from the landfill is shown on Figure 1. The estimated travel time to the discharge point is approximately 43.3 years.

Table 1a					
Hydraulic Parameters					
Zone	K_h [m/d]	σ_{Y_h}	K_v [m/d]	σ_{Y_v}	Porosity
Sand	8.25	0.013	0.825	0.013	0.3
Till	0.0039	0.25	0.00039	0.25	0.3
Landfill	2.55	0.25	2.55	0.25	0.3

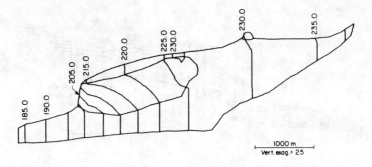

Figure 3: Estimated hydraulic head distribution in meters A.M.S.L.

Table 1b		
Infiltration Parameters		
Zone	Rate [cm/yr]	σ[cm/yr]
1	9.50	3.65
2	22.9	3.65

The uncertainty in the estimated travel time for groundwater flowing from the landfill to a point below A in Figure 1 was calculated using both the moment method described in a preceding section and the Latin hypercube method. For the latter, the pdfs for hydraulic conductivity and infiltration were divided into 50 intervals of equal probability. It is assumed that the uncertainty in travel time is a result of uncertainty in hydraulic conductivity and infiltration only. The parameter standard deviations are listed in Table 1, with log normal values being given for hydraulic conductivity. Infiltration is assumed to be uncorrelated from element to element. The hydraulic conductivity values for a zone are independent of both the values in the other zones as well as the infiltration. The magnitude of the selected standard deviations reflect the parameter identification procedure; the uncertainty in the sand hydraulic conductivity and the zone 2 infiltration is small as a result of the high sensitivity of the objective function to these parameters. The lower sensitivity of the objective function to the other parameters resulted in the use of higher estimated standard deviations. The results of the uncertainty analysis are given in Table 2. The two methods yield similar estimates of both mean travel time and the variance of travel time. The magnitude of the estimated variance is consistent with the error bound for the observed travel time. While the Latin hypercube method only yields an estimate of the travel time variance, the contribution of the uncertainty in the various parameters to the uncertainty in the travel time are given in the moment method. These values are also given in Table 2. The results indicate that the hydraulic conductivity for the till and landfill contribute little to the uncertainty of the travel time. The distribution of the contribution to the travel time variance of the surface infiltration uncertainty is illustrated in Figure 4. The abscissa represents distance along the surface with key features being marked. The ordinate is the log of the contribution to the variance in travel time of the uncertainty in infiltration for a given flux element divided by the length of that element (days squared per meter). The results indicate that infiltration west of the local divide above the eastern end of the till zone are relatively unimportant. Infiltration to the areas upgradient of the landfill have a greater impact on travel time. The marginal sensitivity coefficients of travel time to infiltration are negative for regions west of the track end and positive for the eastern portion; the impact of this sign change is evident in Figure 4.

By assuming that the hydraulic conductivity values in an element are independent of the values in other elements, the contribution of the uncertainty in the element value to the travel time variance can be determined (see Figure 5). The hydraulic conductivity values upgradient of the landfill and downgradient of the discharge point of the travel path (see Figure 1) contribute little to the variance in the travel time to point A. As expected, elements along the travel path contribute the most to travel time uncertainty.

Table 2 Travel Time Uncertainty		
	Moment Method	Latin Hypercube
Mean [yrs]	17.47	17.62
Variance [yrs^2]		
Sand	0.176	★
Till	0.064	★
Landfill	0.000	★
Infiltration	0.117	★
Total	0.357	0.373

★ denotes value not obtainable with this method

CONCLUSIONS

A prerequisite for the undertaking of uncertainty analysis of estimated system responses in porous media is parameter optimization. The sensitivity coefficients used in the moment method are local derivatives, for many problems they may have a limited range of applicability. For direct sampling techniques such as the Latin hypercube method, the use of large standard deviations can result in the generation of physically unreasonable realizations. For both methods, the parameter standard deviations should be small; the moment methods require this for computational reasons, the direct parameter sampling methods for physical reasons. When

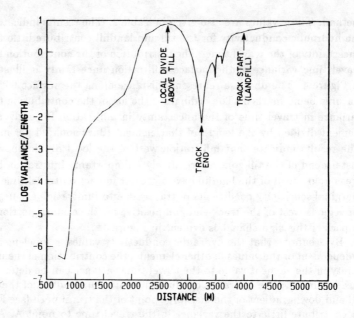

Figure 4: Contribution to travel time variance due to the variance in surface infiltration.

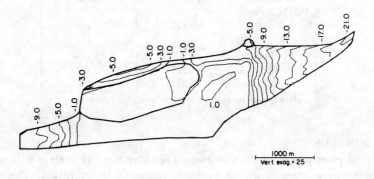

Figure 5: Contribution to travel time variance due to the variance in elemental hydraulic conductivity values (log of days^2).

the parameter standard deviations are small, the results of the two methods compare quite favourably. However, the computational burden of the two methods are significantly different: the assessment of the forward problem, track determination and the calculation of travel time sensitivity and uncertainty required 11 minutes execution time on a Microvax II computer, the 50 realizations of the Latin hypercube method required 301 minutes of execution time.

REFERENCES

1. Yeh, W. (1986), Review of parameter identification procedures in groundwater hydrology: The inverse problem, Water Resour. Res., 22(2), pp. 95-108.

2. Cooley, R. L. (1977), A method of estimating and assessong reliability for models of steady state groundwater flow 1. Theory and numerical properties, Water Resour. Res., 13(2), pp. 318-324.

3. Cooley, R. L. (1982), Incorporation of prior information on parameters into nonlinear regression groundwater flow models 1. Theory, Water Resour. Res., 18(4), pp. 965-976.

4. Carrera, J. and S.P. Neuman (1986), Estimation of aquifer parameters under transient and steady state conditions 1. Application to synthetic and field conditions, Water Resour. Res., 22(2), pp. 228-242.

5. Dettinger, M.D., and J.L. Wilson (1981), First order analysis of uncertainty in numerical models of groundwater flow. Part 1. Mathematical development, Water Resour. Res., 17(1), pp. 149-161.

6. Iman, R.L. and W.J. Conover (1980), Small sample sensitivity analysis techniques for computer models, with an application to risk assessment, Commun. Statist.-Theor. Meth., A9(7), pp. 1749-1842.

7. Sykes, J.F. (1987), GWPGM3 documentation and user guide, University of Waterloo, Waterloo, Ontario, 38pp.

8. Neuman, S. P. (1973), Calibration of distributed parameter groundwater flow models viewed as a multiple-objective decision process under uncertainty, Water Resour. Res., 9(4), pp. 1006-1021.

9. Neuman, S. P. (1980). A statistical approach to the inverse problem of aquifre hydrology, 3, Improved solution method and added perspective. Water Resour. Res., 16(2), pp. 331-346.

10. Sykes, J.F., J.L. Wilson and R.W. Andrews (1985), Sensitivity analysis for steady state groundwater flow using adjoint operators, Water Resour. Res., 21(3), pp. 359-371.

Modeling of Highly Advective Flow Problems

M.F. Wheeler

Department of Mathematical Sciences, Rice University, POB 1892, Houston, TX 77251, and Department of Mathematics, University of Houston, Houston, TX 77004, USA

ABSTRACT

Numerical methods appropriate for modeling highly advective flow problems in porous media are considered. Emphasis is placed on fruitful approaches for emerging parallel computer architectures; these include operator splitting and domain decomposition techniques.

INTRODUCTION

In this paper we consider numerical methods for highly advective flow problems with particular application to simulation of flow in porous media. We emphasize the discretization by operator splitting methods since such an approach provides a computationally efficient way of decoupling various difficulties, namely advection, diffusion, reaction, incompressibility and nonlinearities. In operator time-splitting one "breaks up" the spatial operator into two or more terms and solves each piece sequentially using a numerical technique particularly suited for that operator. For example, for the partial differential equation

$$u_t = L_R u + L_A u + L_D u, \tag{1}$$

one could sequentially solve the problems

$$u_t = L_R u, \tag{2}$$

$$u_t = L_A u, \tag{3}$$

$$u_t = L_D u. \tag{4}$$

Here L_R, L_A and L_D denote the reaction, advection and diffusion operators respectively. This splitting is extremely useful if the time scales are quite different for each of the three operators.

The paper is divided into four additional sections. In Section 2 numerical algorithms for the advection operator which we feel are accurate, robust and computationally efficient are described. In particular we examine the modified method of characteristics or one of several higher-order techniques developed in recent years for modeling problems in gas dynamics. A discussion of schemes for adding physical diffusion and dispersion are included in Section 3. In Section 4 a brief description of an operator-splitting algorithm formulated by Dawson and the author for application to microbial biodegradation of hydrocarbons in groundwater is presented. A section on various parallel computing aspects arising from operator splitting follows.

ADVECTION-DOMINATED PARABOLIC PROBLEMS

We shall consider the following finite element and finite difference procedures for treating the advection-dominated flow problems:

(i) modified method of characteristics or Euler characteristic Galerkin

(ii) higher-order Godunov

(iii) filtering and smoothing techniques

The modified method of characteristics procedure was first proposed and analyzed by Douglas and Russell [11] for linear advection-dominated problems in one space variable. Extensions of this work can be found in [24,25,26,27, 28,29,31,33,8,9,10]. A similar procedure was analyzed by Pironneau [22] for the Navier-Stokes equations. In the modified method of characteristics one combines the time derivative and the advection term as a a directional derivative; that is one takes time steps in the direction of flow, along the characterisitics of the velocity field of the total fluid. Physical diffusion or dispersion is usually treated using a Galerkin finite element method. Since the time-stepping direction is along a characteristic where the unknowns are changed slowly, large time steps can be taken without introducing serious time truncation errors. The characteristic approach also minimizes certain overshoot difficulties which accompany many finite element methods for the problem with sharp interfaces. Moreover this method involves solving large sparse symmetric systems instead of nonsymmetic equations. We remark that this procedure can be viewed as a time-splitting algorithm.

In [13,14] Ewing, Russell and Wheeler describe numerical results for modeling incompressible miscible displacement in porous media. They combine the modified method of characteristics for treating the concentration equations and a mixed finite element procedure for approximating the Darcy velocity. They report that this combined numerical scheme eliminates numerical dispersion and greatly reduces grid orientation and material balance problems. Vectorization and parallelization of these algorithms as well as generalizations to contaminant problems has been treated by Wheeler and Dawson [9,10,31,33]. Moeissis, Miller, and Wheeler [18] have also applied these schemes to the studying of viscous fingering.

The modified method of characteristics (MMOC) is numerically unstable without the presence of diffusion or dispersion or unless exact quadrature is performed (see [8] and [29]). Recently Dawson, Russell and Wheeler [8] established optimal $L^2(L^2)$ error estimates for MMOC applied to linear advection problems. These results involve both optimal rates of convergence as well as optimal norms on the solution. Preliminary results indicate results extend to more general settings such as contaminant transport systems.

The modified method of characteristics is not a conservative scheme, even though as mentioned above the material balance errors are small. A similar scheme to MMOC is the Euler characteristic Galerkin method proposed by Morton [19, 20]. This method is conservative, and results by Morton appear quite promising. Computationally however it is more complicated than MMOC.

Schemes (ii) and (iii) have been defined for hyperbolic conservation laws. A type of time-splitting would involve alternating solving convection and diffusion problems. The advective terms would be treated explicitly using small time steps followed by a large implicit time step in which the physical diffusion-dispersion would be added.

The higher order Godunov procedure [30] and in particular MUSCL [5,6] has proven to be a very effective finite difference procedure for approximating solutions of problems in gas dynamics. For one-dimensional scalar conservation laws, the higher-order Godunov procedure involves an integral formulation of the equation, a discontinuous linear approximation of the solution and incorporation of the nonlinear wave structure of the problem. The latter is treated by solving Riemann problems. A slope-limiting step is also used to prevent nonphysical behavior of the numerical solution. In particular the slope limiting is a numerically diffusive step. Bell et. al. and Chavent and Jaffre [1,2,4] have entended these techniques to problems in reservoir simulation using finite difference and finite element formulations respectively. Bell et. al. report improvement in the resolution of fronts and essentially no grid orientation dependence. In his Ph.D. thesis [7], Dawson has derived a priori error estimates for the finite difference schemes in one space variable. Numerical results for these schemes can be found in Wheeler, Dawson, and Kinton [32] for the Buckley-Leverett problem with capillary pressure.

Another numerical approach gaining attention for hyperbolic conservation laws is that of filtering or smoothing techniques. These algorithms involve the recovery of accurate functional values at points from oscillatory high order approximations; these approximations typically have have high-order error estimates in Sobolev norms of negative index. The recovery is performed by averaging or convolution. Examples are higher order difference methods, higher order integral formulations, and particle methods. Much of this work is still in a preliminary stage [12]. Although particle methods have been used for for many years by Peaceman [16] and more recently by Farmer, Raviart, and many others [15, 23], no rigorous mathematical treatment for porous media problems has been established, especially when diffusion-dispersion terms are present.

TREATMENT OF DIFFUSION-DISPERSION

In operator-splitting there are many possibilites in the numerical method chosen for adding diffusion-dispersion. One could apply a Galerkin finite element, a mixed finite element method or one of several schemes based on the use of nonconforming approximations spaces: Petrov Galerkin, H^{-1} Galerkin, interior penalty Galerkin, discontinuous H^1 Galerkin or interior penalty collocation [34,35]. It should be noted that these operator-splitting schemes also have great flexibility in the choice of approximating spaces for the advection and diffusion steps, respectively. These spaces need not be the same, see Wheeler et. al. [32].

AN OPERATOR SPLITTING TECHNIQUE FOR MICROBIAL BIODEGRADATION OF HYDROCARBONS IN GROUNDWATER

An operator-splitting algorithm for solving this problem was first formulated by Dawson and Wheeler in [31]. In addition, a detailed theoretical analysis was obtained, as well as extensive calculations and testing. We briefly describe the scheme for modeling of *in situ* biorestoration, in particular, oxygen-limited biorestoration.

The equations describing this process are given as follows

$$R_H \frac{dH}{dt} = -M_t \cdot \bar{k} \cdot (\frac{H}{K_H + H})(\frac{O}{K_O + O}) = R_1(O, H, M_t), \tag{5}$$

$$\frac{dO}{dt} = -M_t \cdot \bar{k} \cdot F \cdot (\frac{H}{K_H + H})(\frac{O}{K_O + O}) = R_2(O, H, M_t), \tag{6}$$

and

$$\begin{aligned} \frac{dM_t}{dt} &= M_t \cdot \bar{k} \cdot Y \cdot (\frac{H}{K_H + H})(\frac{O}{K_O + O}) + K_c \cdot Y \cdot OC - bM_t \\ &= R_3(O, H, M_t), \end{aligned} \tag{7}$$

where H = hydrocarbon concentration, O = oxygen concentration, M_t = total microbial concentration, $\bar{k}$ = maximum hydrocarbon utilization rate per unit mass microorganisms, Y = microbial yield coefficient, K_H = hydrocarbon half saturation constant, K_O = oxygen half saturation constant, b = microbial decay rate, F = ratio of oxygen to hydrocarbon consumed, K_c = first order decay rate of natural organic carbon, and OC = natural organic carbon concentration. If Equations (5) and (6) for oxygen and hydrocarbon removal are combined with the advection-dispersion equation for a solute undergoing linear instantaneous adsorption, the following equations are obtained:

$$\Phi \frac{\partial H}{\partial t} - \frac{1}{R_H} \nabla \cdot (D \nabla H - \mathbf{u} H) = \frac{q}{R_H} \tilde{H} + \frac{\Phi}{R_H} R_1(O, H, M_t), \tag{8}$$

$$\Phi \frac{\partial O}{\partial t} - \nabla \cdot (D \nabla O - \mathbf{u} O) = q \tilde{O} + \Phi R_2(O, H, M_t), \tag{9}$$

$$\mathbf{u} = -\frac{K}{\mu} \nabla p, \tag{10}$$

and

$$\nabla \cdot \mathbf{u} = q,$$

where D = dispersion tensor, $\mathbf{u} = (u_1, u_2)^T$ = incompressible Darcy velocities, p = pressure, K/μ = permeability/viscosity, R_H = retardation factor for hydrocarbons, q = hydraulic source term, ϕ = porosity, $\tilde{H}$ = concentration of hydrocarbon in hydraulic sources, and $\tilde{O}$ = concentration of oxygen in hydraulic sources. Here $\tilde{O}$ is specified at injection wells and $\tilde{O} = O$ at production wells.

The movement of naturally occurring microorganisms will be limited by the tendency of the organisms to grow as microcolonies attached to the formation. In most aquifers, greater than 95 percent of the native organisms are attached, and consequently the affinity of microorganisms for solid surfaces will control the transport of the total population. For simplicity, we assume that the exchange of microorganisms between the solid surface and the free solution will be rapid and satisfy M_s = constant M_t, where M_s = concentration of microbes in solutions. Under this assumption the movement of the microbes can be treated using a simple retardation factor approach, i.e.

$$\Phi \frac{\partial M_s}{\partial t} - \frac{1}{R_m} \nabla \cdot (D \nabla M_s - \mathbf{u} M_s) = \frac{q}{R_m} \tilde{M}_s + \phi R_3(O, H, M_s), \tag{11}$$

where R_m = microbial retardation factor, $R_m = M_t/M_s$ and $\tilde{M}_s$ = concentration of microbes in hydraulic sources. Further details of the equation formulation and parameter selection can be found in Borden and Bedient.

The modeling of these advection-dominated diffusion reaction problems, i.e. (4) – (7), involves the treatment of sharp fronts, heterogeneities, point sources and sinks, phsical dispersion and time scale effects. In particular, reactions generally need to be approximated on a much smaller time scale than advection and dispersion. All of these difficulties must be handled in a robust and computationally efficient fashion. We feel that the approach described below satisfies these requirements. The algorithm we have implemented is as follows:

1. Initially accurate Darcy velocities and pressure are calculated using a higher-order mixed finite element procedure. Steps (2) and (3) are repeated N times where $N\Delta t =$ final time.

2. An initialization of the injected component(s) around wells is calculated via a radial solution.

   Steps (3) and (4) are repeated N times where $N\Delta t =$ final time.

3. A modified method of characteristics is used to treat each of the three advection-dispersion equations in parallel.

4. The reaction terms become a system of nonlinear ordinary differential equations. A second-order explicit Runge-Kutta is employed M times ($M \sim 100$ or 1000).

The approximating spaces used in (1) were the Raviart-Thomas spaces with $r = 1$. That is, the pressure approximations are piecewise discontinuous bilinears on a tensor product rectangular grid. The x-component of velocity (y-component) is approximated by piecewise continuous quadratics in $x(y)$ tensor discontinuous linears in $y(x)$. The dimension of these spaces equals approximately $12N_x N_y$ where N_x and N_y are the number of intervals in the x and y directions respectively; see [13,36]

The spaces we chose for pressure and velocity yield higher order convergence rates than the standard cell-centered finite differences. Moreover, in the mixed method finite element formulation fluid conductivities are treated in an integral formulation which allows for more accurate modeling of highly varying permeabilities. In addition point source terms (wells) are handled correctly, i.e. they are treated as distributions. A complete analysis of the mixed formulation and superconvergence results can be found in [36].

As an initialization step for steps (3) and (4), the injected concentration is approximated by calculating a radial solution for a small time period. This is an inexpensive way of obtaining accurate approximations at early time on a coarse two-dimensional computational grid. This approach is justified by approximation theory, and has been substantiated by asymptotic analysis of Peaceman [13].

The time-splitting algorithm given by steps (3) and (4) for solving the systems of equations given by (4),(5), and (7) is most conveniently described by considering a single component. We time-split the equation

$$\Phi\frac{\partial C}{\partial t} - \nabla \cdot D\nabla C + \mathbf{u} \cdot \nabla C = q(\tilde{C} - C) + \Phi R(C) \tag{12}$$

as

$$\frac{dC^*}{dt} = R(C^*) \tag{13}$$

and

$$\Phi\frac{\partial \bar{C}}{\partial t} + u \cdot \nabla \bar{C} = \sqrt{\Phi^2 + |u|^2}\frac{\partial \bar{C}}{\partial \tau} = \nabla \cdot D\nabla \bar{C} + (\tilde{C} - \bar{C})q, \tag{14}$$

where $\frac{\partial \bar{C}}{\partial \tau}$ is a directional derivative. Solutions to 8 and 9 are numerically approximated in a sequential fashion. Advection-dispersion (step 3) is handled using a finite element modified method of characteristics with continuous bilinear approximating spaces.

Let $(\cdot, \cdot)$ denote the usual L^2 inner product and M a finite dimensional subspace of $H^1(\Omega)$. Let Δt and $t^n = n\Delta t$. Denote by $\overline{C^n} \in M$ the approxiamtion to $\bar{C}$ at $t = t^n$. For $n \geq 0$, we define $\overline{C^{n+1}}$ by

$$(\Phi \frac{\overline{C^{n+1}} - \overline{\tilde{C}}^n}{\Delta t}, v) + (D\nabla \overline{C^{n+1}}, \nabla v) = (q(\tilde{C} - \overline{C})^{n+1}, v), v \in M, \quad (15)$$

where

$$\overline{\tilde{C}}^n(\mathbf{x}) = C^*(\mathbf{x} - \frac{\mathbf{u}}{\Phi}\Delta t, t^{n+1}). \quad (16)$$

We wish to remark that the physical dispersion terms, i.e. longitudinal and tranverse dispersion, are handled in a straightforward fashion by integration by parts, which yields a symmetric bilinear form as indicated in (10). This is not true for schemes such as C^1 collocation or the standard method of characteristics. In addition, the use of C^0 approximating spaces does not involve adding artificial dispersion.

Step (4), in which we approximate reactions (i.e. Equations 1-3) is performed locally at each node, using a small time step Δt_s, which is 100 to 1000 times smaller than the advection-dispersion time step.

PARALLEL COMPUTING

For the most part simulations of coupled systems of nonlinear advection-diffusion-reaction problems has been restricted to two spatial variables and two or three components. Recent technological advances in parallel computing offer great promise in developing more physically realistic computer codes. In particular, the development of parallel algorithms is essential if one is to use grid discretizations in three dimensions that are necessary to obtain accurate numerical solutions. We discuss here the parallelizability of the operator splitting methods discussed above as well as the parallel features of some of the physical problems arising in flow in porous media.

The modified method of characteristics is ideally suited for implementation on parallel machines. Each computational point (quadrature point in a Galerkin implementation) must be tracked backward in time from the current time step to tbe preceding time step along the integral curves of the velocity vector field. A procedure using micro-time steps of size depending on the curvature of these characteristics has been developed for this purpose [13, 14]. The tracking scheme is quite successful, but it does not vectorize well because it involves different steps in different parts of the domain according to variations in the velocity. However it parallelizes naturally; the tracking from each point is independent of the tracking from all other points. Parallel computing also provided the flexibility of using different quadrature in different parts of the domain as dictated by the physics.

Much of the computational burden in higher-order Godunov methods is in the solution of Riemann problems; each of these is independent of all others, so this part of these schemes is readily parallelizable. Filtering and smoothing techniques involve local calculations which parallelize easily.

Decomposition methods are a very important concept for solving large applied problems modeled by partial differential equations. Among them domain decomposition techniques play a crucial role; one of the main reasons being that these methods provide algorithms well suited for taking advantage of the specific architectures provided by recent vector and parallel machines.

The basic idea in domain decomposition for a boundary value problem is to break up the domain into subdomains, solve (exactly or approximately) each subdomain problem separately, and then match the local solutions together.

Evidence of the importance of domain decomposition methods clearly follows from noting that intensive research efforts are presently underway in both the United States and the USSR as well as in Europe and Japan, (see for example the Proceedings of the First International on Domain Decomposition, Paris, 1987, edited by SIAM).

Glowinski and Wheeler [11] have formulated domain decomposition techniques for mixed finite element methods. These schemes first involve the use of a coarse grid calculation to obtain either flux or pressure boundary value guesses, followed by projection of these initializations into a finer grid. Subsequently, subdomain solutions are obtained by solving local problems with these boundary conditions. A conjugate gradient technique is applied to a symmetric bilinear form to update the appropriate boundary conditions. Convergence reults for these schemes has been established. Preliminary computational results are quite promising and indicate that the algorithms are robust. Problems with noflow boundary conditions and with rough coefficients and right-hand sides (Dirac measures) have been tried with success. Presently these algorithms are being modified to run on the Cray XMP4 and the Intel Hypercube.

We wish to remark that the Glowinski-Wheeler schemes are also applicable to cell-centered five-point finite difference discretizations, the most commonly used finite difference schemes for elliptic and parabolic p.d.e.'s. These methods have been shown by Russell and Wheeler [27] to be mixed finite element methods with special numerical quadrature formulae.

In addition, for advection-diffusion-reaction problems, domain decomposition techniques can be used in the operator-splitting to treat the diffusion problem for each component. The latter is the only piece of the splitting that is not naturally parallalizable.

In the time-splitting application to microbial biodegradation it is clear that this approach is ideal for parallel computation. The advection-dispersion step can be solved simultaneously on three different processors. Furthermore, the reaction ordinary differential equations can be done in parallel over as many processors as one desires.

The porous-media problems will benefit greatly from parallelism. An aspect that has received much attention in recent years is phase behavior, in which one must determine how a collection of chemical components of varying amounts will partition itself into flowing phases. The nonlinear equations modeling this process in each cell of a computational grid take different forms depending on the presence or absence of possible phases, making vectorization difficult (though recent efforts have shown improvements) and parallelization natural. Similar observations apply to the evaluation of nonlinear constitutive relationships from tables of data and to the determination of source and sink terms at wells.

REFERENCES

1. Bell, J.B., Dawson, C. and Shubin, G.R. "An Unsplit Higher Order Godunov Method for Scalar Conservation Laws in Higher Dimensions," to appear in *Journal of Computational Physics*.

2. Bell, J.B. and Shubin, G.R. "Higher-order Godunov Methods for Reducing Numerical Dispersion in Reservoir Simulation," *SPE Symposium on Reservoir Simulation*, Dallas (Feb. 1985).

3. Borden, R.C and P.B. Bedient, "Transport of Dissolved Hydrocarbons Influenced by Reaeration and Oxygen Limited Biodegradation." To appear in *Water Resource Research*.

4. Chavent, G. and Jaffre, J. *Mathematical Models and Finite Elements for Reservoir Simulation*, Elsevier Science Publisher, New York, 1987.

5. Collela, P. "A Direct Eulerian MUSCL Scheme for Gas Dynamics," LBL-14104, Lawrence Berkeley Laboratory (July 1982).

6. Collela, P. and Woodward, P.R. "The Piecewise Parabolic Method (PPM) for Gasdynamics Simulations," LBL-14661, Lawrence Berkeley Laboratory (July 1982).

7. Dawson, C.N. "Error Estimates for Godunov Mixed Method for Nonlinear Parabolic Equations," Ph.D. Thesis, Rice University, in preparation.

8. Dawson, C.N., T.F. Russell, and M.F. Wheeler, "Some Improved Error Estimates for the Modified Method of Characteristics," to appear.

9. Dawson, C.N., M.F. Wheeler and R.C. Borden, "Numerical Simulation of Microbial Biodegradation of Hydrocarbons in Groundwater," *Proceedings of the Sixth International Symposium on Finite Element Methods in Flow Problems*, INRIA, Antibes, France, (1986), pp. 353-358.

10. Dawson, C.N., M.F. Wheeler, T.M. Nguyen and S.W. Poole, "Simulation of Hydrocarbon Biodegradation in Groundwater," *Cray Channels 8*, 3, (1986), pp. 14-19.

11. Douglas, J., Jr. and Russell, T.F. "Numerical methods for Convection-Dominated Diffusion Problems based on Combining the Method of Characteristics with Finite Element or Finite Difference Procedures," *SIAM J. Numer. Anal.* (1982) pp. 871-885.

12. Engquist, B. "Filtering Methods for Scalar Conservation Laws," presentation at *Meeting on Nonlinear PDE's*, Mathematics Department, University of Arizona, Tucson, AZ (Jan. 1986).

13. Ewing, R.E., Russell, T.F. and Wheeler, M.F. "Simulation of Miscible Displacement Using Mixed Methods and a Modified Method of Characteristics," *Computer Methods in Applied Mechanics and Engineering*, 47 (1984), pp. 73-92.

14. Ewing, R.E., Russell, T.F. and Wheeler, M.F. 1984. "Simulation of Miscible Displacement Using Mixed Methods and a Modified Method of Characteristics," SPE 12241, paper presented at SPE Symposium on Reservoir Simulation, San Francisco, CA (Nov. 1983).

15. Farmer, C.L., "Moving Point Techniques," NATA Advanced Meeting Deleware, (1985).

16. Garder, Jr., A.D., Peaceman, D.W., and Pozzi Jr., A.L., "Numerical Calculation of Multi-dimensional Miscible Displacement by the Method of Characteristics," *Society of Petroleum Engineers Journal*, (March 1964), pp. 26-36.

17. Glowinski, R. and Wheeler, M.F. "Domain Decomposition and Mixed Finite Element Methods for Elliptic Problems," to appear in *Proceedings of First International Congress of Domain Decomposition*, (1987), SIAM.

18. Moeissis, D.E., C.A. Miller, and M.F. Wheeler, "A Parametric Study of Viscous Fingering in Miscible Displacement by Numerical Simulation," *Numerical Simulation in Oil Recovery* (M.F. Wheeler editor), Springer-Verlag, (1988), pp. 227-248.

19. Morton, K.W. and A. Priestly. "On Characteristics and Lagrange Galerkin Methods," *Putnam Research Notes in Mathematics*, Series, (eds. D.F. Griffiths and G.A. Watson), Longman Scientific and Technical, Harlow, (1986).

20. Morton, K.W. *Generalized Galerkin Methods for Steady and Unsteady Problems*, Academic Press, London, (1982), pp. 1-32.

21. _____. "Shock Capturing, Fitting and Recovery," *Numerical Analysis Report* 5/82, Department of Mathematics, University of Reading, and presented at the 9th International Conference on Numerical Methods in Fluid Dynamics, Aachen (June 28-July 2, 1982).

22. Peaceman, D.W. and D.H.E. Tang, "New Analytical and Numerical Solutions for the Radial Convection-Dispersion Problem," *SPE Reservoir Engineering* 2,3 (1987), pp. 343-359.

23. Pironneau, O. "On the Transport-Diffusion Algorithm and Its Application to the Navier-Stokes Equations," *Numeri. Math.* 38, (1982), pp. 309-332.

24. Raviart, P.A., "An Analysis of Particle Methods," *C.I.M-E. Course in Numerical Methods in Fluids*, Springer Lecture Notes, 1983.

25. Russell, T.F. "An Incompletely Iterated Characteristic Finite Element Method for a Miscible Displacement Problem," Ph.D. Thesis, University of Chicago, (1980).

26. _____. "Finite Elements with Characteristics for Two-Component Incompressible Miscible Displacement," Paper 10500, presented at the 6th SPE Symposium on Reservoir Simulation, New Orleans, LA (Feb. 1982), pp. 123-135.

27. _____. "Time-Stepping Along Characteristics with Incomplete Iteration for a Galerkin Approximation of Miscible Displacement in Porous Media," *SIAM J. Numer. Anal.* 22, (1985), pp. 970-1013.

28. _____ and Wheeler, M.F. "Finite Element and Finite Difference Methods for Continuous Flows in Porous Media," in Ewing, R.E. (ed.), *Mathematics of Reservoir Simulation*, SIAM Publications, Philadelphia, (1983), pp. 35-106.

29. _____, Wheeler, M.F. and Chiang, C. "Large-Scale Simulation of Miscible Displacement by Mixed and Characteristic Finite Element Methods," *Proceedings of the SEG/SIAM SPE Conference* on Mathematical and Computing Methods in Seismic Exploration and Reservoir Modeling, SIAM, (1985), pp. 85-107.

30. Süli, E. "Convergence Analysis of the Lagrange-Galerkin method for the Navier-Stokes Equations," Oxford University Computing Laboratory, Numerical Analysis Group 8613.

31. van Leer, B. "Towards the Ultimate Conservative Difference Scheme, V.A. Second-Order Sequel to Godunov's Method," *Journal of Computational Physics*, (1979), pp. 101-136.

32. Wheeler, M.F. and Dawson, C.N. "An Operator-Splitting Method for Advection-Diffusion-Reaction Problems," to appear in *MAFELAP Proceedings*, (1987).

33. Wheeler, M.F., Kinton, W. and Dawson, C.N. "Time-Splitting for Advection-Dominated Parabolic Problems in One-Space Variable," to appear.

34. Wheeler, M.F., C.N. Dawson, P.B. Bedient, C.Y. Chiang, R.C. Borden, and S. Rafai. "Numerical Simulation of Microbial Biodegradation of Hydrocarbons in Groundwater". *Proceedings of National Water Wells Conference on Solving Groundwater Problems with Models*, Colorado, Feb 10-12, 1987.

35. Wheeler, M.F. and P. Percell, "A Local Residual Finite Element Procedure for Elliptic Equations," *SIAM J. Numer. Anal.* 15 (1978), pp. 705-714.

36. Wheeler, M.F. "An Elliptic Collocation Finite Element Method with Interior Penalties," *SIAM J. Numer. Anal.* 15 (1978), pp. 152-161.

37. Wheeler, M.F., A. Weiser, and M. Nakata, "Some Superconvergence Results for Mixed Finite Element Methods on Rectangular Domains," *MAFELAP Proceedings*, 1985.

SECTION 2 - MODELING FLOW IN POROUS MEDIA

SECTION 2A - SATURATED FLOW

Cross-Borehole Packer Tests as an Aid in Modelling Ground-Water Recharge

J.F. Botha and J.P. Verwey
Institute for Ground-Water Studies, University of the Orange Free State, Bloemfontein, South Africa

INTRODUCTION

The town of Atlantis, situated about 40 km north of Cape Town, is the centre of a large industrial development area, which is solely dependent on ground water from an underlying phreatic aquifer for its water supply. At the beginning of the development in 1976, it was decided to use the underlying phreatic aquifer for the supply of water, until surface water could be brought to the area from the Berg River about 100 km north. The rapid growth in population during the initial stages of the development led the authority in control to recharge the central part of the aquifer with storm- and purified sewerage water discharged into a natural depression. Although the recharged water is generally of a good quality, it does contain some hazardous materials from accidental and intentional industrial spills. The result is that these materials are now encroaching on some of the nearby production boreholes and spreading towards the basement of the aquifer.

As the Atlantis aquifer is a coastal aquifer, the quality of its water may also be adversely affected by sea-water intrusion caused by the over-utilization of the aquifer. One possibility to prevent this and to purify the sewerage and storm-water effluents naturally, would be to discharge the effluents along the coast. However, before this project can be implemented, the Department of Health has to be satisfied that the process would not jeopardize public health. To achieve this, requires that a detailed investigation has to be conducted of the permeability of the aquifer and its ability to adsorb unwanted hazardous materials.

An investigation of this nature is obviously a long-term project with a strong emphasis on modelling artificial recharge. This raises the question of what model to use, particularly in this case where the recharge will take place close to the contact zone of the sand deposits and the bed rock. The present paper discusses the investigation of this zone with cross-borehole packer tests as developed by Hsieh et al [2] and the numerical methods used in analysing the results of the tests.

NATURE OF THE AQUIFER

The Atlantis aquifer is situated in the sand dunes which stretch along the

West Coast of South Africa from Cape Town to Saldanha 100 km further north-west. Previous investigations (Van der Merwe [4]) have shown that the aquifer is underlain by an impermeable layer of clay, a few metres thick, formed by the weathering of greywacke (Malmesbury formation), which underlies the sand deposits. A drilling program, conducted as part of the present investigation, revealed, however, that another aquifer is present in the Malmesbury formation and that there is an interaction between this partially confined aquifer and the one in the sand, particularly along the coast. As the Malmesbury pinched out towards the coast, this aquifer presents a particular challenge for the implementation of the fresh-water curtain. For example, if the Malmesbury is conducive to fracture flow, the proposed scheme may not be as effective as originally envisaged. This is indeed a distinct possibility as outcrops of the Malmesbury along the coast display a highly fractured structure. It was, for this reason, that the present investigation was conducted.

CROSS-BOREHOLE PACKER TESTS

Of primary importance in the present investigation is obviously the nature of flow in the Malmesbury. There are basically two methods that can be used to investigate the nature of ground-water flow. The first method involves the detailed analysis of ordinary pumping test data (Botha [1]). While this method can distinguish between porous and fracture flow, the information derived from it is of a general nature. It cannot, for example, yield information on the tensorial character of the hydraulic conductivity or its variation with depth. A method that can supply this type of information is the cross-borehole packer tests devised by Hsieh *et al* [2].

In a cross-borehole packer test, parts of two neighbouring boreholes are sealed off with packers. Water is then injected at a constant rate into the sealed off section (R) in one borehole, and the pressure in the sealed off section (S) of the other borehole continuously monitored (see Figure 1).

If the length of the sections R and S in Figure 1 are short in relation to the depth of the borehole and the boreholes are not too closely spaced, the sections may be considered as a point source and sink respectively. If it is further assumed that the flow is porous, it can be shown (Hsieh *et al* [2]) that the pressure increase in B must satisfy the equation

$$\Delta h = [Q/\{4\pi G^{1/2}\}] \operatorname{erfc}(S_0 G/4Dt)^{1/2}, \quad (1)$$

where,

$$\operatorname{erfc}(u) = [2/\pi^{1/2}] \int_u^\infty \exp(-s^2)\, ds = 1 - \operatorname{erf}(u), \quad (2)$$

is the complementary error function,

$$D = \|\mathbf{K}\| = K_{11}K_{22}K_{33} + 2K_{12}K_{23}K_{13} - K_{11}(K_{23})^2$$
$$- K_{22}(K_{13})^2 - K_{33}(K_{12})^2, \quad (3)$$

the determinant of **K**, and G a quadratic expression of the form

$$G = \mathbf{x}^T \mathbf{A} \mathbf{x} = x_1^2 A_{11} + x_2^2 A_{22} + x_3^2 A_{33} + 2x_1 x_2 A_{12}$$
$$+ 2x_2 x_3 A_{23} + 2x_1 x_3 A_{13} \quad (4)$$

with (x_1, x_2, x_3), the components of a cartesian coordinate system centred at the midpoint of the interval R, Q the constant recharge rate applied at R and $A_{i,j}$, the components of the symmetric matrix adjunct to K, given explicitly by

$$A_{11} = K_{22}K_{33} - K_{23}^2, \ A_{22} = K_{11}K_{33} - K_{13}^2, \ A_{33} = K_{11}K_{22} - K_{12}^2 \quad (5)$$

$$A_{12} = K_{13}K_{23} - K_{12}K_{33}, \ A_{23} = K_{12}K_{13} - K_{23}K_{11},$$

$$A_{13} = K_{12}K_{23} - K_{13}K_{22}.$$

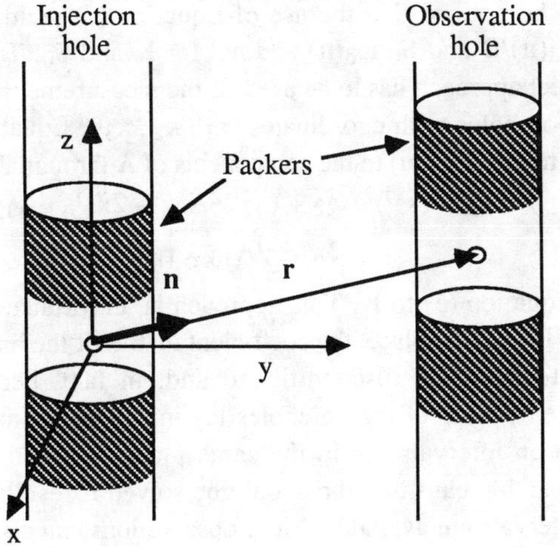

Figure 1   Graphical representation of a cross-borehole packer test.

The previous expressions can be simplified to a certain extent. Let **r** be the radius vector from the midpoint of R to that of S with magnitude r. If **n** is a unit vector in the direction of **r** (see Figure 1), then **r** can also be expressed as

$$\mathbf{r} = r\mathbf{n}.$$

Substitution of this expression into Equation (4) yields (Hsieh et al [2])

$$G = r^2(\mathbf{n} \bullet \mathbf{A} \bullet \mathbf{n}) = r^2 D[\mathbf{n} \bullet (\mathbf{K}^{-1}) \bullet \mathbf{n}] = r^2 D / K_d(\mathbf{n}) \quad (6)$$

where

$$K_d(\mathbf{n}) = \mathbf{n} \bullet \mathbf{K} \bullet \mathbf{n}$$

is the *directed hydraulic conductivity*. With these substitutions, Equation (1) becomes

$$\Delta h = (Q/4\pi r) \, [K_d(\mathbf{n})/D]^{1/2} \, \text{erfc} \, [r^2 S_0 / 4K_d(\mathbf{n}) t]^{1/2} \quad (7)$$

DATA REDUCTION

The expression for Δh in Equation (7) contains the two unknown parameters [$K_d(n)$ and S_0] that can be determined, provided that Δh is measured at two or more time steps. Since Equation (7) is non-linear, special techniques have to be used for its solution. In their original work Hsieh et al [2] used a graphical technique. In practice, Δh is usually measured at a large number of time steps. For this reason, the more objective non-linear least squares technique was used in the present investigation. Although any suitable method can be used for this purpose, the structure of Equation (7) allows one to use the simple Newton-Raphson method for two unknowns. Starting values for the method can be obtained from the graphical procedure described by Hsieh et al [2].

It should be noticed that the use of Equation (7) yields values for the parameters $K_d(n)/D$ and $S_0/K_d(n)$ and not for **K** and S_0. To determine the latter, an indirect approach has to be used. If the measurements are repeated in a number of boreholes with coordinates (x_1^i, x_2^i, x_3^i), Equations (4) and (6) can be used to relate $D/K_d(n)$ to the components of **A** through the equation

$$(x_1^i)^2 A_{11} + (x_2^i)^2 A_{22} + (x_3^i)^2 A_{33} + 2x_1^i x_2^i A_{12} + 2x_2^i x_3^i A_{23} + 2x_{1i} x_3^i A_{13} = r_i^2 D/K_d(n_i) \qquad (8)$$

and through Equation (5) to **K**. This approach is, unfortunately, not without its difficulties. In the first place, the coefficient matrix of the linear equations in Equation (8) tends to be ill-conditioned and, in fact, becomes singular, whenever three or more of the boreholes lay in a straight line, or more than three observation intervals lay in the same plane. Equation (8) contains 6 coefficients, which means that it can only be solved if results from at least 6 observation intervals are available. More observations can also be handled by again applying the method of least squares. In fact, the experience gained at Atlantis, indicates that, if the boreholes are situated close to one another, the ill-conditioning of the coefficient matrix increases with a decreasing number of observation intervals. To circumvent this, requires either a large number of observation points, or that the observation points be spaced not too closely. The latter condition will, unfortunately, require a more refined observation system.

RESULTS

The results of the cross-borehole packer tests to be described here were obtained at a site 26 mamsl and 150 m from the coast. The reasons for choosing this site were twofold: first the site corresponds to a position where a previous investigation (Müller and Botha [3]) has shown that sea-water intrusion into the aquifer is a possibility and secondly because of its easy access.

The basic principle of the cross-borehole packer test is that the observed pressure changes should satisfy the solution of Equation (1). The simplest method to check this, is to draw a graph of $\Delta hr/Q$ against t/r^2. Examples of both porous and fracture flow observed in the Atlantis aquifer are shown in Figure 2. The porous flow was observed at the contact between the sand deposits and the Malmesbury, at a depth of approximately 20 m (mbmsl), and the fractured flow at a depth of 50 (mbml); are shown in Figure 2. There is of course a possibility that presure leakage may influence the form of the curve if special precautions are not taken in constructing the apparatus.(Verwey [5]).

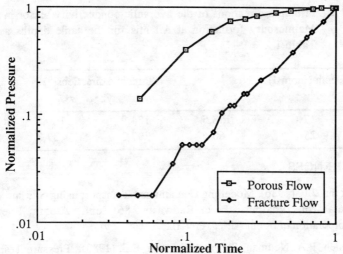

Figure 2 Behaviour of the pressure difference in the observation interval during a cross-borehole packer test for (a) porous and (b) fracture flow

Examples of the hydraulic tensor, calculated by the procedure discussed above, are given in Tables 1 and 2 for different depths. These results indicate that the Malmesbury formation is so fractured up to a depth of approximately 25 m below mean sea level, that it must be considered as an anisotropic porous medium. After this depth, the fracture density decreases rapidly and the flow changes completely from porous to fracture flow.

A very interesting result revealed by these figures is that at the contact of the sand and Malmesbury formation, the main principle component of the hydraulic conductivity is almost vertical, while the second principle is directed perpendicular to the coast line. These results show, not only that the conditions existing along the coast at Atlantis are conducive for artificial recharge, but also that one can use a porous flow model with confidence when modelling artificial recharge into the Atlantis aquifer.

Table 1 Principal components of the hydraulic conductivity tensor and storativity in the Malmesbury formation at Atlantis for a few selected depths

Depth (mbmsl)	Principle Components of the Hydraulic conductivity tensor (m/s)			Storativity (m^{-1})
17	$2,0 \times 10^{-6}$	$4,3 \times 10^{-6}$	$9,9 \times 10^{-5}$	$7,0 \times 10^{-7}$
25	$9,0 \times 10^{-6}$	$2,1 \times 10^{-6}$	$1,4 \times 10^{-5}$	$1,1 \times 10^{-5}$

Tabel 2 Principle directions of the hydraulic conductivity tensor in the Malmesbury formation at Atlantis for the same depths as in Table 1

Depth (mbmsl)	Principle directions(°)		
	xy	yz	zx
17	158,5	174,9	160,5
25	60,6	297,3	37,3

REFERENCES

1. Botha, J.F. (1986). Modelling Contaminant Transport using Site specific Data from Vaalputs. Proc. Radwaste '86 Conf., Atomic Energy Corporation of South Africa, Pretoria.

2. Hsieh, P.A. Neuman, S.P. and Simpson, E.S. (1983). Pressure Testing of Fractured Rocks − a Methodology employing Three-dimensional Cross-hole tests. U.S. Nuclear Regulatory Research Commission, Report NUREG/CR-3213, Washington D.C.

3. Müller, J.L. and Botha, J.F. (1986). A Preliminary Investigation of Modelling the Atlantis Aquifer, Bull. 15, Institute for Ground-water Studies, Bloemfontein.

4. Van der Merwe, A.J. (1983). Exploration, Development and Evalution of Ground-Water in the Sand Deposits in the Atlantis Area for Water Supply to the Atlantis Development Area. Unpublished M.Sc. Thesis, University of the Orange Free State, Bloemfontein.

5 Verwey, J.P. (1987). A Practical Guide for Cross-borehole Packer Tests. Unpublished Report, Institute for Ground-Water Studies, Bloemfontein.

The Boundary Element Method (Green Function Solution) for Unsteady Flow to a Well System in a Confined Aquifer*

Xie Chunhong
Department of Mathematics, Nanjing University, Nanjing, China
Zhu Xueyu
Department of Earth Sciences, Nanjing University, Nanjing, China

M·Radojkovic and J·Pecaric(1984) gave the BEM solution for steady flow to well system[1]. In this paper we consider both unsteady state and flow to well system using the boundary element method (Green function solution).

Let Ω be a two-dimensional domain and $\Gamma, s_1, s_2, \cdots, s_m$ its boundaries (See Figure 1). Where s_j ($j=1, \cdots, m$) are circles of wells with radius r_{0j} ($j=1, \cdots, m$), and $\Gamma = \Gamma_1 + \Gamma_2$.

Consider the parabolic equation:

$$T\left(\frac{\partial^2 H}{\partial x^2} + \frac{\partial^2 H}{\partial y^2}\right) - S\frac{\partial H}{\partial t} = Q \tag{1}$$

with boundary conditions:

$$H = H_1 \quad \text{on } \Gamma_1 \tag{2}$$

$$q = T\frac{\partial H}{\partial n} \quad \text{on } \Gamma_2 \tag{3}$$

and

$$T\frac{\partial H}{\partial n}\bigg|_{s_j} = q_j \quad \text{or} \quad \int_{s_j} q_j \, ds_j = Q_j \quad j=1, 2, \cdots, m \tag{4}$$

$$H(x, y, 0) = H^0(x, y) \quad (x, y) \in \Omega \tag{5}$$

The boundary integral formulation of the problem above can be derived from Green's second identity. It reads[2]:

---
*Project is supported by National Science Fund of China

$$\lambda H_{M_0} = T \int_0^t \int_\Gamma [G\frac{\partial H}{\partial n} - H\frac{\partial G}{\partial n}] d\Gamma d\tau + T\sum_{j=1}^m \int_0^t \int_{S_j} [G\frac{\partial H}{\partial n} - H\frac{\partial G}{\partial n}] d\Gamma d\tau$$

$$+ \int_\Omega SH^0 G(r,t) d\Omega - \int_0^t \int_\Omega GQ \, d\Omega d\tau \qquad (6)$$

where
$$\lambda = \begin{cases} 1 & \text{if } M_0(x_0, y_0) \text{ is in } \Omega. \\ 1/2 & \text{if } M_0(x_0, y_0) \text{ is on } \Gamma \text{ and } \Gamma \text{ is smooth.} \\ \theta/2\pi & \text{if } M_0(x_0, y_0) \text{ is on } \Gamma \text{ and } \Gamma \text{ is not smooth.} \end{cases}$$

θ is the interior angle, G is the fundamental solution of equation (1). It is given by

$$G(r, t-\tau) = \frac{1}{4\pi T(t-\tau)} \exp[\frac{-r^2 S}{4T(t-\tau)}] \qquad (7)$$

$$r = [(x-x_0)^2 + (y-y_0)^2]^{1/2} \qquad (8)$$

We first rewrite equation(6) in following form[2]

$$\lambda H_{M_0} = \int_\Gamma \int_0^{t_{k+1}} Hq^* d\tau d\Gamma - \int_\Gamma \int_0^{t_{k+1}} Gq d\tau d\Gamma + \sum_{j=1}^m \int_{S_j} \int_0^{t_{k+1}} Hq^* d\tau d\Gamma$$

$$- \sum_{j=1}^m \int_{S_j} \int_0^{t_{k+1}} Gq d\tau d\Gamma + S\int_\Omega H^0 G(r, t_{k+1}) d\Omega - \int_\Omega \int_0^{t_{k+1}} GQ d\tau d\Omega \qquad (9)$$

where q and q^* are the actual and associated outward normal fluxes, respectively. They are given by

$$q = -T\frac{\partial H}{\partial n}, \qquad q^* = -T\frac{\partial G}{\partial n} \qquad (10)$$

and
$$H_{M_0}^{k+1} = H(x_0, y_0, t_{k+1})$$

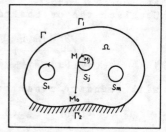

Figure 1. Flow domain

Because the expression of G is given by equation(7), q^* can be obtained as follows

$$q^* = -T\frac{\partial G}{\partial n} = \frac{S\eta}{8\pi T(t-\tau)^2} \exp(-\frac{r^2 S}{4\pi(t-\tau)}) \qquad (11)$$

where
$$\eta = (x-x_0)\cos(n,x) + (y-y_0)\cos(n,y).$$

The integral value of the first term on the right-hand of equation(9) can be evaluated as

$$\int_\Gamma \int_0^{t_{k+1}} Hq^* d\tau d\Gamma = \int_\Gamma \{\sum_{l=0}^k \int_{t_l}^{t_{l+1}} \frac{S\eta H}{8\pi T(t_{k+1}-\tau)^2} \exp(\frac{-r^2 S}{4T(t_{k+1}-\tau)}) d\tau\} d\Gamma$$

$$= -\sum_{l=0}^{k} \frac{1}{2\pi} \int_\Gamma \langle H \rangle \frac{\eta}{r^2} \{ \exp[\frac{-r^2 S}{4T(t_{k+1}-t_{l+1})}] - \exp[\frac{-r^2 S}{4T(t_{k+1}-t_l)}] \} d\Gamma \quad (12)$$

where $t_0 = 0$, $\langle H \rangle = (H^l + H^{l+1})/2$ and H^l is the value of H at time level l. The integral of the third term on the right side of equation (9) can be expressed as

$$\int_{S_J} \int_0^{t_{k+1}} Hq^* d\tau d\Gamma \approx -\sum_{l=0}^{k} \frac{1}{2\pi r_{M_0 M_i}} \exp[-\frac{r^2_{M_0 M_i} S}{4T(t_{k+1}-\tau)}]_{t_l}^{t_{l+1}} \int_{S_J} \cos(\overline{MM_0},$$

$$\overline{MM_J}) d\Gamma_M = 0 \quad (13)$$

The calculation of the second time integral in equation (9) is accomplished by making following changes of variables:

$$u^* = \frac{r^2 S}{4T(t_{k+1}-\tau)}, \quad du^* = \frac{r^2 S}{4T(t_{k+1}-\tau)^2} d\tau$$

We obtain

$$\int_\Gamma \int_0^{t_{k+1}} Gq d\tau d\Gamma \approx \frac{1}{4\pi T} \int_\Gamma q(-E_i(\frac{-r^2_{M_0 M_i} S}{4T t_{k+1}})) d\Gamma_M \quad (14)$$

where $-E_i(-u)$ is expotential integral given by

$$-E_i(-u) = \int_u^\infty \frac{1}{u} \exp(-u) du \quad (15)$$

if q is a constant. As the same we obtain

$$\int_{S_J} \int_0^{t_{k+1}} Gq d\tau d\Gamma = \frac{1}{4\pi T} Q_j(-E_i(-\frac{r^2_{M_0 M_i} S}{4T t_{k+1}})) \quad (16)$$

Next, we evaluate the remaining time integral in the equation (9). The result is given by

$$\int_\Omega \int_0^{t_{k+1}} GQ d\tau d\Omega = \frac{1}{4\pi T} \int_\Omega Q(-E_i(-\frac{r^2 S}{4T t_{k+1}})) d\Omega \quad (17)$$

where, for the sake of simplicity, it is assumed that Q is invariant with time.
Substituting Eqs. (12)-(17) into Eq. (9), We obtain

$$\lambda H_{M_0}^{k+1} = -\frac{1}{2\pi} \sum_{l=0}^{k} \int_\Gamma \langle H \rangle \frac{\eta}{r^2} \{ \exp[\frac{-r^2_{M M_0} S}{4T(t_{k+1}-t_{l+1})}] - \exp[\frac{-r^2_{M M_0} S}{4T(t_{k+1}-t_l)}] \} d\Gamma$$

$$- \frac{1}{4\pi T} \int_\Gamma q(-E_i(\frac{-r^2_{M M_0} S}{4T t_{k+1}})) d\Gamma_M + \sum_{j=1}^{m} \frac{Q_J}{4\pi T} (-E_i(\frac{-r^2_{M_0 M_i} S}{4T t_{k+1}}))$$

$$- \frac{1}{4\pi T} \int_\Omega Q(-E_i(\frac{-r^2_{M M_0} S}{4T t_{k+1}})) d\Omega_M + S \int_\Omega H^0 G(r, t_{k+1}) d\Omega \quad (18)$$

The spatial discretization of (18) can be performed. It is convenient to evaluate the boundary integrals using Gauss quadrature. Because t appears both as a limit of integration and in the integrand, we cannot derive a recurrence relation between the value of the integration and that in the integrand, we cannot derive a recurrence relation between the value of the integrals at the current and at previous time levels also. In order to circumvent the problem related to the time integration, Brebbia and Wrobel(1980)[3] employed a stepwise integration procedure. It is given by

$$\lambda H_{Mo}^{l+1} = \int_\Gamma \int_{t_l}^{t_{l+1}} Hq^* d\tau d\Gamma + \sum_{j=1}^{m} \int_{S_j} \int_{t_l}^{t_{l+1}} Hq^* d\tau d\Gamma - \int_\Gamma \int_{t_l}^{t_{l+1}} Gq d\tau d\Gamma$$

$$-\sum_{j=1}^{m} \int_{S_j} \int_{t_l}^{t_{l+1}} Gq_J d\tau d\Gamma + S \int_\Omega H^l G(r, t_{l+1}-t_l) d\Omega$$

$$-\int_\Omega \int_{t_l}^{t_{l+1}} GQ d\tau d\Omega \tag{21}$$

where G, q^*, and $G(r, t_{l+1}-t_l)$ are given by

$$G = \frac{1}{4\pi T(t_{l+1}-\tau)} \exp\left(\frac{-r^2 S}{4T(t_{l+1}-\tau)}\right) \tag{22}$$

$$q^* = \frac{S\eta}{8\pi T(t_{l+1}-\tau)^2} \exp\left(\frac{-r^2 S}{4T(t_{l+1}-\tau)}\right) \tag{23}$$

and

$$G(r, t_{l+1}-t_l) = \frac{1}{4\pi T(t_{l+1}-\tau)^2} \exp\left(\frac{-r^2 S}{4T(t_{l+1}-\tau)}\right) \tag{24}$$

substituting Eqs.(22)-(24) into equation(21), finally we obtain

$$\lambda H_{Mo}^{l+1} = \int_\Gamma \langle H \rangle \frac{\eta}{2\pi} \frac{1}{r^2_{MMo}} \exp\left(-\frac{-r^2_{MMo} S}{4T(t_{l+1}-t_l)}\right) d\Gamma_M$$

$$-\int_\Gamma \frac{\langle q \rangle}{4\pi T} \left(-E_1\left(\frac{-r^2_{MMo} S}{4T(t_{l+1}-t_l)}\right)\right) d\Gamma_M$$

$$+\sum_{j=1}^{m} \frac{\langle Q_J \rangle}{4\pi T} \left(-E_1\left(\frac{-r^2_{MoMj} S}{4T(t_{l+1}-t_l)}\right)\right)$$

$$+S\int_\Omega H^l \frac{1}{4\pi T(t_{l+1}-t_l)} \exp\left(\frac{-r^2_{MMo} S}{4T(t_{l+1}-t_l)}\right) d\Omega$$

$$-\frac{1}{4\pi T}\int_\Omega \langle Q\rangle(-E_1(\frac{-r_{MM_0}^2 S}{4T(t_{l+1}-t_l)}))d\Gamma_M \qquad (25)$$

where
$$\langle H\rangle = (H^l + H^{l+1})/2 \qquad (26)$$
$$\langle q\rangle = (q^l + q^{l+1})/2 \qquad (27)$$
$$\langle Q\rangle = (Q^l + Q^{l+1})/2 \qquad (28)$$
$$\langle Q_j\rangle = (Q_j^l + Q_j^{l+1})/2 \qquad (29)$$

We begin to solve the problem at first time step by use of Eq.(25) where $l=0$ and H^0 is the initial condition. At the end of the time step the nodal values H^l are determined and used as the initial condition for the second time step. Substituting Eqs.(26)-(29) into eq.(25) and discritizing it, then we place M_0 at point M_i and rewrite the equation as follows

$$\lambda H_i^{l+1} - \sum_{j=1}^{N}(-E_{ij}+F_{ij}\xi_{j+1})H_j^{l+1} - \sum_{j=1}^{N}(E_{ij}-F_{ij}\xi_j)H_{j+1}^{l+1}$$

$$+\sum_{j=1}^{N}(-G_{ij}+H_{ij}\xi_{j+1})\left[\frac{\partial H}{\partial n}\right]_j^{l+1} + \sum_{j=1}^{N}(G_{ij}-H_{ij}\xi_j)\left[\frac{\partial H}{\partial n}\right]_{j+1}^{l+1}$$

$$= A - B + C + D - R \qquad i=1,2,\cdots,N \qquad (30)$$

where
$$A = \sum_{j=1}^{N}\frac{H^l}{4\pi}\int_{\xi_j}^{\xi_{j+1}}\frac{\eta}{\xi^2+\eta^2}\exp(-u)d\xi \qquad (31)$$

$$B = \sum_{j=1}^{N}\frac{1}{8\pi}\left[\frac{\partial H}{\partial n}\right]_j^l \int_{\xi_j}^{\xi_{j+1}}(-E_1(-u))d\xi \qquad (32)$$

$$C = \sum_{j=1}^{m}\frac{\langle Q_j\rangle}{4\pi T}(-E_1(-u_1)) \qquad (33)$$

$$D = S\int_\Omega H^l \frac{1}{4\pi T(t_{l+1}-t_l)}\exp(-u_2)d\Omega \qquad (34)$$

$$R = \frac{1}{4\pi T}\int_\Omega \langle Q\rangle(-E_1(-u_2))d\Omega \qquad (35)$$

$$E_{ij} = \frac{1}{\xi_{j+1}-\xi_j}\int_{\xi_j}^{\xi_{j+1}}\frac{\eta\xi}{4\pi}\frac{1}{(\xi^2+\eta^2)}\exp(-u)d\xi \qquad (36)$$

$$F_{ij} = \frac{1}{\xi_{j+1}-\xi_j}\int_{\xi_j}^{\xi_{j+1}}\frac{\eta}{4\pi}\frac{1}{(\xi^2+\eta^2)}\exp(-u)d\xi \qquad (37)$$

$$G_{ij} = \frac{1}{8\pi(\xi_{j+1}-\xi_j)}\int_{\xi_j}^{\xi_{j+1}}\xi(-E_1(-u))d\xi \qquad (38)$$

$$H_{ij} = \frac{1}{8\pi(\xi_{j+1} - \xi_j)} \int_{\xi_j}^{\xi_{j+1}} (-E_i(-u)) d\xi \qquad (39)$$

$$u = \frac{(\xi^2 + \eta^2) S}{4T(t_{l+1} - t_l)}, \quad u_1 = \frac{r_{M_kM_i}^2 S}{4T(t_{l+1} - t_l)}, \quad u_2 = \frac{r_{MM_l}^2 S}{4T(t_{l+1} - t_l)}$$

In the equations above N is symbolizing the number of nodes and ξ, η are local coordinate.
The solution procedure is then repeated until the final time step is reached.
In order to compare the results of BEM with analytic solutions we give an example. The flow domain is a $4 \times 4 km^2$ square. It is shown in Figure 2. The parameters of the aquifer $S = 0.001$, $T = 400 m^2/d$. The analytic results are based on Theis solution. The comparison between the BEM results and analytic solutions are shown in Figure 3. There is a good agreement between them.

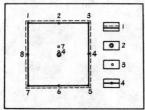

Figure 2. The example
1. Boundary, $H = H_0$
2. Pumping well
3. Observation hole
4. Node

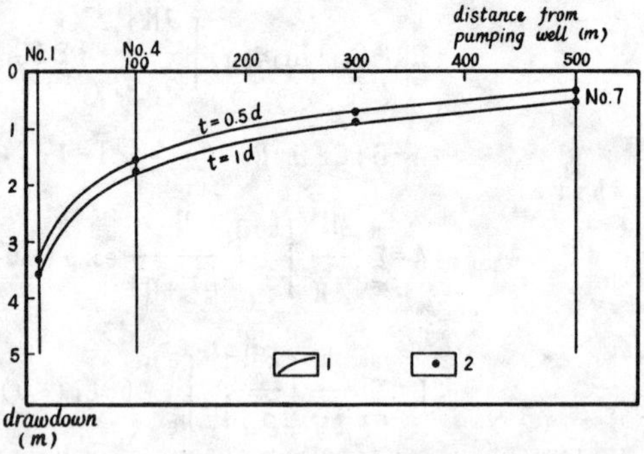

Figure 3. The comparison of BEM results with analytic solutions
1. Analytic results
2. BEM results

REFERENCES

1. Radojkovic, Miodrag and Pecaric, Josip, Boundary Element Analysis of Flow in Aquifers, (Ed. Gray W·G· et. al) PP. 723 to 736, Proceedings of The Fifth International Conference on Finite Elements in Water Resources, CMC, Southampton, 1984
2. Huyakorn, Peter S· and Pinder, George F·, (1983) Computational Methods in Subsurface Flow, Academic Press, New York, PP. 299 to 339
3. Brebbia, C·A· and Wrobel·L·C, (1980), Steady and unsteady potential problem using the boundary element method, Chapter I, Recent Advances in Numerical Methods in Fluids, (ed. Taylor and Morgan), Pineridge Press, U·K.

Finite Element Solution of Groundwater Flow Problems by Lanczos Algorithm

A.L.G.A. Coutinho, L.C. Wrobel and L. Landau

COPPE/UFRJ, Civil Engineering Department, Caixa Postal 68506 21945, Rio de Janeiro, RJ, Brazil

INTRODUCTION

For flow of groundwater that obeys Darcy's law, the governing equation can be written as

$$\nabla \cdot (k \nabla H) = 0 \tag{1}$$

where k is permeability, H is piezometric head, equal to $p\gamma + z$; p is pore water pressure, γ is specific weight of water and z is elevation above datum.

Applying a finite element discretization, we arrive at a system of algebraic equations which can be expressed as,

$$\underline{A}\,\underline{u} = \underline{b} \tag{2}$$

where $\underline{A}$ is a nxn symmetric banded positive definite matrix, $\underline{b}$ is the known independent vector arising from boundary conditions and $\underline{u}$ is the solution sought.

Iterative methods for solving the algebraic system of equations that arises from the finite element discretization of self-adjoint boundary value problems are currently under intensive research (Axelsson and Barker, 1984). The basic iterative method (Hageman and Young, 1981) computes an approximate solution $\underline{u}_{j+1}$ by the recursive formula,

$$\underline{u}_{j+1} = \underline{u}_j + \underline{C}^{-1}(\underline{A}\underline{u}_j - \underline{b}) \tag{3}$$

where the subscript denotes the iteration and $\underline{C}$ is called a splitting matrix.

Certainly, the most popular method to accelerate the basic iterative procedure is the conjugate gradient method (CG) introduced by Hestenes and Stiefel (1952). This method provides an elegant means to treat the sparsity of matrix $\underline{A}$, since the

only requirement is the evaluation of a matrix-vector product
throughout the iterations. Further, the convergence rate of CG
can be dramatically improved by using a preconditioning technique.

On the other hand, in the same year of the CG publication,
Lanczos (1952) showed that his method of minimized iterations
(Lanczos, 1950) for solving the matrix eigenproblem could also be
used for the solution of an algebraic system of equations. It is
also known that Lanczos' algorithm and CG share the same
properties in infinite precision arithmetic. However on using
Lanczos, unlike CG, there is no need to compute the approximate
solution in each iteration.

In the present work, the Lanczos algorithm is employed in
the finite element solution of groundwater flow problems.
Results obtained using regular meshes in a model
problem are compared with conjugate gradients, in order to
assess the efficiency of the numerical procedures.

LANCZOS' ALGORITHM

The Lanczos algorithm (LAN) is an oblique projection method
that constructs weak forms of equation (2) in the subspaces
spanned by the Krylov sequence. The projection of matrix $\underline{A}$ in
each Krylov subspace is a tridiagonal matrix $\underline{T}$. The theoretical
details of LAN can be found elsewhere (Golub, 1983). The Lanczos
algorithm can be summarized as follows:

<u>Step 1</u>   Initialize

$$j = 0 \tag{4}$$
$$\underline{w}_0 = \underline{0} \tag{5}$$
$$\underline{q}_0 = \underline{r}_0 = \underline{b} - \underline{A}\,\underline{u}_0 \tag{6}$$
$$\beta_1 = \|\underline{r}_0\|, \tag{7}$$

<u>Step 2</u>   Iterations

for $j = 1, 2, \ldots$, until convergence do

(a) Compute and store Lanczos vectors and matrix $\underline{T}$

$$\underline{q}_j = \underline{q}_{j-1}/\beta_j \quad,\text{ save Lanczos vector } \underline{q}_j \tag{8}$$
$$\underline{w}_j = \beta_j \underline{w}_{j-1} \tag{9}$$
$$\underline{w}_j = \underline{w}_j + \underline{A}\,\underline{q}_j \tag{10}$$

swap $\underline{q}_j, \underline{w}_j$

$$\alpha_j = \underline{w}_j^T \underline{q}_j \tag{11}$$
$$\underline{q}_{j+1} = \underline{q}_j - \alpha_j \underline{w}_j \tag{12}$$
$$\beta_{j+1} = \|\underline{q}_{j+1}\| \tag{13}$$

(b) Update last component of $\underline{x}_j$ ($\underline{T}_j \underline{x}_j = \beta_1 \underline{e}_1^j$)

where $\underline{T} \equiv$ tridiag $(\beta_j, \alpha_j, \beta_{j+1})$

(c) Update residual norm

$$\| \underline{r}_j \| = \beta_{j+1} x_j \qquad (14)$$

(d) Check convergence

if $\| \underline{r}_j \| \leq$ RTOL $\| \underline{r}_0 \|$ then stop $\qquad (15)$

else continue

Step 3   Solve tridiagonal system

$$\underline{T}_j \underline{x}_j = \beta_1 \underline{e}^j \qquad (16)$$

Step 4   Compute solution

$$\underline{u} = \underline{Q}_j \underline{x}_j, \quad \underline{Q}_j = \left[\underline{q}_1, \underline{q}_2, \ldots, \underline{q}_j \right] \qquad (17)$$

Remark 1: The computation of the last component of vector $\underline{x}_j$ can be performed without solving the tridiagonal system at each iteration by a scalar recurrence formula based on Givens rotations (Simon, 1982). It can be shown that the residual norm is proportional to this value (see Eq. (14)).

Remark 2: With respect to storage demands, LAN needs only 2 n-dimensional arrays plus matrix $\underline{A}$. Like CG methods, it is necessary to compute the matrix-vector product $\underline{A} \underline{w}_j$ which represents the main computational task of LAN. This operation can be computed at element level, through the decomposition,

$$\underline{A} \underline{w}_j = \sum_{i=1}^{Nel} \hat{\underline{A}}^i \hat{\underline{w}}_j^i \qquad (18)$$

where $\hat{\underline{A}}^i$ is the i-th element matrix, $\hat{\underline{w}}_k^i$ is the component of vector $\underline{w}_k$ related to this element and Nel is the total number of elements. This decomposition characterizes the element-by-element (EBE) solution scheme (Coutinho et al., 1987a). For the practical implementation of EBE schemes, a special data structure for the gather/scatter operations to transfer information from element level to the global dimensions is needed.

Remark 3: LAN and CG share the same properties in infinite precision arithmetic as shown by Simon (1982), Cullum and Willoughby (1985). The CG method computes implicitly the factorization of $\underline{T}_j$,

$$\underline{T}_j = \underline{L}_j \underline{D}_j \underline{L}_j^T \qquad (19)$$

while LAN performs it explicitly. Therefore, the convergence rate of LAN and CG will strongly depend on the eigenvalue distribution of $\underline{A}$. Obviously, this rate can be improved by adopting a preconditioning technique. In this work, a Jacobi regularization is employed and the LAN algorithm applied to the transformed system of equations. This choice retains the same advantages of the Modified CG method (Axelsson and Barker, 1984).

MODEL PROBLEM

The seepage below a concrete dam under an impervious foundation presented by Cheng (1984) was used as the model problem. The dam is resting on a layer of unit thickness and permeability, and has a basis of unit width. Owing to symmetry, only one half of the domain is solved. For convenience of numerical solution the left side of the domain is truncated at a distance of -10, where the no-flux boundary condition is prescribed. Therefore, the problem is posed on a rectangular domain and its solution presents a singularity on the edge of the dam.

Table 1 compares the results obtained using 4-noded isoparametric elements with 2x2 Gauss integration, for two meshes comprising 200 (10x20), MESH-1, and 800 (20x40), MESH-2, elements, totalizing respectively 220 and 840 equations. In this table, the prefix J refers to Jacobi regularization and M stands for the Modified CG with Jacobi regularization. The analyses were carried out in a Burroughs A-10 machine from the Computer Centre of the Federal University of Rio de Janeiro and the algorithms coded in FORTRAN. In the column STORAGE, the first number is the required amount of storage, in double precision words, for the main loop of each algorithm, and the second, the area needed for the upper triangle of the element matrices.

METHOD	ITERATIONS	CPU (s)	I/O (s)	STORAGE
CG	86	73	15	1100 + 2000 = 3100
	234	670	91	4200 + 8000 = 12200
J-CG	33	38	9	1320 + 2000 = 3320
	63	224	35	5040 + 8000 = 13040
M-CG	26	37	8	1100 + 2000 = 3100
	51	204	31	4200 + 8000 = 12200
LAN	110	93	21	440 + 2000 = 2440
	314	903	122	1680 + 8000 = 9680
J-LAN	27	38	9	440 + 2000 = 2440
	46	193	31	1680 + 8000 = 9680

Table 1 - Results for 4-noded elements
(Upper row: MESH-1; Lower row: MESH-2)

Convergence was achieved within a tolerance of 10^{-3}. It can be noted that J-LAN becomes more efficient than CG-based procedures as the mesh is refined, requiring also less storage area.

Table 2 shows a similar comparison for constant triangular elements, for equivalent meshes of 400 (MESH-3) and 1600 (MESH-4) elements. The same conclusions previously observed also apply to this case.

METHOD	ITERATIONS	CPU (s)	I/O (s)	STORAGE
CG	93	128	23	1100 + 4000 = 5100
	253	1176	138	4200 + 16000 = 20200
J-CG	39	62	13	1320 + 4000 = 5320
	75	365	51	5040 + 16000 = 21040
M-CG	28	55	12	1100 + 4000 = 5100
	55	303	47	4200 + 16000 = 20200
LAN	131	180	34	440 + 4000 = 4440
	374	1748	198	1680 + 16000 = 17680
J-LAN	28	55	13	440 + 4000 = 4440
	48	279	45	1680 + 16000 = 17680

Table 2 - Results for constant triangles
(Upper row: MESH-3; Lower row: MESH-4)

Finally, it should be pointed out that the results obtained were in good agreement with the analytical solution given in Cheng (1984), and improved as the mesh was refined.

CONCLUSIONS

It was shown in this work that Lanczos algorithm can be efficiently aplied to the solution of groundwater flow problems. However, it is essential that a preconditioner be employed to accelerate convergence.

Nonlinear problems can also be analysed using LAN as an iterative driver in Newton-like algorithms. Some results obtained using such scheme are reported by Coutinho et al. (1987b).

REFERENCES

O. Axelsson and V.A. Barker (1984), Finite Element Solution of Boundary Value Problems, Academic Press, New York.

A.H.D. Cheng (1984), Darcy's Flow With Variable Permeability: A Boundary Integral Solution, Water Resources Research, Vol. 20, pp. 980-984.

A.L.G.A. Coutinho, J.L.D. Alves, L. Landau, E.C.P. Lima and N.F.F. Ebecken (1987a), On the Application of an Element-by-Element Lanczos Solver to Large Offshore Structural Engineering Problems, Computers and Structures, Vol. 27, pp.27-37.

A.L.G.A. Coutinho, L.C. Wrobel and L. Landau (1987b), Comparison between Lanczos and Conjugate Gradient Algorithms for the Finite Element Solution of Heat Conduction Problems, submitted for publication in Int. J. Num. Meth. Engng.

J.K. Cullum and R.A. Willoughby (1985), Lanczos Algorithms for Large Symmetric Eigenvalue Computations, Vol. I: Theory, Vol. II: Programs, Birkhäuser, Boston.

G.H. Golub and C.F. Van Loan (1983), Matrix Computations, The John Hopkins University Press, Baltimore.

L.A. Hageman and D.M. Young (1981), Applied Iterative Methods, Academic Press, New York.

M.R. Hestenes and E. Stiefel (1952), Methods of Conjugate Gradients for Solving Linear Systems, J. Research Nat. Bureau Standards, Vol. 49, pp. 409-436.

C. Lanczos (1950), An Iteration Method for the Solution of the Eigenvalue Problems of Linear Differential and Integral Operators, J. Research Nat. Bureau Standards, Vol. 45, pp. 255-282.

C. Lanczos (1952), Solution of Systems of Linear Equations by Minimized Iteration, J. Research Nat. Bureau Standards, Vol. 49, pp. 33-53.

H.D. Simon (1982), The Lanczos Algorithm for Solving Symmetric Linear Systems, PhD Thesis, Dept. of Mathematics, University of California, Berkeley.

Finite Element Model of Fracture Flow
R. Deuell, I.P.E. Kinnmark and S. Silliman
Department of Civil Engineering, University of Notre Dame, Notre Dame, IN 46556, USA

INTRODUCTION

The first attempts to model flow in fractures assumed laminar flow, with fracture walls represented by flat, parallel plates. This simplification was motivated by the fact that it resembles a fracture, and that an analytic solution is known. Many textbooks in hydraulics, such as Lamb[1] and Langlois[2], include the solution to this problem, quite often called Poiseuille or Hele–Shaw flow. When the governing equations, the Navier–Stokes equation of motion and the continuity equation, are solved for flat parallel boundaries and slow viscous flow, a parabolic velocity distribution is obtained. The average velocity, v_{avg}, is given by

$$v_{avg} = -\frac{b^2 g \Delta H}{12 \nu L} \quad (1)$$

where b is the plate separation (fracture aperture), L is length parallel to flow, ΔH is the change in head over length L, ν is kinematic viscosity, and g is gravitational acceleration. For unit width, perpendicular to flow and parallel to the plates, the flow quantity Q, is given by

$$Q = -\frac{b^3 g \Delta H}{12 \nu L} \quad (2)$$

which is known as the "cubic law". Langlois[2] states that this equation is valid when b is variable, provided the change in aperture per unit length is small.

As with most natural processes, fracture geometries are complex. It is obvious that the simplifying assumptions necessary to allow derivation of an analytical solution do not truly represent the hydraulic conditions present in the rock fracture. Fracture walls are not smooth, they have associated roughness, similar to that defined in pipe flow, which impedes flow through the fracture. Neglecting roughness, fracture separation is variable, both perpendicular and parallel to flow, making it necessary to determine which aperture to use in calculations.

This study will examine fracture variation in the direction of flow. The answers to two questions will be sought. First, both roughness and non-parallel fracture walls can be considered as variations in fracture aperture. At what scale do these variations exhibit characteristics associated with roughness and when do variations exhibit characteristics associated with nonparallel plates? Second, when fracture walls are nonparallel, what value for the effective aperture is appropriate for use in the cubic law?

MODEL

To answer these questions different fracture wall geometries were envisioned. These geometries are symmetric perpendicular to flow and repeating parallel to flow. For brevity only the triangular shaped boundary is presented here (Figure 1). Simplified geometries, of this sort, allow variations in scale factors to be addressed for the given problem. To determine flow in fractures with these geometries the governing equations are again simplified using the typical assumptions associated with slow viscous flow (i.e. neglect inertial forces, incompressible, homogeneous etc.). For ease of computation and interpretation, the equations of motion and continuity are put in dimensionless form relating to the cubic law using the minimum aperture. Define

b_{min} = minimum aperture

v_{avg} = average velocity using cubic law and b_{min}

L = length of system parallel to flow

p = pressure

Δp = $\nabla p \cdot L$

x = coordinate in direction of flow

y = coordinate perpendicular to flow

ξ = x/b_{min}

η = y/b_{min}

u = v/v_{avg}

$$\phi = p/\Delta p$$

$$a = L/b_{min}$$

Using these definitions and the typical slow viscous flow assumptions the governing equations reduce to

$$0 = -\nabla\phi + \frac{1}{12a}\nabla^2 \mathbf{u} \qquad (3)$$

$$\nabla \cdot \mathbf{u} = 0 \qquad (4)$$

These equations were solved, in the desired region, using finite elements. However, the numerical problem can be simplified by utilization of the penalty method. The principle idea of the method is to relate pressure and velocity through

$$p = -\gamma \nabla \cdot \mathbf{v} \qquad (5)$$

where γ is a large constant known as the penalty parameter. The importance of (5) is more clearly shown when rearranged

$$p/\gamma = -\nabla \cdot \mathbf{v} \qquad (6)$$

Clearly in the limit as $\gamma \to \infty$, $\nabla \cdot \mathbf{v}$ must go to zero. Since $\nabla \cdot \mathbf{v}$ is a measure of volume change large values of γ represent a fluid that is slightly compressible. Therefore (5) replaces (4) as the continuity equation. Substituting (5) into the weak form of (3) yields

$$\frac{1}{12a}\int_\Omega 2\mathbf{D}\cdot\nabla\delta\mathbf{u}d\Omega + \frac{\gamma}{12a\mu}\int_\Omega (\nabla\cdot\mathbf{u})(\nabla\cdot\delta\mathbf{v})d\Omega = \int_\Gamma \left[\frac{2\mathbf{D}}{12a}+p\mathbf{I}\right]\mathbf{n}\cdot\delta\mathbf{v}d\Gamma \qquad (7)$$

where

$$\mathbf{D} = 1/2(\nabla\mathbf{u} + \nabla\mathbf{u}^T) \qquad (8)$$

and $\delta\mathbf{v}$ is a weighting function.

The problem was discretized as shown in Figure 1. Only half of the system was solved since it is symmetric and flow in both halves are the same. Now the scale factors can be varied. The length scale 'a' corresponds to the frequency of aperture variation while ϵ is the magnitude of the variation.

NONPARALLEL FRACTURE WALLS

Nonparallel fracture wall characteristics can be described as a small rate of change in aperture variation. In the experiments performed this is represented when 'a' is large, as compared to ϵ/b_{min}. As 'a' is increased relative to ϵ/b_{min}, the extreme values of 'a' should provide flow characteristics that are only influenced by nonparallel plate effects.

As is shown in Figures 2 and 3, increasing 'a' to extremes produces limiting values of Q, which are inconsistent with equations (2) using b_{avg} or the average aperture. Through analogy with flow perpendicular to layers in porous media, an argument for using the harmonic mean of the aperture was developed by Silliman. In porous media it is convenient to find a single effective value of K to use in Darcy's law. In a system with flow perpendicular to layers of different thicknesses and permeabilities, it has been shown that the effective conductivity, K, is equivalent to the harmonic mean of the individual conductivities, K_i, weighted proportional to thickness.

Using a similar approach, this concept is developed for fractures varying in aperture along the flow path. For aperture variation of discrete magnitude and length the effective aperture is given by

$$b_{eff}^3 = \frac{L}{\sum_{i=1}^{n} \frac{L_i}{b_i^3}} \qquad (9)$$

In the fractures represented herein, flow is in the x direction, parallel to the aperture variation. For the triangular and sinusoidal boundaries, the variation is continuous. Thus, the magnitude of the aperture is a continuous distribution. The harmonic mean for a continuous distribution would be given by

$$H.M. = \left[\int_{-\infty}^{+\infty} \frac{1}{x} f(x) dx \right]^{-1} \qquad (10)$$

where $f(x)$ is the distribution function of x. For the triangular boundaries this becomes

$$b_{eff}^3 = \frac{4\epsilon}{\left[\frac{1}{b_{min}}\right]^2 - \left[\frac{1}{b_{min} + 2\epsilon}\right]^2} \qquad (11)$$

As seen in Figures 2 and 3 the flow calculated numerically agrees with that predicted by the cubic law using the harmonic mean of the aperture when the value of 'a' is large compared to ϵ/b_{min}.

FRACTURE ROUGHNESS

In contrast to nonparallel plate behavior, roughness is associated with a high frequency of aperture variation. In this case the spatial variance of aperture is such that parabolic flow does not become established completely. Eddy flows are established near the minimum and maximum apertures.

Smaller values of 'a' increase the eddy flow, which increases the energy required to overcome the viscous forces associated with this circulation.

If the magnitude of the variations is large, compared to the total aperture size, this eddy influence can dominate flow. It is at this point where it is questionable whether the inertial terms of the Navier–Stokes equation can be neglected. When the magnitude of the aperture variations is small, compared to the total aperture, the eddy influences are minimized. Parabolic flow is essentially established in the fracture, except for deviation at the boundary.

Figures 2 through 3 shows that as 'a' becomes smaller, flow volume deviates from that predicted by use of the harmonic mean. In fact, flows less than those expected using b_{min} in the cubic law are observed. It would be expected that this deviation is even greater when inertial forces are not neglected. Restraint must be used in relying on the numerical results as 'a' becomes smaller than ϵ/b_{min}. At this point the distortion of the elements becomes severe for triangular and sinusoidal boundaries, decreasing the accuracy of calculations.

These figures also show a trend which makes it possible to quantify geometric conditions, under which parabolic flow can be assumed. For linear and sinusoidal distributions, actual flow values are 93–99% of those predicted by the harmonic mean when $a = 10 \times \epsilon/b_{min}$. As 'a' is increased to $\epsilon/b_{min} \times 100$, flow is as predicted by the harmonic mean. This same trend was shown for other fracture wall geometries examined. Below these limits it is necessary to consider aperture variations as roughness and include a friction factor.

CONCLUSIONS

Through numerical simulations it has been possible to better characterize flow in fractures when the aperture varies in the direction of flow. Two distinct types of flow were discussed. First, the aperture deviations can exhibit characteristics associated with nonparallel plates, when the rate of change in aperture is sufficiently small and parabolic velocity profiles are established. Second, when the rate of aperture change is large enough, the aperture variations show characteristics associated with roughness, causing eddies to form. It has been possible to better identify the aspects of fracture wall geometry that control these regions of flow. These aspects are:

Flow governed by nonparallel plate behavior:

1. Flow is predicted by the cubic law, using the harmonic mean of the aperture cubed.

2. This type of flow is essentially present when the ratio of the length of aperture change to the magnitude of aperture change is greater than 10. It is completely established when this ratio is 100 or larger.

REFERENCES

1. Lamb, H. (1946). Hydrodynamics, Dover, New York.

2. Langlois, W.E. (1964). Slow Viscous Flow, MacMillan, New York.

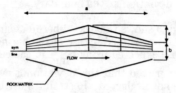

Figure 1. Fracture Geometry and Discretization.

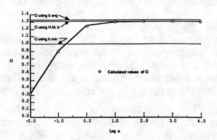

Figure 2. Volumetric Flow Rate (Q) as a Function of the Logarithm of Fracture Period (Log a) for $\varepsilon/b_{min} = 0.1$

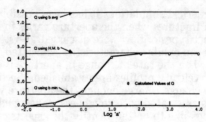

Figure 3. Volumetric Flow Rate (Q) as a Function of the Logarithm of Fracture Period (Log a) for $\varepsilon/b_{min} = 1$.

Finite Element Modeling of the Rurscholle Multi-Aquifer Groundwater System
H-W. Dorgarten
Institute for Hydraulic Engineering and Water Resources Development, Aachen University of Technology, Federal Republic of Germany

INTRODUCTION

With an extension of 5000 km² the West German region between Aachen and Cologne is one of the largest reservoirs of lignite in Europe (Fig. 1). The lignite that lies several hundred metres below ground surface has been gained by open-pit mining for more than 30 years. Today's planning reaches up to the year 2100 with pits of 600 m depth. As mining technology requires dry working conditions, the groundwater level has to be lowered to the bottom of the pits by pumpage rates of more than 1000 million m³ per year (Eckschlag and Dorgarten, 1987).

It is obvious that the groundwater system of a great area is strongly affected by the resulting overdraft. The drawdown raises e.g. the following problems:

- restrictions for the surrounding waterworks,
- land-subsidence,
- ecological changes in rivers, brooks and wetlands,
- dryout of springs and brooks.

Planning of new pits requires the prediction of the future groundwater situation and investigations on possible countermeasures in order to protect groundwater reservoirs. A complex finite element multi-aquifer model was developed for this purpose. The project area, called 'Rurscholle', is one of six major groundwater systems in the lignite region, which are almost disconnected from each other by a complex system of dislocations. The geological formation is vertically structured in six aquifers separated by layers of clay and lignite.

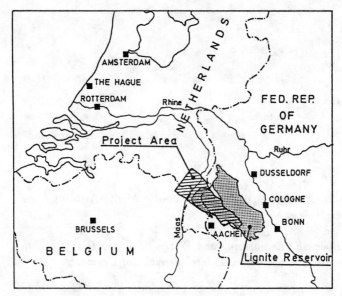

Fig. 1: Location of the Rurscholle project area

NUMERICAL MODEL

The application of mathematical modeling for transient phenomena in multi-aquifer systems leads to problems that involve four independent variables. Although three-dimensional models of groundwater flow systems are available, when considering economic aspects, quasi-three-dimensional models are generally preferred. Their advantages are considerable reductions in computer-time as well as in storage requirements, when compared with fully three-dimensional models (Herrera, 1983).

Governing Equations

In the quasi-three-dimensional model the groundwater system is composed of aquifers and aquitards. Flow is assumed to be horizontal in the aquifers and vertical in the aquitards. Under these assumptions groundwater flow is governed by the two-dimensional equation

$$S(h) \frac{\partial h}{\partial t} - \frac{\partial}{\partial x_i} \left[T_{ij}(h) \frac{\partial h}{\partial x_j} \right] + q_L + q_S = 0 \tag{1}$$

for the aquifers and the one-dimensional equation

$$S'_s \frac{\partial h'}{\partial t} - \frac{\partial}{\partial z} \left[K'_z \frac{\partial h'}{\partial z} \right] = 0 \tag{2}$$

for the aquitards. Coupling of eqs. (1) and (2) is given by the leakage-flux

$$q_L = \left[K_z' \frac{\partial h'}{\partial z}\right]_{bottom} - \left[K_z' \frac{\partial h'}{\partial z}\right]_{top} \qquad (3)$$

across the bottom and top aquifer-aquitard interfaces and by the identity of the hydraulic heads

$$h_{aquifer} = h'_{bottom} = h'_{top} \qquad (4)$$

at the interfaces (e.g. Chorley and Frind, 1978; Sartori and Peverieri, 1983).

Numerical Solution

The finite element solution of eqs. (1) and (2) is formulated by employing the well-known Galerkin method. The aquifers are represented by two-dimensional triangular elements, while the aquitards are discretized using one-dimensional prismatic elements (Chorley and Frind, 1978). For simplicity linear shape functions were chosen for both element types.

The strong coupling of the aquifers and aquitards didn't allow an iterative solution. Thus, in opposition to other multi-aquifer models, the matrix system was solved simultaneously for all aquifers and aquitards. The finite element equations of both element types are assembled in a global matrix system

$$\left[S_{mn} + S'_{mn}\right]\left\{\frac{\partial h_n}{\partial t}\right\} + \left[D_{mn} + D'_{mn}\right]\{h_n\} = \{R_m\} \,. \qquad (5)$$

The coefficient matrices are

$$S_{mn} = \sum_e \int_A \phi_m \, S \, \phi_n \, dA \qquad (5a)$$

$$D_{mn} = \sum_e \int_A \frac{\partial \phi_m}{\partial x_i} T_{ij} \frac{\partial \phi_n}{\partial x_j} \, dA \qquad (5b)$$

for the aquifers and

$$S'_{mn} = \sum_{e'} A' \int_{L'} \phi_m \, S'_s \, \phi_n \, dL' \qquad (5c)$$

$$D'_{mn} = \sum_{e'} A' \int_{L'} \frac{\partial \phi_m}{\partial z} K'_z \frac{\partial \phi_n}{\partial z} \, dL' \qquad (5d)$$

for the aquitards (Rouvé and Dorgarten, 1987).

Program Structure

The complexity of the Rurscholle groundwater system resulted in transient simulations of about 160 time-steps with a finite element discretization of about 28,000 elements and more than 10,000 nodes (Fig. 2). Obviously, the calculations required not only a powerful super-computer (CRAY X-MP), but also an optimized program structure.

Basing on the experience of vectorizing two-dimensional transmissivity models (Pelka and Peters, 1986), a highly vectorized code of the multi-aquifer model was developed. The computation of the element matrices as the most time-consuming part of the program was completely vectorized and, additionally, pre-calculated as far as possible.

For the solution of the linear equations system as the second time-consuming part, both direct and iterative solvers were tested. The iterative solvers showed bad convergence or even none at all, probably due to extreme differences in element volumes. Thus, a direct-solving Cholesky-algorithm was vectorized. As the matrix size didn't allow an in-core solution, the equation system was solved out-of-core making use of the CRAY high-speed buffer.

Simulation of Excavations

For the simulation of open-pit mining the development of algorithms is required to describe the progressing excavation and refilling of the pits. The obvious idea to eliminate the respective nodes and elements proves to be unqualified, as this would disturb vectorization. Therefore the pits are treated as special material types with time-dependent material parameters. After excavation, both aquifer and aquitard elements are represented by extremely low hydraulic conductivities to attain flux values near to zero. After refilling, the hydraulic conductivities of both element types are changed to a uniform value, representing the dump material. Thus, although there is no longer a physical differenciation between aquifers and aquitards, the layers are still simulated separately.

PROJECT RESULTS

The simulation period of the Rurscholle model was about 125 years, using time-steps between 3 months and 2.5 years. The basic calculations described future groundwater management according to today's planning and technology. Results show a wide-ranging decline of the phreatic groundwater table by more than one metre, that may cause several ecological damages. Simultaneously, depressions in the lower aquifers are about two hundred metres near the pits and still several metres far beyond the German border in the Netherlands.

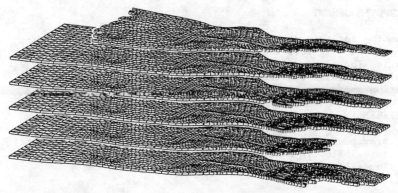

Fig. 2: Schematic view of the Rurscholle multi-aquifer discretization

Based on these results, several countermeasures were simulated with the model, e.g.:

1. Reducing the pits to smaller volumes could lead to an earlier refilling of the groundwater reservoir. Furthermore, nearby rivers would be considerably saved.

2. By building injection walls down to the bottom of the first aquifer ecologically valuable areas could be protected from groundwater decline.

3. By artificial groundwater recharge the water table could be raised to the undisturbed level. As parts of the infiltrated water would flow back to the pits, this measure, however, would cause higher pumpage rates.

4. When closing waterworks and supplying the required quantity of water from the pits, the groundwater reservoir could be saved in certain areas.

The simulation results demonstrate the effectivity and the consequences of the respective measures. Furthermore, they are the basis for future developments and combinations in order to attain an optimal management of the intensely used groundwater reservoir of the Rurscholle area.

CONCLUSIONS

In order to simulate the impact of open-pit mining, a complex finite element multi-aquifer groundwater model was developed. Special algorithms were included to describe the progressing excavation and refilling of the pits. As the groundwater drawdown causes both wide-ranging depressions and steep gradients, extended project-areas and relatively small elements are required, resulting in extremely large meshes. Thus, simulations can only be performed by highly vectorized models on today's supercomputers.

ACKNOWLEDGEMENTS

Funding for this study was provided by the Ministry for Environment and Agriculture (MURL) of the state of Northrhine-Westfalia. The author wishes to thank the authorities of Northrhine-Westfalia and the Netherlands for their continuous cooperation and support.

REFERENCES

1. Bredehoeft J.D. and Pinder G.F. (1970), Digital Analysis of Aereal Flow in Multiaquifer Groundwater Systems: A Quasi-Three-Dimensional Model, Water Res. Res., Vol. 6, pp. 883-888.

2. Chorley D.W. and Frind E.O. (1978), An Iterative Quasi-Three-Dimensional Finite Element Model for Heterogeneous Multiaquifer Systems, Water Res. Res., Vol. 14, pp. 943-952.

3. CRAY Research, Inc. (1984), CRAY-1 and CRAY X-MP Computer Systems, Fortran Reference Manual SR-0009 and Library Reference Manual SR-0014.

4. Eckschlag N. and Dorgarten H.-W. (1987), Das Grundwassermodell Rurscholle, Mitt. Inst. für Wasserbau und Wasserwirtschaft, RWTH Aachen, Vol. 66, pp. 23-53.

5. Herrera I., Hennart J.P. and Yates R. (1983), A Critical Discussion of Numerical Models for Multiaquifer Systems, in: Flow through Porous Media, Recent Developments (Ed. Pinder G.F.), CML Publications, Southampton, U.K., pp. 51-55.

6. Pelka W. and Peters A. (1986), Implementation of Finite Element Groundwater Models on Vector and Parallel Computers, in: Finite Elements in Water Resources (Ed. Sá da Costa, A. et al.), pp. 301-312, Proc. of the 6th Int. Conf., Lisboa, Portugal, 1986, Springer-Verlag, Berlin-Heidelberg-New York-Tokyo and Computational Mechanics Publications, Southampton.

7. Rouvé G. and Dorgarten H.-W. (1987), Grundwassermodell Rurscholle, Schlußbericht, final project report (unpublished).

8. Sartori L. and Peverieri G. (1983), A Frontal Method Based Solution of the Quasi-Three-Dimensional Finite Element Model for Interconnected Aquifer Systems, Int. Journal for Numerical Methods in Fluids, Vol. 3, pp. 445-479.

On the Computation of Flow Through a Composite Porous Domain

J.P. du Plessis
Department of Mechanical Engineering, University of Pretoria, 0002 Pretoria, South Africa

ABSTRACT

Recent developments in the continuum theory of flow through porous media permit the implementation of one general equation, which governs the flow in all parts of a composite medium. Only two parameters, the porosity and a microscopic length scale, are needed to describe the hydrodynamic permeability locally within each control volume of the flow domain. This unified approach can be used to predict flow characteristics through a composite domain consisting of sections of porous media with different permeabilities. Some aspects of the computational application of this method are discussed with reference to a finite volume numerical procedure. It is shown that an internal jump in permeability, which defines the interface between different porous media, can be treated without the explicit matching of flow properties on either side of the interface. Special attention is given to the large source term introduced by the microscopic shear terms.

NOMENCLATURE

a	finite difference coefficient
b	finite difference source term not including microscopic shear
d	microscopic characteristic length
F	microscopic shear factor
g	gravitational constant
n	porosity (void fraction)
p	phase average pressure
S	finite difference source term relating to microscopic shear
T	tortuosity
u	component of $\underline{v}$ in X-direction
v	component of $\underline{v}$ in Y-direction
$\underline{v}$	phase average velocity
x	coordinate
y	coordinate

μ   fluid dynamic viscosity
ρ   fluid mass density
φ   computational variable

Subscripts
C   constant part
nb neighbouring nodes
x   component in X-direction
P   present node/ variable part of S
y   component in Y-direction

INTRODUCTION

The modelling of flow through porous media has for decades been done by utilizing the well-known Darcy equation[1] or some empirically generalized form of it. Such equations normally have the limitation of being unable to provide the correct limits for either or both free flow through void regions and no flow in a solid. In case of an internal boundary, defining the interface between media of different hydrodynamic permeability, a matching procedure has therefore to be followed to bring flow properties on either side of the interface into hydrodynamic equilibrium. If void space and a solid are considered porous media in the general sense, most engineering problems in fluid flow entail a composition of different porous media each of which is governed by different fluid transport equations.

The computation of flow through and in the vicinity of porous materials has thus been hampered by the lack of a unified set of equations, applicable over an entire computational domain. Such equations, theoretically based on volume-averaging concepts[1], were recently introduced by du Plessis and Masliyah[2]. Only two parameters, namely the porosity (void fraction) and a characteristic length of the porous microstructure, are needed to describe the influence of a porous medium on a traversing fluid. A composite domain can therefore be described by two functions defining these two parameters over the domain. In case of the finite decomposition of such a domain, as is done for a discrete numerical analysis, two matrices will be sufficient to fix the permeability in each control volume. In this paper some of the computational aspects of such numerical modelling will be discussed. The emphasis will be on the modifications needed when this unified approach is used together with the finite volume method SIMPLEC[3,4].

MATHEMATICAL MODEL

Following Bachmat and Bear[1], equations governing the momentum transport of flow through porous media can be derived by the volumetric averaging of free flow equations in conjunction with the permeability characteristics of the solid matrix. In this paper the equations are given in two-dimensional form although they can be similarly expressed in three dimensions.

The continuity equation, expressing the conservation of fluid mass, can be written in terms of the components of the phase average velocity $\underline{v}$ as:

$$\frac{\partial u}{\partial x} + \frac{\partial v}{\partial y} = 0. \tag{1}$$

Similarly the momentum transport equation may be represented component-wise by

$$\rho\frac{\partial}{\partial x}(uu/n) + \rho\frac{\partial}{\partial y}(vu/n) - \mu\frac{\partial^2 u}{\partial x^2} - \mu\frac{\partial^2 u}{\partial y^2} + \frac{\partial p}{\partial x} + \rho g_x + \mu F u = 0, \tag{2}$$

and

$$\rho\frac{\partial}{\partial x}(uv/n) + \rho\frac{\partial}{\partial y}(vv/n) - \mu\frac{\partial^2 v}{\partial x^2} - \mu\frac{\partial^2 v}{\partial y^2} + \frac{\partial p}{\partial y} + \rho g_y + \mu F v = 0. \tag{3}$$

The stress influence of the solid material on the traversing fluid is controlled by the factor F. Du Plessis and Masliyah[2] have recently introduced a simple geometrical model, which takes into account the three-dimensionality of the flow within the pores, although the average movement of fluid is locally only one-dimensional. According to this model, which eliminates the need for empirically derived coefficients, it follows for consolidated isotropic porous media that:

$$F = \frac{42.69(1-T)}{n^2 d^2 T^2}\sqrt{1 + \frac{0.117T}{1-T}\cdot\frac{\rho n d}{\mu}\cdot\sqrt{u^2 + v^2}}, \tag{4}$$

The porosity and the tortuosity of the solid matrix are (still for the same model) related by the equation:

$$4nT^3 = (3T - 1)^2. \tag{5}$$

Equations (1) through (5) present a set of strongly coupled equations governing the flow of single phase flow through a composite porous domain. Equation (5) uniquely determines the porosity in terms of the tortuosity and vice versa, although in the latter case an iterative measure is needed. The momentum equations (2) and (3) are strongly coupled through the pressure variable, the convective terms and the factor F, while the latter is a function of the total velocity magnitude.

FINITE DIFFERENCE PROCEDURE

The numerical solution of the set of equations (1) through (3) may, amongst others, be effected by means of a finite volume procedure such as SIMPLE[3] or the improvement to it SIMPLEC[4]. These procedures require the momentum transport equations (2) and (3) to be written in the following discretized form:

$$\phi_P = \sum_{nb} a_{nb}\phi_{nb} + b + S_C + \phi_P S_P. \tag{6}$$

with S_P non-positive. In the present work the terms in S denotes the shear stress on the fluid exerted locally within the control volume by the solid matrix. The other source terms like gravity are grouped together under the symbol b.

Spatially variable permeability

A numerical porosity matrix may be applied which fixes the locally mean porosity for each finite difference control volume. Similarly, a matrix supplying the locally mean characteristic length of the microstructure within each control volume may be defined. At each control volume these two values determine the constant part of the hydrodynamic permeability according to the factor before the outer root in equation (4). The part of F which depends on the velocity has to be updated after each numerical adjustment of the velocity field.

Initialization

At the outset of the numerical computation an initial velocity distribution, which preferably satisfies local and global continuity, has to be supplied. This could be effected most simply by a velocity field which conforms to the external velocity boundary conditions, regardless of non-uniformities in the permeability in the interior of the computational domain. In many cases a totally uniform velocity field will do.

As soon as the permeability is changed by application of the porosity and characteristic length matrices to the momentum transport equations, the F-terms may produce unphysically high shear rates, which in turn lead to overpowering large corrective measures for the velocities in locations of low permeability. This may destabilize the numerical procedure considerably. A very effective remedy is the proper linearization of the source term $S_C + S_P \phi_P$, which is the cause of the trouble. By splitting the effect equally over the two components as follows

$$- S_C = \frac{1}{2}\mu F \phi_P^o \tag{7}$$

and

$$- S_P = \frac{1}{2}\mu F. \tag{8}$$

the numerical procedure is in no way upset by too large velocity corrections. The relaxation provided by this measure was found very effective in all cases studied. The internal composition of the porous domain may therefore be altered without the need to adjust the initial velocity field accordingly.

Solid islands
Within a solid the porosity is zero and in their present form
the momentum transport equations entail division by zero. The
easiest way around this problem is to enforce the no-flow condition in a solid control volume by employment of a zero value
for S_C and a very large negative value for S_P according to the
suggestion by Patankar[3]. In this manner the numerical values
of n and d become irrelevant and a zero velocity component is
enforced within the control volume comprising solid material
only.

Free flow
In regions of free flow where there is no solid material present
the porosity is set to unity. This will correctly effect a zero
value for F and therefore yield the normal flow conditions without impedance by a porous matrix.

Jump in permeability
The treatment of internal jumps in permeability between areas
comprising different media is usually effected automatically
if careful interpolation is done during the finite difference
discretization of the partial differential equations. It is
also true that, physically, any jump could occur at the worst
piece-wise linear over length intervals of the same magnitude
as the characteristic lengths on either side of the interface.
If the control volumes at such an interface are tailored to be
less than or at most equal to these lengths, the interpolation
at the discretization stage should be sufficient towards trust-
worthy numerical results.

If the conditions mentioned above are adhered to, arbitrary
isotropic composite domains can be specified by the two matrices
for porosity and characteristic length.

DISCUSSION

The procedure outlined in this paper has already been applied in
practice and the preliminary results are very promising. Flow
past a porous obstruction (n = 0.976 and d = 0.001) in a straight
tube was modelled[5] to investigate the effect of the so-called
inertial terms of F in the non-Darcy region of intermediate
Reynolds number flows. Also, preliminary results for flow past
a porous sphere yield flow patterns almost identical to that
given by experiemential flow visualization. The simplistic method mentioned here thus seems capable of capturing the essential
features of flow through composite porous media.

The advantages of the present approach are the relative
ease by which consisting codes for free flow can be altered to
handle composite porous flows, the elimination of matching problems at internal boundaries or interfaces and the consistency
of the equations at extremal values of the hydrodynamic permeability.

In this paper the two-dimensional application of the model is discussed. It is, however, also directly applicable to three-dimensional cases of porous flows. Since the equations are a generalization of the full Navier-Stokes equation, recirculation of the the locally average flow present no problem. The straight-forward introduction of time-dependent terms into the momentum transport equations will also render this work applicable to transient flows.

The numerical application of the unified model described in this paper forms a powerful means to illustrate the difference between the normal macroscopic inertial terms and the microscopic inertial terms which are due to the fact that the flow within the pore sections may not be fully developed. Failure to observe the inherent difference between these two inertial effects has led to much speculation and confusion in the past.

REFERENCES

1. Bachmat Y. and Bear J. (1986), Macroscopic Modelling of Transport Phenomena in Porous Media 1: The Continuum Approach, Transport in Porous Media, Vol.1, pp. 213-240.

2. Du Plessis J.P. and Masliyah J.H. (1988), Mathematical Modelling of FLow Through Isotropic Consolidated Porous Media, submitted.

3. Patankar S.V. (1980), Numerical Heat Transfer and Fluid Flow. Hemisphere Publishers, Washington, D.C.

4. Van Doormaal J.P. and Raithby G.D. (1984), Enhancements of the SIMPLE Method for Predicting Incompressible Flows, Numerical Heat Transfer, Vol.7, pp. 147-163.

5. Du Plessis J.P. and Masliyah J.H. (1987), Flow in a Tube with a Porous Obstruction, Proceedings of the 30th Heat Transfer and Fluid Mechanics Institute, Sacramento, CA, pp. 89-96.

Two Perturbation Boundary Element Codes for Steady Groundwater Flow in Heterogeneous Aquifers

O.E. Lafe and O. Owoputi
Hydraulics Research Unit, University of Lagos, Lagos, Nigeria
A.H-D. Cheng
Department of Civil Engineering, University of Delaware, Newark, Delaware, USA

Introduction

The boundary element method (BEM) has received increased attention as a method capable of solving problems containing material heterogeneity. See for example Cheng[2] for a review. The search for the efficient codes continues. Lafe and Cheng[4] proposed a perturbation-based BEM for steady state groundwater flow in heterogeneous aquifers. The new technique expands the potential into a perturbation series and solves the resultant Laplace and Poisson equations using the existing BEM codes. The method is highly efficient for slow to moderately varying hydraulic conductivities. For problems with hydraulic conductivity varying over several orders of magnitude within the domain, however, this technique failed to provide a converged solution.

This paper presents two BEM codes that seek to circumvent the convergence problems in the original formulation. The first method adheres to the earlier formulation[4], but now incorporates a zoning concept. This is to permit the use of slow varying functions for the hydraulic conductivity within each zone while the variability throughout the entire flow field is large. In this approach the numerical process involves domain integrations. As stated in the original paper such integrations do not require representation of the interior unknowns as in the finite element method. Extensive numerical experiments are performed to show the effect of zoning on the convergence of the asymptotic series. Exact solutions to the perturbed equations are also presented for comparison.

In the second method the fundamental solution is perturbed into an asymptotic series. The solution process neither requires domain integration nor any iteration. The main difficulty with this approach is that the Green's function is difficult to derive for an arbitrarily prescribed heterogeneity.

In this paper we propose combining zoning with the use of simple hydraulic conductivity fields for which the fundamental solution can be more readily constructed. The result is an efficient BEM code for aquifers exhibiting complex heterogeneity. Other noteworthy attributes of the developed codes is the ease of data preparation and in cases where area integrations are required the area elements are automatically generated from the boundary elements.

Governing Equation

We are concerned about the Darcy flow of an incompressible fluid through a rigid, fully saturated, but heterogeneous porous formation. The governing equation is:

$$\nabla \cdot (K \nabla \Phi) = 0 \qquad (1)$$

in which Φ = the piezometric head; and $K = K(x, y, z)$ = the hydraulic conductivity. We assume that the boundary Γ consists of two parts, Γ_1 and Γ_2, with the potential and flux type boundary conditions prescribed: $\Phi = H$ on Γ_1 and $-K \partial \Phi / \partial n = q$ on Γ_2, where n is the outward normal to the boundary. For the zoned problem there is an interzonal boundary, Γ_3, on which the compatibility relations for flux and piezometric head must be enforced.

Boundary Integral Formulations

The perturbed potential approach

The piezometric head is expanded into an asymptotic series $\Phi = \sum_{i=0}^{\infty} \Phi_i$ and it is demanded that $\nabla^2 \Phi_0 = 0$. When these are used in equation (1) the following integral equations are obtained:

$$\alpha \Phi_0 = \int_{\Gamma} (\Phi_0 \frac{\partial g}{\partial n} - g \frac{\partial \Phi_0}{\partial n}) \, d\Gamma$$

$$\alpha \Phi_i = \int_{\Gamma_2} \Phi_i \frac{\partial g}{\partial n} \, d\Gamma - \int_{\Gamma_1} g \frac{\partial \Phi_i}{\partial n} \, d\Gamma - \int_{\Omega} g \nabla \ln K \cdot \nabla \Phi_{i-1} \, d\Omega; \quad i = 1, 2, 3 \ldots \quad (2)$$

where α = a coefficient obtained from the Cauchy principal value integration of the kernel singularity; $d\Gamma$ denotes the boundary integral; $d\Omega$ the domain integral; and g = free space Green's function of the Laplace equation, which is

$$g = \begin{cases} \ln r / 2\pi & \text{for two-dimensional problems} \\ -1/4\pi r & \text{for three-dimensional problems} \end{cases} \qquad (3)$$

where r = radial distance from the base point to a field point.

The detailed derivation of the above and the general solution procedure of the integral equations (2) have been given in Lafe and Cheng[4], therefore will not be elaborated herein. The zoning approach is identical to that of Lafe[3] in which the final global matrix has a blocked structure, hence allowing the use of an efficient blocked equation solver.

The perturbed Green's function approach

Following Cheng[1] we convert the governing equation (1) into the boundary integral equation:

$$\alpha \Phi = \int_{\Gamma} (K \Phi \frac{\partial G}{\partial n} - K G \frac{\partial \Phi}{\partial n}) \, d\Gamma \qquad (4)$$

where $G = G(\bar{x}, \bar{x}')$ is the fundamental solution defined by:

$$\nabla \cdot (K \nabla G) = \delta(\bar{x}, \bar{x}') \qquad (5)$$

in which δ is the Dirac delta function introduced at the point $\bar{x}'$ and felt at point $\bar{x}$. We assume G can be expanded into a perturbation series:

$$G = G_0 + G_1 + G_2 + G_3 + \cdots \quad (6)$$

and require that G_0 satisfy:

$$K\nabla^2 G_0 = \delta(\bar{x}, \bar{x}') \quad (7)$$

That is:

$$G_0 = \begin{cases} \ln r / 2\pi K(\bar{x}') & \text{for two-dimensional problems} \\ -1/4\pi K(\bar{x}')r & \text{for three-dimensional problems} \end{cases} \quad (8)$$

Using equations (6) and (8) in (5) and equating terms of the same order in the perturbation quantities we obtain:

$$\nabla^2 G_i = -\nabla(\ln K) \cdot \nabla G_{i-1}; \quad i = 1, 2, 3 \ldots \quad (9)$$

For a two-dimensional problem for which the permeability field is log-linear we write: $\ln K = ax + by + c$ where a, b, c are constants and x, y are the Cartesian coordinates. It can be shown that:

$$\begin{aligned} G_1 &= \frac{r}{4\pi K(x_o, y_o)}[-\theta(a\sin\theta - b\cos\theta) - 2(\ln r - 1)(a\cos\theta + b\sin\theta)] \\ G_2 &= \frac{r^2}{16\pi K(x_o, y_o)}[\theta(a^2 - b^2)\sin 2\theta - 2ab\theta \cos 2\theta - (a^2 + b^2)] \end{aligned} \quad (10)$$

in which $\theta = \tan^{-1}\frac{(y-y_o)}{(x-x_o)}$. Higher order terms also have been derived.

Numerical Experiments

The model problem is the flow in a 2×1 rectangular region with boundary condition shown in Figure 1. The exact solution is:

$$\Phi = 2 - \frac{\int_1^x \frac{dz}{K}}{\int_1^3 \frac{dz}{K}} \quad (11)$$

Perturbed potential results

Two types of permeability functions have been tested. The first is in the power form:

$$K = (1+x)^\lambda \quad (12)$$

The second has the exponential form:

$$K = e^{\lambda x} \quad (13)$$

Up to 10 terms in the perturbed series have been used ($i = 0$ to 9 in equation (2)). In general we found that convergence (where it exists) is normally evident by the fourth term. Adding more terms does not help for the diverged results.

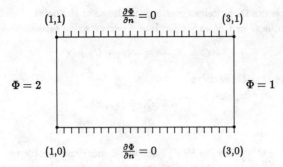

Figure 1: Problem Definition

The domain is subdivided into 1 to 4 zones for the observation of improvement due to zoning. The experiments involved λ values ranging between 0 and 10. This is indeed the most stringent test since the variation of permeability in the extreme case is of the order 10^8. During the experiment we found that the solutions are quite sensitive to the manner in which the hydraulic conductivity function is interpolated for the area integrations. We present two algorithms in this paper. In the first one the exact values of K are obtained for the area integration. The second one involves the use of a linearly fitted K-field within the subzones. Integrations for both algorithms make use of the Liggett[5] cubature formula.

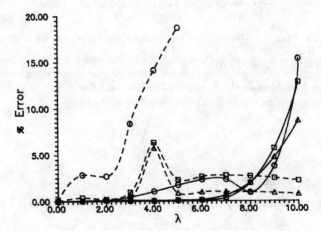

Figure 2: Error analysis for power K field. (Legends: — algorithm 1, - - - algorithm 2, ○ 1 zone, □ 2 zones, △ 4 zones)

The error plots for the power and exponential fields are respectively shown in Figures 2 and 3 for the various λ values. Error calculations are based on the calculated values of Φ at $x = 2$ which is at the center of the region. The one zone results are generally poor. The solutions diverge for $\lambda > 2$. By using up to

4 zones we are able to carry the calculation to larger values of λ, but all of them eventually diverge. We also observe that convergence is generally more rapid using the second (dashed lines) than the first (solid lines) algorithm. Further details of the test are reported by Owoputi[6].

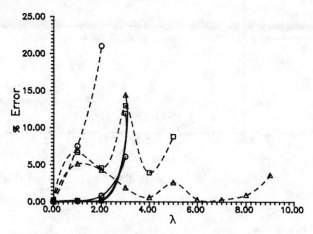

Figure 3: Error analysis for exponential K field. (Legends: — algorithm 1, - - - algorithm 2, ○ 1 zone, □ 2 zones, △ 4 zones)

Perturbed Green's function results

Only the 1-zone, exponential K-field case has been simulated using the perturbed Green's function method since the improvement by zoning has already been highlighted above. Three terms in the perturbations have been used in obtaining the results summarized on Figure 4. The results for $\lambda < 0.8$ are satisfactory but experiences convergence problems as λ increases. We expect the improvement of solution by adding more terms in the perturbed Green's function (6). As demonstrated in the previous example, zoning should also improve the solution.

Conclusions

Two boundary element methods have been presented in this paper for the simulation of flows in heterogeneous formations. The combination of a zoning approach with simple interpolation functions for K allows the simulation of problems that exhibit large variability in hydraulic conductivity. The perturbed potential technique presents less algebraic difficulties than the perturbed Green's function approach since the successful use of the latter depends on the ability to obtain expressions for the perturbed Green's function. The second method however has the computational advantage of not requiring domain integrations.

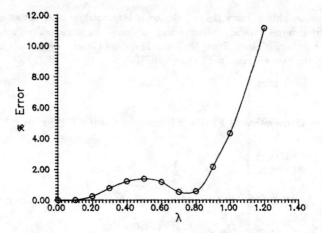

Figure 4: Error analysis for exponential K field: Perturbed Green's function approach.

References

[1] Cheng, A.H-D., "Darcy's flow with variable permeability—a boundary integral solution," Water Resour. Res., Vol. 20, No. 7, pp. 980-984 (1984).

[2] Cheng, A.H-D., "Heterogeneities in flows through porous media by the boundary element method", Chapter 6 in *Topics in Boundary Element Research, Vol. 4: Applications in Geomechanics*, ed. C.A. Brebbia, Springer-Verlag, pp. 129–144 (1987).

[3] Lafe, O. E., "The simulation of two-dimensional confined flows in zoned porous media—A boundary integral approach," Master thesis, Cornell University (1980).

[4] Lafe, O. E. and Cheng, A.H-D., "A Perturbation boundary element code for groundwater flow in heterogeneous aquifers," Water Resour. Res., Vol. 23, No. 6, 1079-1084 (1987).

[5] Liggett, J. A., "Singular cubature over triangles," Int. J. Numer. Meth. Engng., Vol. 18, pp. 1375-1384 (1982).

[6] Owoputi, L.O., "A perturbation boundary element code for groundwater flow in zoned heterogeneous aquifers," Master thesis, University of Lagos (1987).

A Three-Dimensional Finite Element - Finite Difference Model for Simulating Confined and Unconfined Groundwater Flow

A.S. Mayer and C.T. Miller
Department of Environmental Sciences and Engineering, CB# 7400, 105 Rosenau Hall, University of North Carolina, Chapel Hill, NC 27599-7400, USA

INTRODUCTION

Numerical models for the simulation of groundwater flow and contaminant transport are well established. It is clear that accurate simulation of many contaminant transport problems requires a three-dimensional groundwater flow solution (Burnett and Frind[1]). Some three-dimensional groundwater flow models have been developed and tested. Application of these models often has been constrained by core storage and CPU time requirements, especially for unconfined flow problems. Babu et al.[2] and later Huyakorn et al.[3] developed an algorithm that combines finite element discretization in the horizontal direction and finite difference discretization in the vertical direction. The algorithm is known as the ALALS algorithm (Alternate subLayer And Line Sweep). This algorithm is more efficient than pure finite element algorithms, while giving accurate solution to three-dimensional, confined aquifer problems (Huyakorn et al.[3]). The objective here is to extend the ALALS algorithm to unconfined flow.

BACKGROUND

Groundwater flow through saturated porous media is described by

$$L(h) = \frac{\partial}{\partial x}\left(K_x \frac{\partial h}{\partial x}\right) + \frac{\partial}{\partial y}\left(K_y \frac{\partial h}{\partial y}\right) + \frac{\partial}{\partial z}\left(K_z \frac{\partial h}{\partial z}\right) - S_s \frac{\partial h}{\partial t} = -\Gamma \quad (1)$$

where $L(h)$ is a differential operator; h is hydraulic head (L); K_x, K_y, and K_z are the principal components of the hydraulic conductivity tensor in the x, y, and z directions, respectively (LT^{-1}); S_s is specific storage (L^{-1}); Γ describes sources or sinks (T^{-1}); and t is time (T).

The ALALS algorithm describes head with the trial function

$$\hat{h}(x,y,z,t) = \sum_{j=1}^{n_{xy}} N_j(x,y) h_j(z,t) \approx h(x,y,z,t) \qquad (2)$$

where n_{xy} equals the number of nodes in each horizontal layer, and N_j are two-dimensional basis functions in the x-y plane.

Applying the Galerkin approximation to Equation (1) in the x-y plane gives

$$\iint_{\mathcal{D}} N_i \left[L(h) + \Gamma \right] \, dx \, dy = 0 \qquad \text{for } i = 1, ..., n_{xy} \qquad (3)$$

where $\mathcal{D}$ is the x-y problem domain, which is assumed constant for each layer k. Application of Green's formula to Equation (3) reduces the order of the highest derivatives, giving

$$\iint_{\mathcal{D}} \left(K_x \frac{\partial N_i}{\partial x} \frac{\partial \hat{h}}{\partial x} + K_y \frac{\partial N_i}{\partial y} \frac{\partial \hat{h}}{\partial y} \right) dx \, dy + \iint_{\mathcal{D}} N_i S_s \frac{\partial \hat{h}}{\partial t} \, dx \, dy$$
$$= \iint_{\mathcal{D}} N_i \frac{\partial}{\partial z} \left(K_z \frac{\partial \hat{h}}{\partial z} \right) dx \, dy + \iint_{\mathcal{D}} N_i \Gamma \, dx \, dy + \int_{\mathcal{B}} N_i \left(K_n \frac{\partial \hat{h}}{\partial n} \right) d\mathcal{B} \qquad (4)$$
for $i = 1, ..., n_{xy}$

where $\mathcal{B}$ is the vertical boundary of $\mathcal{D}$, $\partial \hat{h}/\partial n$ is the outward normal derivative on $\mathcal{B}$, and K_n is the normal component of hydraulic conductivity on $\mathcal{B}$.

The ALALS algorithm applies a central finite difference approximation to the z-derivative terms in Equation (4), and a two-step, finite difference approximation to resolve the time-dependent derivative. In the first step, the z-component terms are evaluated at the previous time step l—reducing the original set of three-dimensional equations to n_z (number of layers or nodes in the z direction) subsets of equations, each of which contains n_{xy} equations, resulting in

$$[K]_k \{h\}_k^{(l+1)^*} + \frac{[S_s]_k}{\Delta t} \left(\{h\}_k^{(l+1)^*} - \{h\}_k^l \right) = \{F\}_k^{(l+1)^*} + \sum_{m=k-1}^{k+1} [K_v]_m \{h\}_m^l$$
for $k = 1, ..., n_z$

$$(5)$$

where $[K]$ represents the horizontal conductance terms, $[S_s]$ represents the specific storage terms, $\{F\}$ represents the sum of source and sink terms and the boundary flux, and $[K_v]$ represents the vertical conductance terms. In the second step, hydraulic heads associated with the z-component and specific storage terms are assumed unknown and all other terms are taken as known functions of the estimated head, giving

$$[K]_k\{h\}_k^{(l+1)^*} + \frac{[S_s]_k}{\Delta t}\left(\{h\}_k^{(l+1)} - \{h\}_k^l\right) = \{F\}_k^{(l+1)^*} + \sum_{m=k-1}^{k+1}[K_v]_m\{h\}_m^{l+1}$$

for $k = 1, ..., n_z$

(6)

An efficient algorithm results if $[S_s]$ and $[K_v]$ are lump diagonalized, and if Equation (6) is replaced by the difference between Equations (6) and (7)—giving n_{xy} tridiagonal systems of equations each with n_z unknowns (Babu et al.[2]).

UNCONFINED FLOW

The confined ALALS procedure is extended to include unconfined conditions by vertically averaging Equation (4) for each layer giving

$$\iint_D \left(T_{xk}\frac{\partial N_i}{\partial x}\frac{\partial \hat{h}}{\partial x} + T_{yk}\frac{\partial N_i}{\partial y}\frac{\partial \hat{h}}{\partial y}\right) dx\,dy + \iint_D N_i S_k \frac{\partial \hat{h}}{\partial t} dx\,dy$$
$$= \iint_D \Delta z_k N_i \frac{\partial}{\partial z}\left(K_z \frac{\partial \hat{h}}{\partial z}\right) dx\,dy + \iint_D N_i \Gamma_k' dx\,dy + \int_B N_i \left(T_{nk}\frac{\partial \hat{h}}{\partial n}\right) dB$$

for $i = 1, ..., n_{xy}$

(7)

where T_x, T_y, and T_n are components of transmissivity in the x, y, and normal directions, respectively, for layer k (L^2 T^{-1}); S is the storage coefficient; and Γ' is a vertically averaged source or sink term (LT^{-1}). Equation (7) can be developed into a two-step solution procedure similar to that shown in Equations (5) and (6) for the confined case.

The resulting equations are no longer linear for the unconfined problem, because $T_{xk} = f(h)$ and $T_{yk} = f(h)$. For the unconfined case, one unconfined node will exist at each of the n_{xy} locations in the horizontal plane. Various iterative schemes are available to solve this type of quasilinear problem, such as Picard iteration, the Newton-Raphson method, or chord-slope iteration (Huyakorn and Pinder[4]). The Newton-Raphson method and chord-slope iteration often have the advantage of more robust behavior and convergence in fewer iterations, but more operations are required per iteration compared to Picard iteration. Picard iteration was found to converge quickly for the unconfined ALALS procedure, and so alternate schemes were not used.

A second-order head predictor was included to increase the convergence rate of solution by estimating the solution prior to each new time step, using

$$h_n^{(l+1,m=0)} = h_n^l + \frac{(h_n^l - h_n^{(l-1)})^2}{h_n^{(l-1)} - h_n^{(l-2)}} \qquad \text{for } n = 1, ..., n_n \qquad (8)$$

for a constant time-step increment, where m is the iteration level and n_n is the number of nodes in the system.

In most multiple-layer, unconfined flow problems, the simulated system consists of both unconfined and confined elements. For confined elements, the coefficient matrices $[K], [S]$, and $[K_v]$ do not change between iterations—allowing coefficient matrices to be selectively updated. A selective update is performed for each unconfined element at each iteration level by summing the changes in each coefficient matrix caused by successive estimates of $\{h\}$.

Elements that become completely drained by pumping—or conversely, previously drained elements that become refilled by recharge—can cause numerical difficulties. A procedure was introduced that effectively removes the influence of the drained element and the associated nodes, but also allows for refilling. The procedure consists of imposing Dirichlet conditions on the associated drained nodes, equal to the head from the layer immediately below. If a source or a sink term exists at a drained node, the location of the source or sink is temporarily removed from the drained node and the quantity of the source or sink is allocated to the remaining layers below.

VALIDATION

Results from the numerical model described above were tested with flow problems amenable to analytical solution. The first flow problem consisted of an transient unconfined flow simulation using a single layer. The approximate analytical solution used for comparison was the Theis equation with specific yield (S_y) as the storage term, adjusted for non-constant transmissivity with the Jacob correction (Bear[5]). The input parameters are listed and simulation results are compared to the analytical solution in Figure 1—where Q is volumetric rate of pumpage ($L^3 T^{-1}$), and b is the aquifer thickness (L).

The second validation consisted of a transient, radially convergent, confined flow problem with a partially penetrating well. The simulation was performed using 29 layers with a stress located only in the bottom 10 layers. The input parameters are listed and simulation results are compared to Hantush's analytical solution (Bear[5]) in Figure 2.

EVALUATION

The USGS McDonald-Harbaugh model (McDonald and Harbaugh[6]) is a popular three-dimensional unconfined/confined flow model. This model was used to provide a benchmark for accuracy and efficiency. Two groundwater sys-

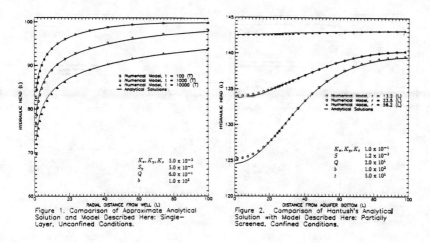

Figure 1. Comparison of Approximate Analytical Solution and Model Described Here: Single-Layer, Unconfined Conditions.

Figure 2. Comparison of Hantush's Analytical Solution with Model Described Here: Partially Screened, Confined Conditions.

tems were chosen for comparison between the McDonald-Harbaugh model and the model described here: a confined aquifer with a single stress, and an unconfined aquifer in which the applied stress resulted in the development of numerous dry nodes. The data sets for all systems had identical discretization in both space and time, while stress was applied over elements or blocks, rather than nodes. The input parameters are listed and results of the simulations are compared with an approximate analytical solution (Hantush[7]) in Figures 3 and 4. Comparison of CPU times for a variety of equivalent discretization runs showed that the model described here is generally more efficient than the McDonald-Harbaugh model for confined conditions, whereas the converse is true for unconfined conditions.

CONCLUSIONS

The three-dimensional ALALS algorithm can be used to simulate accurately unconfined groundwater flow. The model described here compares favorably with the USGS McDonald-Harbaugh model in terms of computational efficiency, accuracy, and in the ability to handle difficult conditions such as the draining and refilling of model elements.

ACKNOWLEDGEMENTS

Although the research described in this article has been supported in part by the U.S. Environmental Protection Agency through assistance agreement CR-814625 to the University of North Carolina, it has not been subjected to Agency review and therefore does not necessarily reflect the views of the Agency and no official endorsement should be inferred.

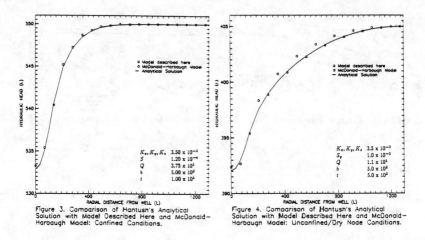

Figure 3. Comparison of Hantush's Analytical Solution with Model Described Here and McDonald–Harbaugh Model: Confined Conditions.

Figure 4. Comparison of Hantush's Analytical Solution with Model Described Here and McDonald–Harbaugh Model: Unconfined/Dry Node Conditions.

REFERENCES

1. Burnett R. D. and Frind E. O. (1987), Simulation of Contaminant Transport in Three Dimensions I. The Alternating Direction Galerkin Technique, Water Resources Research, Vol. 23, pp. 683–694.

2. Babu D.K. Pinder G.F. and Sunada D.K. (1982), A Three Dimensional Hybrid Finite Element-Finite Difference Scheme for Groundwater Simulation, in 10th IMACS World Congress, National Association for Mathematics and Computers, pp. 292–294.

3. Huyakorn P.S. Jones B.G. and Andersen P.F. (1986), Finite Element Algorithms for Simulating Three-Dimensional Groundwater Flow and Solute Transport in Multilayer Systems, Water Resources Research, Vol. 22, pp. 361–374.

4. Huyakorn P.S. and Pinder G.F. (1983). Computational Methods in Subsurface Flow, Academic Press. New York, New York.

5. Bear J. (1979). Hydraulics of Groundwater, McGraw-Hill Inc. New York, New York.

6. McDonald M.G. and Harbaugh A.W. (1984). A Modular Three-Dimensional Finite Difference Groundwater-Flow Model. U.S. Geological Survey, Scientific Publications Co. Washington, D.C.

7. Hantush M.S. (1967), Growth and Decay of Groundwater-Mounds in Response to Uniform Percolation, Water Resources Research, Vol. 1, pp. 227–234.

Galerkin Finite Element Model to Simulate the Response of Multilayer Aquifers when Subjected to Pumping Stresses

A. Pandit and J. Abi-Aoun
Department of Civil Engineering, Florida Institute of Technology, Melbourne, FL 32901, USA

INTRODUCTION

Recently several researchers Chorley and Frind[1], and Huyakorn et al[2] to name a few have used the finite element method to examine the response of multilayer aquifers to pumping stresses. In this paper a finite element model is introduced which can solve many types of complex radial-flow problems. Six example problems are selected to check the accuracy of the model by comparing the results with available analytical and numerical results and to present new solutions. An effort is made to check the effect of selected mesh sizes. Finally, the concept of effective radius is used to increase the computational efficiency of the model.

THEORY AND FINITE ELEMENT FORMULATION

The two-dimensional radial flow of ground water can be described by the partial differential equation

$$\frac{\partial}{\partial r}\left(K_r r \frac{\partial s}{\partial r}\right) + \frac{\partial}{\partial z}\left(K_z r \frac{\partial s}{\partial z}\right) = S_s r \frac{\partial s}{\partial t} \quad (1)$$

in which r and z are the radial and vertical distances, respectively, K_r and K_z are the respective radial and vertial hydraulic conductivities, s is drawdown, S_s is specific storage and t is time.

The finite element formulation for radial flow to wells is fully described by Pinder and Gray[3] and will not be provided here.

DESCRIPTION OF EXAMPLES USED FOR MODELING

The six example problems selected for modeling are:
1. Case 1: Flow to a fully penetrating well in a homogeneous and isotropic confined aquifer.
2. Case 2: Flow to a fully penetrating well in a dual permeability confined aquifer.
3. Case 3: Flow to a partially penetrating well in a homogeneous confined aquifer.
4. Case 4: Flow to a partially penetrating well in a dual permeability confined aquifer.
5. Case 5: Flow to a fully penetrating well in a three layer system which consists of two confined aquifers separated by an aquitard.
6. Case 6: Flow to a fully penetrating well with multiple screens.

The medium was assumed to be isotropic for each case.

Description of selected aquifer-aquitard parameters

The aquifer and aquitard properties selected for the six examples are shown in Table 1. Cases 1, 3 and 5d were selected to compare the model results with available analytical or numerical results. Results for Case 3 have also been obtained by Huyakorn et al.[2] using a convolution-finite element solution. The parameters selected for Case 3 are, therefore, identical to those selected by Huyakorn et al.[2]. Results for Case 5 were obtained with the following four assumptions; the aquitard is completely impermeable (Case 5a), the aquitard transmits water but has no storage capacity (Case 5b), the aquitard can transmit as well as store water but there is no drawdown in the unpumped aquifer (Case 5c), and, finally the aquitard can store and transmit water and there is drawdown in the unpumped aquifer (Case 5d). Numerical results for Case 5d have been obtained by Chorley and Frind[1] and also by Huyakorn et al.[2]. The parameters for Case 5d are identical to those used by these authors. Case 2 was solved for three different values of K_2. Case 6 was solved to study the effect of multiple screens and

TABLE 1. AQUIFER AND AQUITARD PROPERTIES

Case	Pumped Aquifer Properties				Aquitard Properties			Unpumped Aquifer Properties			Screen Length (m)
	S_{s1} (m^{-1})	K_1 (m/s)	K_2 (m/s)	Depth (m)	S_{s2} (m^{-1})	K_2 (m/s)	Depth (m)	S_{s3} (m^{-1})	K_3 (m/s)	Depth (m)	
1	1E-04	2E-05	--	20	--	--	--	--	--	--	20
2a	1E-04	2E-05	2E-04	20	--	--	--	--	--	--	20
2b	1E-04	2E-05	1E-05	20	--	--	--	--	--	--	20
2c	1E-04	2E-05	2E-06	20	--	--	--	--	--	--	20
3	1E-04	2E-05	--	20	--	--	--	--	--	--	5
4	1E-04	2E-05	1E-05	20	--	--	--	--	--	--	10
5a	1E-04	2E-05	--	5	0	0	10	1E-04	2E-05	5	5
5b	1E-04	2E-05	--	5	0	8E-08	10	1E-04	2E-05	5	5
5c	1E-04	2E-05	--	5	8E-04	8E-08	10	1E-04	2E-05	5	5
5d	1E-04	2E-05	--	5	8E-04	8E-08	10	1E-04	2E-05	5	5
6	1E-04	2E-05	--	20	--	--	--	--	--	--	5

TABLE 2. VARIOUS PARAMETERS SELECTED FOR MODELING

Case	Re (m.)	l (m.)	L (m)	L_z (m.)	# of Elements	# of Nodes	Δt (min.)	NTS	Total Time (min.)
1	109	0.5	127	1.43	392	225	0.5	2	1
							10	10	100
							100	10	1000
2a	118	0.5	254	1.43	448	255	10	10	100
2b	254	0.5	254	1.43	448	255	10	10	100
2c	118	0.5	254	1.43	448	255	10	10	100
3a	110	0.5	127	1.25	448	255	5	9	45
3b	421	1	510	1.25	512	289	148.33	10	1483.33
3c	2362	10	2540	1.25	448	255	4666.7	10	46667
4	118	1	254	1.43	392	225	10	10	100
5a	960	2	1020	0.312	512	289	30	2	60
							120	5	600
							1200	5	6000
5b	960	2	1020	0.937	512	289	Same as 5a		6000
5c	960	2	1020	0.937	512	289	Same as 5a		6000
5d	960	2	1020	1.25	512	289	Same as 5a		6000
	9600	3	12282	1.25	792	437	12000	5	60000
							120000	5	600000
6	110	0.5	127	1&3	280	165	5	9	45

the total screen length for this case 5m is the same as in Case 3.

Description of mesh and time-step sizes

The meshes and time-step sizes used to solve the example problems are shown in Table 2. The effective radius, R_e, was calculated assuming a drawdown of 0.01m to be negligible. The pattern of the finite element mesh in the radial direction is selected such that the distance between the nodes doubles after every two nodes. For example, the node locations for Case 1 are 0.5, 1, 2, 3, 5, 7, 11, 15, 23, 31, 47, 63, 94 and 127 meters. In Table 2, l represents the distance of the node closest to the well and L represents the radial extent of the aquifer where the drawdown is assumed to be zero. In each case L is slightly greater than R_e. The last three columns of

Table 2 show the time step scheme (Δt is the time step size, NTS is the number of total time steps) and the total time for which solutions were obtained.

RESULTS AND DISCUSSIONS

The results of all the cases are shown in Figures 2 and 3. The solutions for Cases 1, 3, and 5 are close to the analytical or numerical solutions. The solutions for Cases 2, 4 and 6 are also as expected.

REFERENCES

1. Chorley, D.W. and Frind, E.O. (1978), An Iterative Quasi-Three-Dimensional Finite Element Model for Heterogenous Multiaquifer Systems, Water Resources Research, Vol. 14, No. 5, pp. 943-952.
2. Huyakorn, P.S., Jones, B.G., and Andersen, P.F. (1986), Finite Element Algorithms for Simulating Three-Dimensional Groundwater Flow and Solute Transport in Multilayer Systems, Water Resources Research, Vol. 22, No. 3, pp. 361-374.
3. Pinder, G.F. and Gray, W.G., (1977), Finite Element Simulation in Surface and Subsurface Hydrology, Academic Press, Inc., Orlando and London.

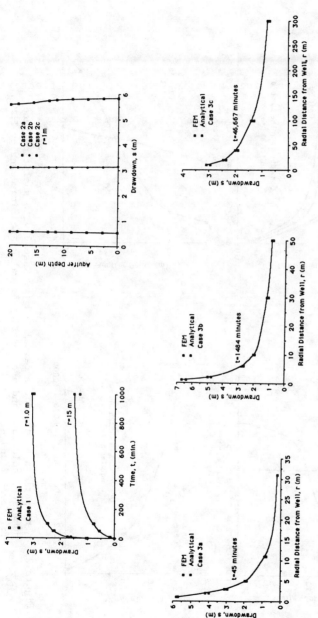

Figure 1. A comparison of drawdowns calculated for Cases 1 and 3 by FEM with analytical solutions and presentation of calculated drawdowns by FEM for Case 2.

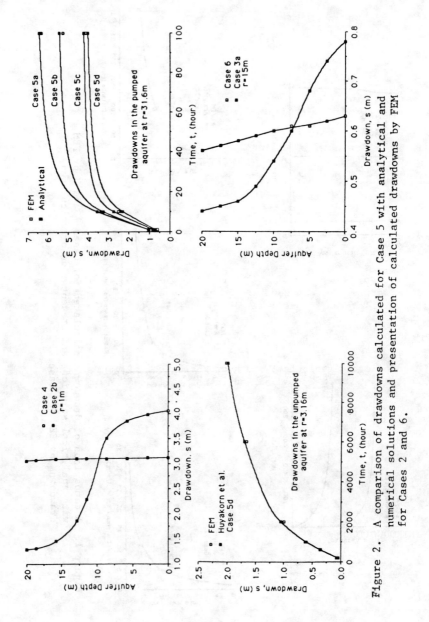

Figure 2. A comparison of drawdowns calculated for Case 5 with analytical and numerical solutions and presentation of calculated drawdowns by FEM for Cases 2 and 6.

Finite Element Based Multi Layer Model of the "Heide Trough" Groundwater Basin
B. Pelka
Institute for Hydraulic Engineering and Water Resources Development, Aachen University of Technology, Federal Republic of Germany

ABSTRACT

As a part of the groundwater management masterplan for Dithmarschen, a region in North Germany near the North Sea, a multi layer model was implemented to optimize and coordinate the development and management policies of seven public and industrial waterworks.

A complicated geological and hydrogeological situation, characterized by several glacial geological processes, had to be abstracted to become input for the numerical multi layer model. The model included large marsh regions, surface drained by distributed patterns of field ditches, which were simulated by an areal leakage boundary condition. A successful calibration of different model parameters has been achieved by performing a stepwise calibration according to the priority of influence.

The model has been used to simulate several management alternatives, and global as well as regional water balances have been evaluated.

INTRODUCTION

The Heide Trough Groundwater Basin is a hydraulic and hydrogeological unit. The geological structure is very complicated. As shown in a typical section (fig. 1) aquifers and aquitards cannot be strictly divided, since they are connected in some parts and the interaction between the aquifers may not be neglected, but also the separation of the two main aquifers is too significant for them to be described by a transmissivity model. That was the reason why a multi layer model was used.

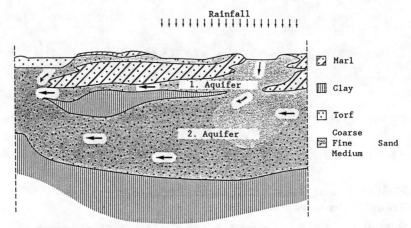

Fig. 1: Typical section of the model area

The model boundaries were enlarged until proper boundary conditions could be defined. The north and south of the model area is bordered by natural channels which are carrying the drainage from the region. The corresponding water levels are the boundary conditions. The western boundary is represented by the outer surface of the lower aquifer. The upper aquifer is defined by non-influenced piezometric heads. The eastern boundary is caused by the end of the lower aquifer. There is a groundwater divide of the upper aquifer. The south-easterly boundary again is represented by non-influenced piezometric heads.

DISCRETIZATION

For discretizing the model region the aquifers were divided into triangular elements, each with an associated permeability, an effective porosity and a storage coefficient.

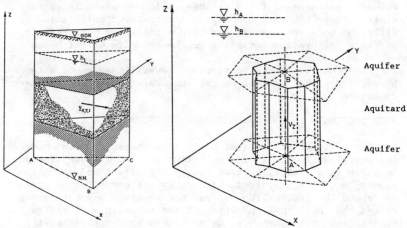

Fig. 2a: Triangular element   Fig. 2b: Prismatic element

Fig. 2a shows a triangular element of the upper aquifer. The local depth-coordinates of the upper and lower aquitards are assigned to the three nodes of the element. The approximative function is chosen in a way that the shape of the piezometric heads is a plane area in the space. For the whole model area the water level of one aquifer is represented by a faceted surface of triangular elements.

The aquitards are described by vertical prismatic elements (fig. 2b). The node in the middle of each prisma is the connection point of one or more elements of the upper and lower aquifers. The area of the prismatic element is composed by the third part of each triangular element which is connected with the centre-node. The flow direction is vertical. Fig. 3a, b and c show the spatial discretization of aquifer and aquitard.

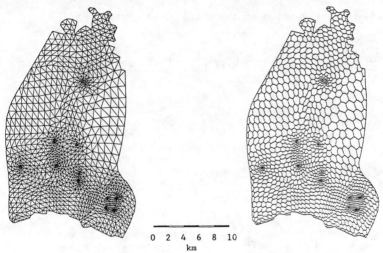

Fig. 3: Discretization of
  a) aquifer   b) aquitard

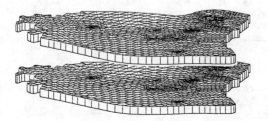

Fig 3c: Three-dimensional view of the aquifersystem

MULTI LAYER SOLUTION

The well known equation for two-dimensional horizontal groundwater flow

$$S_L \frac{\partial h_L}{\partial t} - \frac{\partial}{\partial x_i}\left(T_{ij,L} \frac{\partial h_L}{\partial x_j}\right) - Q_{Q/S,L} = 0$$

and the analogous equation for one-dimensional vertical groundwater flow

$$S_N \frac{\partial h_N}{\partial t} - K_{zz,N} \left(\frac{\partial h_N}{\partial z}\right)^2 - Q_{Q/S,N} = 0$$

leads to the equation of the multi layer system

$$\sum_{L=1}^{\max L} \left[S_L \frac{\partial h_L}{\partial t} - \frac{\partial}{\partial x_i}\left(T_{ij,L} \frac{\partial h_L}{\partial x_j}\right) - Q_{Q/S,L} \right]$$

$$+ \sum_{N=1}^{\max N} \left[S_N \frac{\partial h_N}{\partial t} - K_{zz,N} \left(\frac{\partial h_N}{\partial z}\right)^2 - Q_{Q/S,N} \right] = 0 ,$$

which was solved simultaneously. Max L and max N are the maximum numbers of the layers and non-layers.

CALIBRATION

For simulating several groundwater management policies first the model had to be calibrated. During calibration the model parameters were changed within reasonable limits until the difference between measured data and model results had been minimized.

Obviously the permeability distribution was of minor influence on the groundwater flow pattern. Large marsh regions, surface drained by field ditches and concentrated influx from one aquifer into the other were of more influence. Therefore first the groundwater recharge was calibrated. A subordinate calibration of the permeability followed. The results are shown in fig. 4. The fast and strong convergence already within the calibration of groundwater recharge was dominating.

The results proved to be consistent with regard to the geological and hydrogeological situation.

RESULTS

The model has been used to simulate several management alternatives. It should especially answer the following questions:

- In what way do the waterworks influence each other?
- Is it possible to intensify some local discharges?
- How does the aquifer basin change?
- Is it possible, that the border of fresh and saline groundwater is displacing at the western boundary?

For example there are some interesting results:

First of all there was a fixed basic situation to compare with (fig. 4a). Then the discharge quantities of the seven waterworks were changed. Especially the northern part of the model area reacted upon changes of discharge. For example during the basic situation there was a typical outflow of groundwater in the northwest (fig. 4b). But if there was an intensivation of the northern discharges (fig. 4c) the outflow turned into an inflow, which might lead to a saline influx from coastal regions (in- or outflow in $10^6 m^3/s$). Fig. 4d shows the differences of water level between the basic situation and this studied case. There is a reaction in the whole model area.

CONCLUSIONS

The simulation results inform about the reaction of the groundwater flow and show consequences of possible discharge alternatives. Furthermore they will prevent some management policies, which are of minor economical or ecological profit.

REFERENCES

1. Chorley D.W. and Frind E.O. (1978), An Iterative Quasi-Three-Dimensional Finite Element Model for Heterogeneous Multiaquifer Systems, Water Res. Res., Vol. 14, pp. 943-952.

2. Fujinawa, K. (1977), Finite Element Analysis of Groundwater Flow in Multiaquifer Systems, I. The Behaviour of Hydrogeological Properties in an Aquitard While Being Pumped, II. A Quasi Three-Dimensional Flow Model, Journal of Hydrology, Vol. 33, pp. 59-72 and pp 349-362.

3. Pelka B. (1987), Grundwassermodell Heider Trog - Modellaufbau, Berechnung und Auswertung, Mitt. Institut fuer Wasserbau und Wasserwirtschaft, RWTH Aachen, No. 66, pp. 423-449.

4. Sartori L. and Peverieri G. (1983), A Frontal Method Based Solution of the Quasi-Three-Dimensional Finite Element Model for Interconnected Aquifer Systems, Int. Journal for Numerical Methods in Fluids, Vol. 3, pp. 445-479.

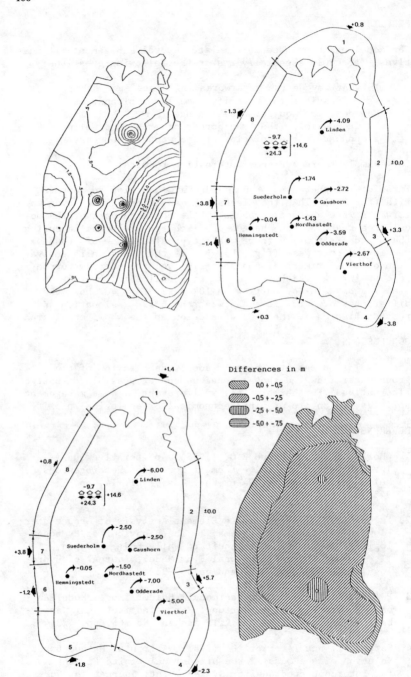

Fig. 4a-d: Model results

Three-Dimensional Finite Element Groundwater Model for the River Rhine Reservoir Kehl/Strasbourg
W. Pelka, H. Arlt and R. Horst
Lahmeyer International GmbH, Consulting Engineers, Environmental Engineering Department, Lyoner Strasse 22, D-6000 Frankfurt/Main 71, Federal Republic of Germany

ABSTRACT

For optimizing the depth of diaphragm walls of the river Rhine reservoir Kehl/Strasbourg a mathematical model has been developed and implemented combining a three-dimensional finite element groundwater flow model and a finite element channel flow model.

INTRODUCTION

River engineering measures at the Upper Rhine from Basel to Strasbourg have been completed by putting the river power plant Strasbourg into operation. To mitigate negative environmental effects, especially to maintain a certain water level in the original river bed and to support the groundwater level, the river Rhine barrage Kehl/Strasbourg has been constructed. Since more than a century, natural detention storage has been eliminated by flood walls and dikes. In connection with the Kehl/Strasbourg barrage large retarding reservoirs and flood storage basins have been built for flood control. To prevent excessive rise of groundwater levels in nearby villages during the retention process, several kilometers of diaphragm walls penetrating the dikes and the aquifer had to be planned.

HYDROGEOLOGICAL AND TECHNICAL SITUATION

For a given length the depth of the diaphragm wall is the most significant cost factor. On one hand the specific costs increase with depth since more sophisticated methods of construction have to be implemented and the overall costs depend on the total area of the wall, which is a linear function of depth. On the other hand the seepage flow bypassing the diaphragm decreases with depth and by this the discharge of the inland drainage system is reduced. Smaller cross-sections and smaller pumping stations would be sufficient to maintain the required depth of water level.

The geological profile (Fig. 1) shows a structure of horizontal layers of different permeability. Several pits caused by gravel dredging in the foreshore lead to inhomogenities of this structure. Because of the thickness of the aquifer (about 150 m), the diaphragm walls can by no means reach the bottom of the aquifer. The hinterland is characterized by complicated surface drainage patterns of open ditches, random field ditches, sheep drains and regional outlet channels.

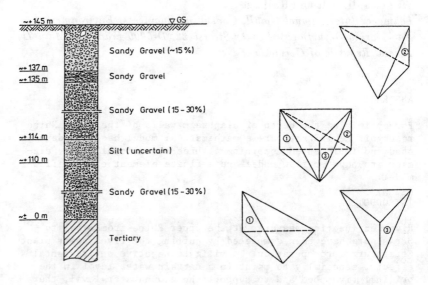

Figure 1:
Geological Structure

Figure 2:
Split-Up of Macro-Elements into Tetrahedrons

THREE-DIMENSIONAL GROUNDWATER MODEL

During the retention process the significant vertical groundwater flow components cannot be neglected, hence the implementation of a regional three-dimensional model became necessary. The finite element method was chosen because of the superior flexibility in the geometrical representation of the model area compared with other methods. Based on regular patterns a realistic spatial discretization of the complex geometry (especially the position of the planned diaphragm as well as the drains and ditches) would not have been possible. The spatial discretization was carried out by triangulization of the horizontal cross-section and defining prisms (macro-elements) in the vertical direction. Each prism is automatically split up into three tetrahedrons (Fig. 2) with the same hydrogeological properties. The position of the free surface of the unconfined aquifer was iteratively determined by cutting the elements at the water level and integrating over the remaining volume (Withum[3]). Overall there were about 3000 nodes and 5000 macro-elements in five layers representing the main hydrogeological structure.

CHANNEL FLOW MODEL

The discharge situation of the surface drainage pattern was simulated by a finite element channel flow model with one-dimensional bar elements representing the channel reaches. Cross-section, bed slope, roughness are assumed to be constant in each element. Boundary conditions are water level, discharge and mean velocity curve. The model is based on the finite element pipe flow model developed by Pelka and Schröder[2].

COUPLED GROUNDWATER-CHANNEL FLOW MODEL

The influence of the water level in the channels on the groundwater flow is introduced by leakage boundary conditions. The reference water level of the surface water results from the channel flow model.

Since the connection of the groundwater and the surface water is quite direct and the cross-sections of the ditches and main outlets are comparatively small, the discharges (and by this the water level) in the channels depend strongly on the drainage of groundwater. Based on the results of the groundwater model, the discharge of groundwater into the channels is computed by the nodal balances and the leakage condition. The coupled process is solved iteratively (Fig. 3), while the models communicate by boundary conditions of related nodes.

MODEL CALIBRATION

Calibration of the model was carried out by means of groundwater levels as well as channel water levels and discharges during a test retention process. The parameters calibrated were permeability and leakage factors for the groundwater model as well as roughness for the channel flow model. As usual the calibration of the three-dimensional model was not without problems of uniquely defined results, especially since in deeper layers the recorded water levels get more and more sparse. Based on the calibration runs additional field tests were designed and carried out to ensure definite permeability distributions. The maximum difference between calculated and measured results was about 25 cm, while the mean derivation was less than 10 cm, a result which by far cannot be generalized for models of this complexity. In many cases even errors two or three times larger will have to be accepted.

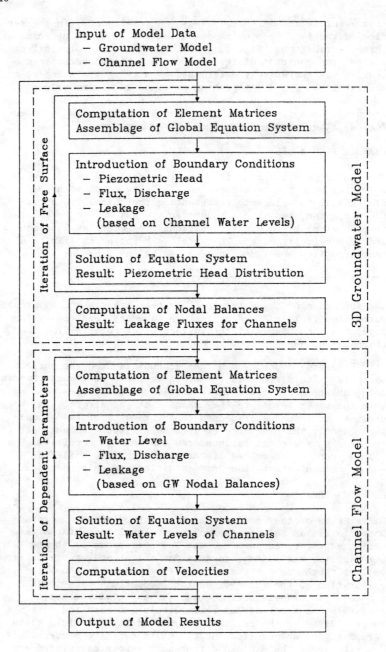

Figure 3: Flow Chart of Combined Model

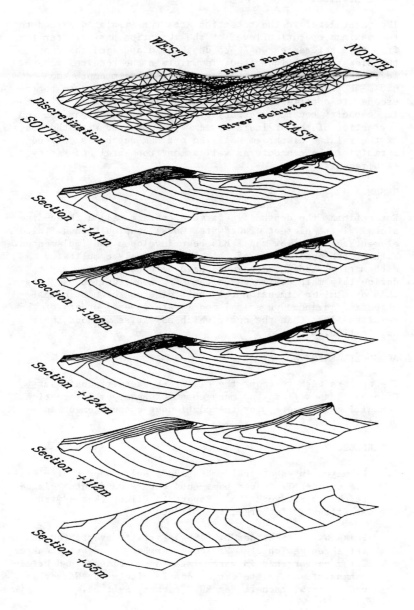

Figure 4: Example of Model Results

MODEL APPLICATION

The water level in the retention area was simulated raised to the maximum operation level of the retention basins. Starting from this reference run (zero depth of diaphragm) the depth of the diaphragm was increased. To maintain the required water level depth in the villages for each diaphragm depth the cross-sections and slope of the drainage system had to be increased to a certain extent and several pump stations of varying capacity became necessary. The costs of the expansion and correction of the drainage system were compared with the costs of the related diaphragm wall. An optimum depth was defined satisfying the economic as well as environmental and safety requirements.

RESUME

For defining the depth of a large diaphragm wall a three-dimensional finite element groundwater model coupled with a finite element channel flow model has been developed and implemented. Without the application of this comparatively complicated mathematical model there would have been no reliable way to define this main design parameter. For construction projects of this or similar dimension the costs of even the most complicated and advanced numerical models are several orders of the magnitude less than the costs which may arise without their application.

ACKNOWLEDGEMENT

The authors wish to thank the client (Wasserwirtschaftsamt Offenburg) as well as the management of Lahmeyer International, Consulting Engineers, for the continuous encouragement and support.

REFERENCES

1. Lahmeyer International GmbH (1988) Kulturwehr Kehl - Optimierung der Tiefe der Damm- und Untergrunddichtung im Bereich des Kieswerks Uhl, Report to Client (Wasserwirtschaftsamt Offenburg), unpublished.

2. Pelka W. and Schröder D. (1984) Variationsverfahren und Verfahren gewichteter Residuen zur Berechnung stationärer Strömungsvorgänge in verzweigten und vermaschten Rohrleitungssystemen, Mitteilungen des Instituts für Wasserbau und Wasserwirtschaft der RWTH Aachen, Heft 54.

3. Withum D. (1967) Elektronische Berechnung ebener und räumlicher Sicker- und Grundwasserströmungen durch beliebig berandete, inhomogene anisotrope Medien, Mitteilungen des Instituts für Wasserwirtschaft und Landwirtschaftlichen Wasserbau der TH Hannover, Heft 10.

SECTION 2B - UNSATURATED FLOW

An Alternating Direction Galerkin Method Combined with Characteristic Technique for Modelling of Saturated-Unsaturated Solute Transport

Kang-Le Huang

Department of Geology and Mining Engineering, Fuzhou University, Fuzhou, China

ABSTRACT

An alternating direction Galerkin method combined with charateristic technique, ADCG, is established for simulation of saturated-unsaturated solute transport in porous media in order to overcome numerical diffusion and oscillation and ensure high accuracy and efficiency of computation. A novel characteristic method, based on single particle tracking of front and high order and linear interpolations near the front and the remaining parts of domain respectively to obtain the 'convective contribution', is also developed. For examination of the model, an idealized contaminant pollution of aquifer and an experiment of wastewater disposal are simulated.

INTRODUCTION

It is well known that the numerical simulation of solute transport in saturated-unsaturated porous media is of paramount importance in groundwater hydrology, agronomy and environmental science. The simulations by conventional FE or FD, however, usually suffer from numerical dispersion and oscillation. Among tries to overcome these difficulties, the Eulerian-Lagrangian approach developed by Neuman[6] is of high efficiency. However, so far it is mainly used for solving fantastical saturated transport problems with simple definited conditions and constant coefficients. Furthermore, it is necessary to track a large number of particles moving in the flow field, and the process is so sophisticated that it can hardly be used for practical computation, especially for 2-D problem. Here, a more effective characteristic method that uses only one particle to track continually forward the front of concentration and high order interpolation for region near the front is developed to solve the 'convective problem'. For 'dispersive problem', the alternating direction Galerkin approach (ADG) (Daus et al[2]) is incorporated. Two examples of saturated-unsaturated transport problems are simulated by the proposed method.

BASIC EQUATION

The saturated-unsaturated solute transport equations considering mass exchange between mobile and immobile zones are given by

$$\partial(\theta_m c)/\partial t + \partial(\theta_{im} s)/\partial t = \nabla \cdot [(\theta_m \underline{D} \nabla c) - (\vec{q} c)] \quad (1)$$

and

$$\partial(\theta_{im} s)/\partial t = \lambda(c-s) \quad (2)$$

where θ_m, θ_{im} represent the water contents in mobile and immobile zones and c, s are concentrations of respective zone. $\underline{D}$, $\vec{q}$, λ and $\nabla = (\partial/\partial x, \partial/\partial y)$ denote dispersive tensor, seepage velocity vector, solute exchange coefficient and operator.

The initial and boundary conditions are

$$c \mid_{t=0} = c_i \tag{3}$$

$$c \mid_{\Gamma_1} = c_0 \tag{4}$$

$$(-\theta_m \underline{D} \nabla c + q c) \cdot \vec{n} \mid_{\Gamma_2} = q_1 c_1 \tag{5}$$

where c_i, c_0 and $c_1 q_1$ are prescribed functions. $\Gamma = \Gamma_1 + \Gamma_2$ is the entire boundary of considered domain and n the unit normal outward.

NUMERICAL MODEL

Using the definition of hydrodynamic derivative

$$Dc/Dt = \partial c/\partial t + \vec{V} \cdot \nabla c \tag{6}$$

and the continuity equation $\partial (\theta_m + \theta_{im})/\partial t = \partial \theta_m/\partial t = -\nabla \cdot \vec{q}$ (here θ_{im} is approximately considered as constant), Eq. (1) can be transformed into a purely parabolic equation

$$\theta_{im} \partial s/\partial t + \theta_m Dc/Dt = \nabla \cdot (\theta_m \underline{D} \nabla c) \tag{7}$$

along the characteristic line described by

$$d\vec{x}/dt = \vec{V} = \vec{q}/\theta_m \tag{8}$$

Assume that the concentration c^k at time t_k is known, c^{k+1} at $t_{k+1} = t_k + \Delta t$ can be solved. In term of Eulerian-Lagrangian approach, $Dc/Dt \approx (c^{k+1} - \bar{c}^k)/\Delta t$ which regards the grid to be solved as the position that a particle will reach at t_{k+1}, and the alternating direction time stepping scheme, Eq. (3) may be decoupled into implicit-explicit one-dimensional advection-dispersion equations in operator notation form:

x in implicit

$$(Lxx+Lyx) c^{k+1/2} + (Lyy+Lxy) c^k = \theta_m^{k+1/2} (c^{k+1/2} - \bar{c}^k)/\Delta t/2 + \theta_{im} (s^{k+1/2} - s^k)/\Delta t/2 \tag{9a}$$

y in implicit

$$(Lxx+Lyx) c^{k+1/2} + (Lyy+Lxy) c^{k+1} = \theta_m^{k+1} (c^{k+1} - \bar{c}^{k+1/2})/\Delta t/2 + \theta_{im} (s^{k+1} - s^{k+1/2})/\Delta t/2 \tag{9b}$$

where $Lxx = \partial (\theta_m Dxx \partial c/\partial x)/\partial x$, $Lyx = \partial (\theta_m Dyx \partial c/\partial x)/\partial y$, etc. definite operators.

For rectangle domain, $\Omega [0, X_L; 0, Y_L]$, upon applying the lumped Galerkin formulation as Daus et al did[2] in each implicit direction with unidirectionally linear basic function $w_i(x)$ and $w_j(y)$ and using appropriate difference discretes, a set of equations about $c^{k+1/2}$ and c^{k+1} may be obtained with the same notations as Ref. (2):

$$[Rx]\{c\}^{k+1/2} = \{Fy\}^k + \{Fxy\}^k + \{Fyx\}^{k+1/2} + \{F_A\}, \quad j=1,\ldots,n \tag{10a}$$

$$[Ry]\{c\}^{k+1} = \{Fx\}^{k+1/2} + \{Fxy\}^{k+1/2} + \{Fyx\}^{k+1} + \{F_B\}, \quad i=1,\ldots,m \tag{10b}$$

where $m = X_L/a$, $n = Y_L/b$; a, b is the spacial step size in x, y direction respectively; $[Rx]$ and $[Ry]$ are tridiagonal coefficient matrices; $\{Fy\}, \{Fx\}$, etc. are flux vectors and $\{F_A\}$, $\{F_B\}$ are dispersive flux boundary terms. The immobile zone concentration s is related with c by Simpson's numerical integration of Eq. (2)

$$s^{k+1/2+p} = \delta (\alpha c^{k+p} + \beta c^{k+1/2+p} + s^{k+p}), \quad p = 0 \text{ or } 1/2$$

and the integration coefficients

$$\alpha = \frac{\gamma \Delta t}{12 l} [1 + 4 \sum_{j=1}^{l} (1 - \frac{2j-1}{2l}) \delta^{\frac{2j-1}{2l}} + 2 \sum_{j=1}^{l-1} \{(1-j/l) \delta^{j/l}\}]$$

$$\beta = \frac{\gamma \Delta t}{12 l} [4 \sum_{j=1}^{l} (\frac{2j-1}{2l}) \delta^{\frac{2j-1}{2l}} + 2 \sum_{j=1}^{l-1} (j/l \delta^{j/l}) + \delta]$$

where $2 \times l$ is the discrete numbers for $[t_k, t_{k+1/2}]$; $\gamma = \lambda/\theta_{im}$ and $\delta = e^{-\gamma \Delta t/2}$.

For elements of constant size $a \times b$, the ith row in $[Rx]$ and the respective components of $\{Fy\}$ etc. of Eq. (10a) are:

$Rx_i = \{ [-b/a(D'_{xx})_{i-1/2,j} + (D'_{yx})_{i,j+1/2} - (D'_{yx})_{i,j-1/2}],$

$[2b(D'_{xx})_{i,j}/a + 2ab(\theta_m)_{i,j}/\Delta t + 2ab\theta_{im}\beta\delta/\Delta t],$

$[-b(D'_{xx})_{i+1/2,j}/a - (D'_{yx})_{i,j+1/2} + (D'_{yx})_{i,j-1/2}] \}$

$Fy^k_i = a/b \times [(D'_{yy})_{i,j+1/2}(c_{i,j+1} - c_{i,j})^k - (D'_{yy})_{i,j-1/2}(c_{i,j} - c_{i,j-1})^k +$

$2ab(\theta_m)_{i,j}\bar{c}^k_{i,j}/\Delta t + 2ab\theta_{im}/\Delta t \times [(1-\delta)s^k_{i,j} - \alpha\delta c^k_{i,j}]$

$Fxy^k_i = \{ (D'_{xy})_{i+1/2,j}(c_{i+1,j+1} - c_{i+1,j-1}) - (D'_{xy})_{i-1/2,j}(c_{i-1,j+1} - c_{i-1,j-1})$

$+ [(D'_{xy})_{i+1/2,j} - (D'_{xy})_{i-1/2,j}](c_{i,j+1} - c_{i,j-1}) \}^k/4$

$Fyx^{k+1/2}_i = [(D'_{yx})_{i,j+1/2}(c_{i+1,j+1} - c_{i-1,j+1}) -$

$(D'_{yx})_{i,j-1/2}(c_{i+1,j-1} - c_{i-1,j-1})]^{k+1/2}/4$

$F_{Ai} = b[D'_{xx} \partial c^{k+1/2}/\partial x \, w_i(x)] \mid_0^{X_1}$

where D'_{xx}, D'_{yy}, etc. are components of dispersion tensor $\underline{D}' = \theta_m \underline{D}$. Upon symmetry, components of $[Ry]$ and $\{Fx\}$ etc. in Eq. (10b) can be attained similarly to Eq. (10a).

CHARACTERISTIC METHOD

In Eqs. (9a) and (9b), $\bar{c}^k$ and $\bar{c}^{k+1/2}$ are 'convective contribution' that can be independently attained by solving the 'hyperbolic problem' constructed by

$D\bar{c}/Dt = 0$ (11)

$\bar{c}\mid_{t=0} = c_i$ (12)

$\bar{c}\mid_{\Gamma_1} = c_0$ or $\vec{q}\,\bar{c}\cdot\vec{n}\mid_{\Gamma_2} = q_1 c_1$ (13)

Neuman[6] and the writer[4] have successfully solved the problem with an adaptive characteristic method that combined the continuous forward particle tracking for steep concentration fronts and the single-step reverse particle tracking away from such fronts. It, unfortunately, is too sophisticated to use and program for most practical problems due to a large number of particles having to be introduced and tracked. Here, a comparatively simple and available scheme is developed as follows.

1. Introduce only one particle initially in the center of each 'source' such as steep front and discharging or leaching boundaries of concentration.
2. Track these particles along the characteristic line and then the fronts are caught.
3. Attain the 'convective contribution' $\bar{c}^k$ and $\bar{c}^{k+1/2}$ with high order Hermittian and linear interpolation for regions near and away from such fronts respectively.

APPLICATION AND CONCLUSION

The proposed method ADCG has been applied to simulate following convective-dispersive problems. Exam.1 An idealized aquifer contamination problem (Frind[3]) with unidirectional flow of constant velocity $V = 0.1 m/d$ with a source is shown in Fig.1. It may be described by Eqs. (1) to (4) when $D_{xy} = D_{yx} = 0$, $\lambda = 0$, $\theta_{im} = 0$. Letting $D_{xx} = 0.1 \, m^2/d$ and $D_{yy} = 8 \times 10^{-5} \, m^2/d$ and using the same space and time steps as those adopted by Frind[3], the solutions of all cases of discretion have errors not large 0.001 (c/c_0) and are of good agreement with analytical ones (Cleary[1]) as shown in Fig.2 even when Pe= 8 and Cu= 4, which would suffer from seriously numerical instability by ADG (Daus et al[2]) or principle direction scheme (Frind[3]).

Exam.2 A Saturated-unsaturated transport experiment on partially ponding wastewater (20g/l NaCl solution) disposal into loam with aggregated structure in a perspex slab 3m in length, 2m in height and 0.3m in thickness under drainage conditions that initially had a water table at depth of 1.3m. It was previously in hydrostatic equilibrium, but was of nonuniform distribution of salt concentration as shown in Fig.3 (t=0). Adopting the parameters supplied by Yang J. -Z.[7], $\lambda = 10^{-4}$, $\theta_{im}=$ 0.04 and dispersities $\alpha_L = 0.2$cm, $\alpha_T = 0.04$cm, the numerical results which are demonstrated in Fig.3, gained by ADCG with $\triangle x= 5$cm, $\triangle y= 2$cm and $\triangle t= 2\sim 30$min are well consistent with the data observed from the salt transducers installed in soil (The flow problem was first solved each time step by ADCI, developed by the writer[5] before concentration problem is simulated). However, nonphysical oscillation and dispersion would seriously occur when $\triangle t > 15$min by ADG with the same spatial steps.

Conclusions may be gained from above applications that thanks to the characteristic method incorporated, the proposed method ADCG is superior to ADG in eliminating numerical difficulties and ensuring high accuracy and efficiency due to the allowable wider ranges of Peclet and Courant numbers. Furthermore, the improved characteristic method greatly simplifies its application in hyperbolic problem and leads to more practical uses of Eulerian-Lagrangian approach for the simulation of general transport problems.

ACKNOWLEDGEMENT

The author gratefully acknowledges the advice and guidance of professor Zhang Wei-Zhen, Department of Irrigation & Drainage Engn., Wuhan Univ. of Hydraulic and Electrical Engn., throughout the writing of the paper.

REFERENCES

1. Cleary, R. W., (1978), Analytical models for groundwater pollution and hydrology, Rep.78-WR-15, Dept. of Civ. Eng., Princeton Univ..
2. Daus, A. D., and E. O. Frind, (1985), An alternating direction Galerkin technique for simulation of contaminant transport in complex groundwater systems, Water Resour. Res., 21 (5) : 653-664.
3. Frind, E. O., (1982), The principal direction technique: A new approach to groundwater contaminant transport modelling, Finite Element in Water Resources, 13.25-42.
4. Huang, K. -L., (1985), A new method for problems of groundwater quality, Hydrology and Engineering Geology, NO.5.
5. Huang, K. -L., (1988) Solution of saturated-unsaturated flow by finite element or finite difference methods combined with characteristic technique, in Proceedings of VII Computational Methods in Water Resources, MIT., June 1988.
6. Neuman, S. P., (1984), Adaptive Eulerian-Lagrangian FEM for advection-dispersion, Int. J. for Num. Meth. in Eng., 20, 321-337.
7. Yang, J. -Z., (1986) Theoretical and experimental studies of saturated-unsaturated water and salt transfer in soil, Ph. D dissertation of Wuhan Univ. of Hydraulic and Electrical Engineering, China.

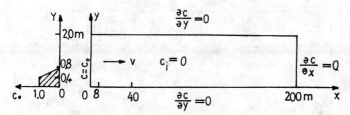

Fig.1 Idealized aquifer contamination: initial and boundary conditions

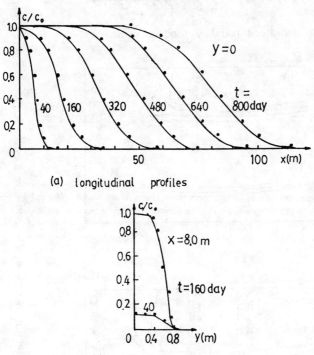

(a) longitudinal profiles

(b) transverse profiles

••• Analytical solution ——— Numerical solution

Fig.2 Idealized aquifer contamination: comparison of numerical and analytical solution

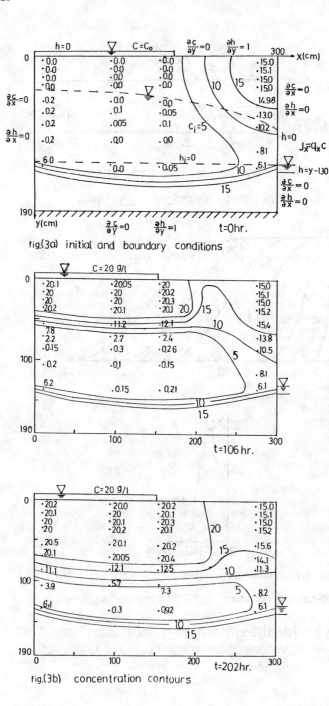

Fig. 3 Concentration profiles of wastewater disposal: numerical solution and experimental data.
—— numerical solutions   ··· observed data

Finite-Element Analysis of the Transport of Water, Heat and Solutes in Frozen Saturated-Unsaturated Soils with Self-Imposed Boundary Conditions

F. Padilla, J.P. Villeneuve and M. Leclerc

Université du Québec, INRS-Eau, 2700 rue Einstein, Ste-Foy, Québec, G1V 4C7, Canada

ABSTRACT

The finite-element method based on a Galerkin technique was used to simulate the one-dimensional transient movement of water, heat and solutes in saturated and partially saturated frozen porous media. Differential equations and physical laws were used to solve simultaneously water pressure, temperature and water chemistry. These unknowns are coupled with phase changes, frost heave and other parameters through several solutions to the discontinuity concept of the Stefan problem. When phase changes occur at the surface of the soil, special boundary conditions need to be considered. The model also uses self-imposed boundary conditions that depend on the solution and are useful for outflow boundaries. Several temporal schemes are available to solve non-linearities, depending on the precision desired and the execution time. The time steps are defined with respect to the diffusion parameter as well as the Courant and Peclet numbers of each equation in order to guarantee stable solutions. However, high gradients in water velocity give occasional monotonic error growth for very small time steps.

INTRODUCTION

The transport of water, heat and contaminants in the unsaturated zone of the soil have received considerable attention in contemporary modeling efforts. However, the deterministic modeling of these phenomena commonly ignore their simultaneity and inter-relationships, as well as the numerical conditions to represent realistic outflow boundaries. Heat and solute transport in unsaturated porous media are of major concern since they affect the water flow pattern in frozen soils to a significant extent (Miller[6]). Temperature also affects the adsorption and degradation of solutes.

In this study, a one-dimensional finite-element model is presented for simulating the simultaneous transport of water, heat and solute in saturated or partially-saturated porous media. Basic processes, namely advection, dispersion, adsorption, degradation, phase changes and frost heave are included. The resulting MELEF-3v model shows more precise results when self-imposed conditions are used for outflow boundaries.

GOVERNING EQUATIONS

The equation for the unsteady flow of water in a saturated-unsaturated soil can be written in one dimension (Laliberté et al.[4]; Milly[7]; Brutsaert and El-Kadi[1]):

$$\rho_w\, m(H,T)\, \frac{\partial H}{\partial t} = \frac{\partial}{\partial z}\left\{\rho_w \frac{k_w(H,T)}{\eta_w(T)} \frac{\partial H}{\partial z}\right\} + Q_w - i_w \qquad (1)$$

where ρ_w is the water density, T is the temperature, H is the water potential (H = P + ρ_w g z), P is the water pressure, k_w is the intrinsic permeability, η_w is the dynamic viscosity, m is the specific moisture capacity, Q_w and i_w are positive values when describing a source function and the rate of freezing.

The convection-dispersion equation for heat transport in saturated or partially saturated porous medium can also be written (De Vries[2]; Menot[5]):

$$c(H) \frac{\partial T}{\partial t} = \frac{\partial}{\partial z}\left\{k_h(H) \frac{\partial T}{\partial z}\right\} - \rho_w c_w \vec{w} \frac{\partial T}{\partial z} + (T^*-T) c_w Q_w + ((c_w-c_i) T + L_f) i_w \quad (2)$$

where c is the calorific capacity by unit volume of the porous medium, c_w and c_i are the specific calorific capacities of water and ice, L_f is the latent heat of fusion, $\vec{w}$ is the Darcy's velocity of the water, k_h is the thermal conductivity of the medium, and T^* is the known temperature of a source water.

Finally, the convection-dispersion equation for the solute transport in saturated-unsaturated soil is, in one dimension (Pickens et al.[12]; Padilla and Gélinas[10]):

$$\{\theta_w(H) + \rho^* K_d(T)\} \frac{\partial C}{\partial t} = \frac{\partial}{\partial z}\left\{\theta_w(H) D_c(H,T) \frac{\partial C}{\partial z}\right\} - \vec{w} \frac{\partial C}{\partial z} + (C^*-C) \frac{Q_w}{\rho_w} + \frac{C}{\rho_w} i_w$$

$$- \{\theta_w(H) + \rho^* K_d(T)\} k(T) C \quad (3)$$

where C is the dissolved solute concentration, C^* is the dissolved solute concentration in a source water, θ_w is the water content, ρ^* is the bulk density of the porous medium, k is the first-order degradation rate, K_d is the adsorption coefficient, and D_c is the hydrodynamic dispersion coefficient.

The parametric non-linear functions of these differential equations are described in the previous references.

TYPES OF BOUNDARY CONDITIONS

The porous media water flow, heat and solute transport equations 1, 2 and 3 have similar mathematical form and therefore can all be represented for example by the form of the solute transport equation. In subsequent explanations, we indicate for eq. 3 the conditions that would apply in the same way to the two other governing equations.

The initial conditions necessary for the solution of eq. 3 are: $C(z,0) = \bar{C}_0(z)$, where $\bar{C}_0$ are prescribed concentrations. The conditions on the boundaries Γ can be Dirichlet, Neumann, Cauchy (transport equations only) expressed as:

$$C(z,t) = \bar{C}(z,t), \text{ on } \Gamma_1, \quad \theta_w D_c \frac{\partial C}{\partial z} = \bar{q}(z,t), \text{ on } \Gamma_2, \quad \theta_w D_c \frac{\partial C}{\partial z} = (C^*-C) \frac{\bar{q}_w}{\rho_w}, \text{ on } \Gamma_3$$

or self-imposed boundary conditions expressed as:

$$\theta_w D_c \frac{\partial C}{\partial z} = \theta_w D_c \frac{\partial C}{\partial z}, \text{ on } \Gamma_4; \quad \text{which is equivalent to: } \theta_w D_c \frac{\partial C}{\partial z} = 0, \text{ at } z = \pm \infty,$$

where $\bar{q}$ is a known function and C^* are prescribed concentrations of the solute in the influx water $\bar{q}_w$. This last condition is only applicable to outflow boundaries.

Another kind of boundary is related to the rate of freezing (i_w). The phase changes in porous media are assumed to be produced on a "freezing front" (Menot[5]). We call "front" a surface where one or several parameters of the model have a discontinuity or a discontinuous derivative. This is pointed out by the Stefan problem. Then, eqs. 1-3 are true in the sense of distributions and they may also be written in terms of "jump". From this point of view, the volumetric rate of freezing can be expressed as:

$$i_w = -\frac{\left[k_h \frac{\partial T}{\partial z}\right]_N}{((c_w - c_i) \overline{T} + L_f) \Delta z} \tag{4}$$

where Δz is the length of spatial discretization and $[\ ]_N$ represents the "jump" or variation between the two sides of the discontinuity (Fig. 1b). Typical discontinuities are the propagation of a frost line in saturated soils (Fremond[3]) and the lenses of ice that grow within forced discontinuities in the frozen soil fabric (O'Neill and Miller[9]). These lenses of ice are responsible for the upward displacement of the soil surface (frost heave).

Notice that the rate of freezing of the solute transport equation became a Cauchy condition at the surface of the soil (where $C^* = 0$ and $\overline{q}_w = \vec{w} \, \rho_w$) if a freezing temperature $\overline{T}$ is prescribed at this boundary and freezing at the surface is also allowed.

FINITE-ELEMENT FORMULATION

The Galerkin technique is used to determine approximate solutions to eqs. 1, 2 and 3 under the appropriate initial and boundary conditions (Fig. 1a). Full details of the rather involved solution procedures will not be given here; however, an outline of our method is presented below to provide some notion of how the nonlinearities were dealt with.

The resulting set of finite-element equations for solute transport can be written in matrix form as:

$$[M] \left\{\frac{\partial C}{\partial t}\right\} + [K] \{C\} = \{F\} \tag{5}$$

where $[M]$ and $[K]$ are $n \times n$ matrices and $\{F\}$ is a vector of length n (the nodal points of the discretized system: $i = 1, 2, \ldots, n$). These are expressed by:

$$M_{ij} = \sum_e \int_{\Omega^e} [(\theta_w + \rho^* K_d)_1 \, N_1 \, N_i] \, dz, \quad \text{for } i=j; \qquad M_{ij} = 0, \quad \text{for } i \neq j;$$

(the nodal time derivatives are weighted averages over the flow region, Pickens et al.[12]),

$$K_{ij} = \sum_e \int_{\Omega^e} \left[(\theta_w D_c)_1 \, N_1 \, \frac{\partial N_i}{\partial z} \frac{\partial N_j}{\partial z} + \vec{w}_1 \, N_1 \, N_i \, \frac{\partial N_j}{\partial z}\right] dz,$$

$$F_i = \sum_e \left\{ -\int_{\Omega^e} ((\theta_w + \rho^* K_d) \, k)_1 \, N_1 \, N_i \, N_j \, C_j \, dz \right.$$

$$+ \oint_{\Gamma^e} (\theta_w D_c)_1 \, N_1 \, N_i \, \frac{\partial N_j}{\partial z} \, C_j \, d\Gamma_4 \qquad \text{Self-imposed boundary}$$

$$+ \oint_{\Gamma^e} N_i \left[\frac{q_w}{\rho_w} C^*\right]_i d\Gamma_3 - \oint_{\Gamma^e} \left[\frac{q_w}{\rho_w}\right]_i N_i \, N_j \, C_j \, d\Gamma_3 \qquad \text{Cauchy boundary}$$

$$\left. + \oint_{\Gamma^e} N_i \, \overline{q}_i \, d\Gamma_2 \right\} \qquad \text{Neumann boundary}$$

where e indicates summation over the elements joining at node i, Ω is the domain of the flow regime, $N_{i,j}$ are spatial quadratic functions, and $(\ )_1 \, N_1$ indicates a functional representation to express a nodal parameter within an element. The vector $\{F\}$ of eqs. 2 and 3 depends on the solutions T_i and C_i. Nonlinear techniques are only needed to solve eq. 1.

SOLUTION TECHNIQUE

The implicit difference scheme has been found to provide good results for the time-dependent nature of eq. 5. Linear systems can then be expressed as:

$$\left[\Delta t \, \alpha \, [K] + [M]\right] \left\{C_{t+\Delta t} - C_t\right\} = \left\{\Delta t \, (\alpha \, \{F_{t+\Delta t}\} + (1-\alpha) \, \{F_t\}) - [K] \, \{C_t\})\right\} \tag{6}$$

When $\alpha = 1$, it is called a fully implicit backward scheme, and when $\alpha = 0.5$, it is the Crank-Nicholson scheme. If $\{F\}$ depends on the solution, it can be expressed as:

$$\{F_t\} = [F']\{C_t\} + \{F''\}, \quad \{F_{t+\Delta t}\} = [F']\{C_{t+\Delta t}\} + \{F''\}; \quad \text{that is,}$$

$$\{F_{t+\Delta t}\} = [F']\{\Delta C_t\} + [F']\{C_t\} + \{F''\} = [F']\{\Delta C_t\} + \{F_t\} \tag{7}$$

Employing this definition in eq. 6 yields:

$$\left[\Delta t \, \alpha \, ([K]-[F']) + [M]\right]\left\{C_{t+\Delta t} - C_t\right\} = \left\{\Delta t \, (\{F_t\} - [K]\{C_t\})\right\} \tag{8}$$

where $[F']$ is a $n \times n$ matrix evaluated on the boundaries, that is for the solute transport:

$$F'_{ij} = \sum_e \left[-\int_{\Omega^e} ((\theta_w + \rho^* K_d) \, k)_l \, N_l \, N_i \, N_j \, dz \right.$$

$$\left. + \oint_{\Gamma^e} (\theta_w D_c)_l \, N_l \, N_i \, \frac{\partial N_j}{\partial z} \, d\Gamma_4 - \oint_{\Gamma^e} \left(\frac{q_w}{\rho_w}\right)_i N_i \, N_j \, d\Gamma_3 \right] \tag{9}$$

Nevertheless, non-linear problems with transient parameters need other methods. Several temporal schemes are available in the MELEF-3v model to solve the non-linearities: the fifth order Runge-Kutta method, evaluation of coefficients at half the time step, as well as some iterative processes to obtain a convergent solution. The time step chosen for each equation is computed for every time by keeping the diffusion parameter as well as the Courant and the Peclet numbers within certain bounds. In general applications give stable solutions. However, high gradients in water velocity produce weak monotonic error growth when transport equations are solved for very small time steps. One method to combat this rare instability is to employ occasional filtering of state variables (temperature and concentration).

MODEL VERIFICATION

The heat and solute transport portions of the model were tested by comparison with the analytical solution proposed by Ogata and Banks[8] for one-dimension convection-dispersion using a step input (C_0 or T_0) at $z = 0$ and a zero gradient ($\partial C/\partial z$ or $\partial T/\partial z$) at $z = \infty$. This last condition can be well represented in a finite-element numerical model by a self-imposed boundary. In consequence, the bottom of the column does not need a specific imposed boundary condition and all the results of the discretized domain are able to closely represent the Ogata and Banks[8] solutions (Fig. 3 and 4). The comparisons are shown for a pore water velocity, soil porosity, longitudinal dispersivity and a thermal conductivity of the solids of 0.24 m/d, 0.4, 0.2 m and 14446 J m^{-1} d^{-1} K^{-1}, respectively.

Testing the accuracy of numerical schemes used to solve the one-dimensional transient non-linear water flow equation is limited by the scarcity of suitable analytical solutions. In order to avoid this difficulty, eq. 1 can be expressed in the Richard's form:

$$\frac{\partial \theta_w}{\partial t} = \frac{\partial}{\partial z}\left\{\frac{k_{wo}}{\eta_w} k_{wr} \frac{\partial P}{\partial \theta_w} \frac{\partial \theta_w}{\partial z}\right\} + \rho_w \, g \, \frac{k_{wo}}{\eta_w} \frac{\partial k_{wr}}{\partial z} \tag{10}$$

where k_{wo} and k_{wr} are the saturated intrinsic and relative permeabilities of water, respectively ($k_w = k_{wo} \, k_{wr}$). Making $k_{wr} = S$, and $P = a \ln S$, yields:

$$\frac{n \, \eta_w}{k_{wo}} \frac{\partial S}{\partial t} = a \frac{\partial^2 S}{\partial z^2} + \rho_w \, g \, \frac{\partial S}{\partial z} \tag{11}$$

where n is the porosity, and a is a constant defined for each type of soil. Eq. 11 has the form of a transport equation and therefore can be compared to the analytical solution of Ogata and Banks[8]. The hypothetical impervious boundary condition at $z = \infty$ can be represented in eq. 1 by making $\partial H/\partial z = \rho_w g$ at the top of the discretized column, far enough of the water table (Fig. 2). Results show nearly steady conditions after 6 hours of water intake to the unsaturated zone of an initially dry soil.

In all of the above cases, the Crank-Nicholson time-dependent scheme shows the most precise results.

SUMMARY AND CONCLUSIONS

A one-dimensional finite-element model, coupling simultaneously moisture, heat and mass transport in saturated-unsaturated or freezing soils, has been developed. The Galerkin technique, in conjunction with several types of boundary and initial conditions, was used to solve also for phase changes, frost heave, adsorption and degradation of a solute. Self-imposed conditions have been developed to take into account the outflow boundaries of more real transport problems. Freezing at the surface of the soil is allowed in the solute transport model when the rate of freezing is replaced by an appropriate Cauchy condition at the boundary.

The heat and solute transport, as well as the flow portions of the resulting MELEF-3v model, compared precisely with analytical solutions. Subsequent applications use occasional filtering of unstable solutions in the transport portions of the model if high gradients in water velocity cause monotonic error growth for very small time steps.

The model presented in this paper also found applications studying the effect of solutes on frozen soils (Padilla and Gélinas[10]) as well as the effect of temperature on the fate of pesticides in the unsaturated zone (Padilla et al.[11]).

REFERENCES

1. Brutsaert W. and El-Kadi A. (1984). The relative importance of compressibility and partial saturation in unconfined ground water flow. Water Resour. Res., Vol.20(3), pp. 400-408.
2. De Vries D.A. (1963). Thermal properties of soils. In: W.R. van Wijk (Editor), The physics of plant environment. North Holland Pub. Co., Amsterdam.
3. Fremond M. (1979). Gel des sols et des chaussées, Problèmes hydrauliques, Méthodes numériques. Ecole Nationale des Ponts et Chaussées et Carleton U., Paris, Ottawa.
4. Laliberté G.E., Corey A.T., Brooks R.H., Corey G.L. and McWhorter D.B. (1967). Similitude for partially saturated flow systems. Soil Moisture, Proceeding of Hydrology Symposium No 6, National Research Council of Canada, pp. 101-117.
5. Menot J.M. (1979). Equations of frost propagation in unsaturated porous media. Engin. Geol., Vol.13, pp. 101-109.
6. Miller R.D. (1980). Freezing phenomena in soils. In: D. Hillel (Editor), Applications of soils physics. Academic Press, New York.
7. Milly P.C.D. (1982). Moisture and heat transport in hysteretic inhomogeneous porous media. A metric head-based formulation and a numerical model. Water Resour. Res., Vol.18(3), pp. 489-498.
8. Ogata A. and Banks R.B. (1961). A solution of the differential equation of longitudinal dispersion in porous media. U.S. Geol. Survey, Professional Paper, 411-A, pp. A1-A7.
9. O'Neill K. and Miller R.D. (1985). Exploration of a rigid ice model of frost heave. Water Resour. Res, Vol.21(3), pp. 281-296.
10. Padilla F. and Gelinas P. (1987). Un modèle pour l'infiltration dans les sols gelés en tenant compte de la qualité de l'eau. Rev. Canadienne Génie Civ. (in press).
11. Padilla F., Lafrance P., Robert C. and Villeneuve J.P. (1987). Modeling the transport and the fate of pesticides in the unsaturated zone considering temperature effects. Ecological Modelling (in press).
12. Pickens J.F., Gillham R.W. and Cameron D.R. (1979). Finite element analysis of the transport of water and solutes in tile-drained soils. J. Hydrology, Vol.40, pp. 243-264.

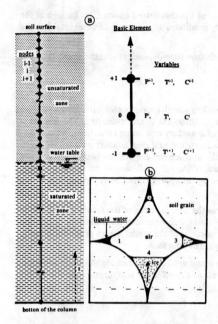

Fig. 1: a) Finite-element vertical discretization of the unsaturated-saturated zone of the soil. b) Four hypothetical configurations of air, water, and ice during the formation of a frozen fringe.

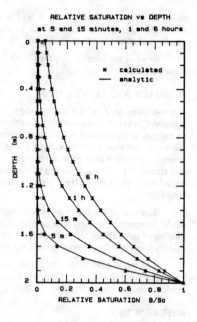

Fig. 2: Comparison of MELEF-3v model and analytical solution results for water intake from a water table 2 m depth to the unsaturated zone of a initially dry soil.

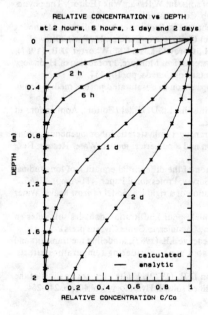

Fig. 3: Comparison of MELEF-3v model and analytical solution results for convection-dispersion of a solute.

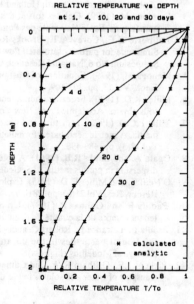

Fig. 4: Comparison of MELEF-3v model and analytical solution results for convection-dispersion of the heat.

A Variably Saturated Finite-Element Model for Hillslope Investigations

S.T. Potter and W.J. Gburek
Northeast Watershed Research Center, University Park, Pennsylvania 16802, USA

INTRODUCTION

The variable-contributing-area (VCA) concept of storm runoff production is based upon the hypothesis that stream response to precipitation is governed by inseparable surface-groundwater interactions. During a storm event in a humid climate, rainfall is initially intercepted by the stream channel, groundwater discharge areas close to the stream (seepage faces), and other wet areas not adjacent to the stream. Waters impinging on these locales result in a well-documented rapid stream response. Immediately upslope of these initial contributing areas, the infiltrating water rapidly fills available storage, saturating the soil and resulting in an expanding area contributing to direct runoff[1,3]. These processes continue over the course of a storm event. The infiltrating water also affects the energy status of the near-stream soil water shortly after a storm's onset[6]. The increased energy of the soil water initiates an increase in discharge of groundwater through prestorm seepage faces as well as through those developed as a direct result of the storm[10]. The groundwater flow regime during these processes is both highly dynamic and strongly nonlinear. The purpose of this study was to develop a state-of-the-art subsurface flow model capable of characterizing this overall response as accurately as possible.

To characterize total hillslope response to a storm event, the model must meet a number of criteria. First, it must properly characterize the several orders of magnitude variation in soil hydraulic properties expected to occur during initial phases of infiltration on a dry soil and which continue to exist at the wetting front. Also, procedures which enable the model to consistently describe abrupt changes in hydraulic conductivity (K) between soil horizons should be used. Boundary conditions at the soil surface must be as realistic as possible. Infiltration rates should vary continuously reflecting the soil's ability to absorb water. Areas immediately upslope of seepage faces are expected to be influent and to produce direct runoff. Finally, the model should accurately define flow line dynamics to discern which portions of the storm hydrograph can be attributed to direct runoff, event water which enters the soil and later re-emerges and contributes to the storm hydrograph, and contributions from pre-event water (base flow).

FINITE-ELEMENT MODEL

To fulfill these needs, a Galerkin finite-element model of Richard's equation was formulated using C^1 continuous Hermitian bases to approximate total head, and serendipity Hermitian bases for hydraulic conductivity and specific moisture capacity (S). These bases were imposed on curved, isoparametric, quadrilateral subspaces restricted to orthogonal corners. The dependent variables were evaluated in normal-tangential coordinates and slope boundary conditions were described using one-dimensional Hermitian bases[4,9]. To eliminate errors associated with integral evaluation, integrations were performed in closed-form using a symbolic formula manipulation compiler (FORMAC)[8], thereby reducing errors to roundoff and truncation. Problems regarding Jacobian discontinuities were handled by constructing the coefficient matrices an element at a time, permitting multiple point values at individual nodes. This formulation in general results in nodal discontinuity of the gradient in local coordinates, but, this does not imply global discontinuity. As a result, this permits formulation and solution of the governing equation in global coordinates. At abrupt changes in hydraulic conductivity, intrinsic discontinuities exist in the hydraulic gradient. To simulate the effect of this condition on infiltration and redistribution, K and S are evaluated on an element-by-element basis, and nodes are

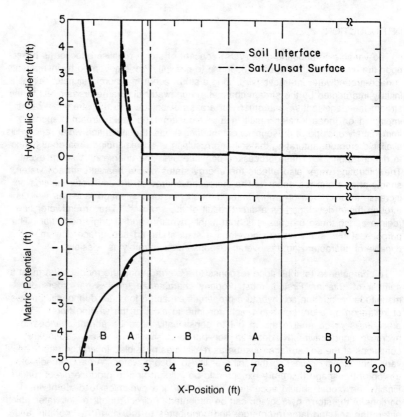

Figure 1. Steady-state analytical solution used for model verification.

always present at soil interfaces. At abrupt changes in K, the specific discharge is continuous. Thus, the dependent variables estimating the hydraulic gradient are mutiplied by K where and when necessary to represent the specific discharge, thus making specific discharge the dependent variable. The time derivative was approximated using a central finite-difference operator.

Due to the complexity of the model, and the unique handling of K discontinuities, model verification can be best achieved using analytic solutions. Steady-state results were evaluated using a solution presented by Potter et al.[7] for flow through a variably-saturated composite soil column. Soil hydraulic properties are characterized using the Campbell[2] equations, and the constants used are in Table 1. Boundary conditions are specified potentials of -20.0 and 0.71 ft at the left and right edges of the column, respectively. As a consequence of the air entry values and the spacings of the soil layers, the soil moisture energy is highly nonlinear and three air water interfaces occur. Figures 1a and 1b show hydraulic gradient and total head, respectively. The solid lines represent the analytic solution, and the dashed lines the numerical solution. Both numerical solutions compare very well to the analytic solutions except at the dry end, where hydraulic gradient is very high (179) and there is a 14-ft increase in potential over only 0.4 feet. Limitations in space preclude presentation of the solution in this region.

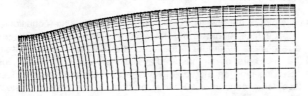

Figure 2. Hypothetical hillslope and finite-element discretization for example problem.

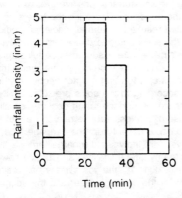

Figure 3. Hyetograph of synthetic two-inch design storm of one-hour duration.

Table 1. Hydrologic parameters in soils used in Figure 1 example.

PARAMETER	SOIL(A)	SOIL(B)
K_S(FT/DAY)	1.00	10.0
Air-entry Potential(FT)	-1.00	-0.25
Exponent	2.50	2.80

EXAMPLE APPLICATION

To test the model's ability to simulate subsurface response to a storm event, a hypothetical hillslope was constructed as shown in Figure 2. The cross-section is 200 ft long, 40 ft thick at the "stream", and 60 ft thick at the "divide". The land surface is S-shaped, with a maximum slope of 27% at 34 ft from the left edge. Groundwater discharges from the section through a constant-potential boundary (P = 0) which corresponds to a surface water body, as well as through any seepage face which may develop upslope. As this is the first test of the model under storm dynamic conditions, the soil was assumed homogeneous and isotropic. Saturated K is 1.0 ft/day, saturated moisture content is 50%, air-entry potential is -0.6 ft, and exponents for K and moisture content in the Campbell equations are 2.71 and 0.237. The groundwater flow system was initially driven to steady state using a recharge value of 0.003 ft/day. This value is characteristic of long-term recharge rates in the Ridge and Valley Region of central PA. Using hydrologic data from the same region, a two-inch design storm of one-hour duration was constructed with a return period of five years[5]. The hyetograph is shown in Figure 3. The model was run using time steps of 10 minutes, and lumping was necessary ensure stability. Due to the length of the flow system and relative low value of saturated K as compared to the simulation time, only the near-stream subsurface flow system played a critical role during the storm. Consequently, only the response of the flow system in the near-stream enviroment is shown in Figure 4.

Fig. 4a is a plot of the specific discharge under steady state conditions prior to the storm. Seepage face length is five feet and groundwater discharge increases downslope reaching a maximum at the toe of the hill. At 10 minutes into the storm (Fig. 4b) the initial burst of rain has filled available storage immediately upslope of the seepage face, initiating an increase in discharge from the seepage face. The seepage face has expanded to 10.9 ft at the completion of this time step. After 20 minutes (Fig. 4c), the seepage face is still the same length, but a fully saturated region has developed upslope of the seepage face which acts as an area of ground water recharge and also produces direct runoff. The third time step of the storm (30 min., Fig. 4d) presents some unique problems in assigning boundary conditions along the right portion of the land surface. When the time step was first solved assuming that rainfall rate was insufficient to saturate the soil surface, positive pressures were predicted along the land surface, implying the opposite. Resolving the time step with the soil surface assumed to be fully saturated resulted in normal gradients predicted in excess of the rainfall rate, implying that the soil wanted to dry. Consequently, the appropriate boundary conditions at the soil surface were assumed to be full saturation with normal gradients equal to the rainfall rate, ultimately producing no runoff from this portion of the hillslope. The fourth time step (40 min.) did not encounter these problems. Infiltration rates were assigned according to the soil's predicted ability to absorb water during the time step. This resulted in surface runoff occuring on all but the far right element of the section. Because of the less intense rain of the fifth time step (50 min., Fig. 4e) runoff has ceased to occur except from the seepage face, and the soil surface has began to dry. However, groundwater discharge through the seepage face continues to increase supported by lateral shallow, subsurface flow. At the completion of the sixth time step (60 min., Fig. 4f), the soil surface has drained everywhere except in the immediate vicinity of the seepage face, and the seepage face has begun to recede.

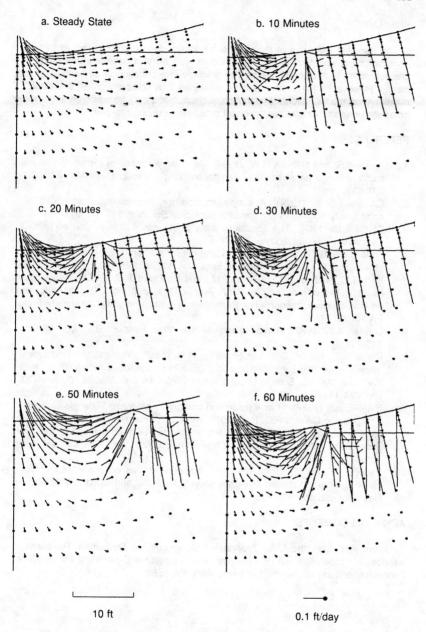

Figure 4. Groundwater response to synthetic storm. Vectors represent specific discharge. Dots represent source of each vector. All vectors longer than 1 inch (q = 0.335 ft/day) have been truncated to 1 inch for clarity.

SUMMARY

The transient response of moisture in a hillslope can be so dynamic as to limit physical characterization in the field. Further, variability in geometry and physical characteristics of the porous media can preclude interpretation of any measurement made. However, if the system can be simulated in a comprehensive, defensible, mathematical model, rational explanation of phenomena may be tendered and a more complete understanding of the important processes may be realized.

REFERENCES

1. Abdul A.S. and Gillham R.W. (1984), Laboratory studies of the effects of the capillary fringe on streamflow generation, Wat. Resour. Res., Vol. 20, pp. 691-698.
2. Campbell G.S. (1974), A simple method for determining unsaturated conductivity from moisture retention data, Soil Sci., Vol. 117, pp. 311-314.
3. Childs E.C. (1960), The nonsteady state of the water table in drained land, J. Geophys. Res., Vol. 65, pp. 780-782.
4. Frind E.O. (1977), An isoparametric Hermitian element for the solution of field problems, Int. J. Num. Methods in Eng., Vol. 11, pp. 945-962.
5. Kerr R.L., McGinnis D.F., Reich B.M., and Rachford T.M. (1970), Analysis of rainfall-duration-frequency for Pennsylvania, Res. pub. no. 70, Institute for res. on land and wat. resour., The Pennsylvania State University, University Park, PA 16802.
6. O'Brien A.L. (1982), Rapid water table rise, Wat. Resour. Bul., Vol. 18, pp. 713-715.
7. Potter S.T., Richie, E.B. and Schnabel R.R. (1987), Verification of numerical models - Variably saturated flow through a heterogeneous media, Proc. Nat. Well Wat. Assoc., Solving Ground Water Problems with Models, Denver CO, pp. 424-443.
8. SHARE Inc. (1986), User's guide and catalog of programs, Triangle University Computation Center, P.O. Box 12076, Research Triangle Park, NC 27709.
9. van Genuchten M.Th. (1983), An Hermitian finite-element solution of the two-dimensional saturated-unsaturated flow equation, Adv. Wat. Resour., Vol. 6, pp. 106-111.
10. Zaltsberg E. (1986), Comment on "Laboratory studies on the effects of the capillary fringe on stream flow generation". by Abdul and Gillham, Wat. Resour. Res., Vol. 22, pp. 837-838.

ACKNOWLEDGMENTS

Contribution from the U.S. Department of Agriculture, Agricultural Research Service, in cooperation with the Pennsylvania Agricultural Experiment Station, The Pennsylvania State University, University Park, PA 16802.

A Subregion Block Iteration to 3-D Finite Element Modeling of Subsurface Flow

G.T. Yeh

Environmental Sciences Division, Oak Ridge National Laboratory, Oak Ridge, Tennessee 37831, USA

ABSTRACT

A subregion block iteration (SBI) technique has been developed in conjunction with finite element approximations of saturated-unsaturated flow equations. The proposed SBI technique is implemented in a three-dimensional finite element saturated-unsaturated flow model. The model is verified with a nonlinear diffusion equation having an analytical solution. It is then applied to a burial trench problem. It is not possible to solve this field problem using the direct elimination finite element method. The SBI technique provides significant improvement over models based on direct band solution methods in both central processing unit (CPU) storage and CPU time.

INTRODUCTION

While iterative schemes for solving large field problems are routinely used in finite difference simulations, direct band solution methods are generally used in finite element approximations because of simplicity in assembling the coefficient matrix in banded form and easy access to band-matrix solver software. To make the application of finite element models to field problems feasible, a number of attempts have been made to incorporate various iterative schemes into finite element formulations (Yeh[1], Huyakorn et al.[2], Yeh[3]). We present here a "subregion block iteration" (SBI) technique for finite element simulations of saturated-unsaturated flow problems using 3DFEMWATER (Yeh[3]).

The SBI technique consists of three steps. First, the grid system is pre-processed to yield two pointer arrays; one is the global stencil array and the other is the local (subregion) stencil array. The pre-processing only has to be performed once unless the spatial discretization is changed during a simulation. Second, the global matrix is assembled and compressed, using the global stencil array, to contain non-zero elements, thereby minimizing the central processing unit (CPU) storage requirements. Third, for each subregion, the sub-matrix equation is formed from the stored global matrix and global vector by the local stencil array and the node-mapping function; the sub-matrix equation is solved with direct elimination, thus resulting in large savings in CPU time because the band-width of the sub-matrix is much smaller than that of the global matrix. The third step is performed for all subregions for each iteration until a convergent solution is obtained. When every nodal point, line, or vertical plane in the region is treated as a subregion, the SBI becomes the point, line, or SSOR iteration method, respectively.

PRE-PROCESSING OF GEOMETRIC DATA

The pre-processing step analyzes the geometric information of a chosen finite element discretization and prepares two pointer arrays. One of the pointer arrays is used for compressing the global matrix to contain non-zero elements and the other is used to recover the actual global position of the rearranged block position. Let us use a simple example (Fig. 1) to illustrate how two pointer arrays are generated. Having a finite element discretization, we first subdivide the region into an arbitrary number of subregions (Fig. 1a). Second, we renumber each subregion so that the band width for the sub-matrix will be minimum. It is noted that intra-block nodes are numbered after all interior nodes have been numbered (Fig. 1b). Third, we input a node-mapping function N(K,I) between global nodes and local nodes for all subregions. Only block interior nodes (those to the left side of the dotted line) have to be input; intra-block nodes (those to the right side of the dotted line) are instead generated based on the input node-mapping function and element connectivity (Fig. 1c). Fourth, NGS(N,I), a global stencil array, is generated based on the element connectivity (Fig. 1d). Fifth, the node-mapping function N(K,I) is filled up for the intra-block nodes based on element connectivity and the partial input of the node-mapping function (Fig. 1c). Finally, NLS(L,I,K), a local stencil array, is generated based on the global stencil NGS(N,I) and the complete node-mapping function N(K,I) (Fig. 1e).

ASSEMBLING OF COMPRESSED GLOBAL MATRIX AND GLOBAL LOAD VECTOR

Since the Galerkin finite element method results in a 27-point formula for three-dimensional problems, the maximum number of non-zero coefficients for any equation is around 27. The question is how to assemble from element matrices the global matrix in the compressed form. This can be easily accomplished by using the pre-processed global stencil array NGS(N,I). For example, if NGS(N,I) = K, then the coefficient corresponding to the K-th unknown in the N-th equation is stored in C(N,I) of the compressed global matrix C. In the meantime, the right-hand side of the N-th equation is stored in

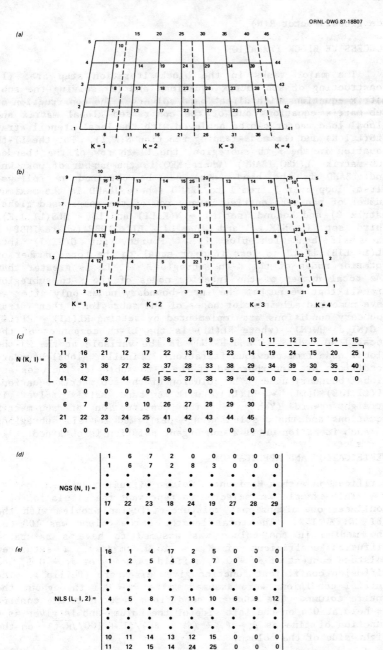

Fig. 1 Pre-processing of global stencil and local stencil arrays, NGS(N,I) and NLS(L,I,K).

the global vector R(N).

SUCCESSIVE BLOCK ITERATION

The major tasks in the block iteration step are: (1) constructing of sub-matrix equations and (2) solving the sub-matrix equations with direct band solvers. The construction of sub-matrix equations out of the compressed global matrix and global load vector can be achieved with the local stencil array NLS(L,I,K) and the node-mapping function N(K,I). For the LI-th equation in the K-th subregion, the LI-th row of the banded sub-matrix CL(NEQ,IBAND) (where NEQ is the number of unknowns and IBAND is the band-width) can be obtained as follows. First, loop over J from 1 to JBAND (where JBAND is the maximum number of non-zero entries in any row of the compressed global matrix C) and second, set NI = N(K.LI) and LJ = NLS(LI,J,K). Third, set NJ = N(K,LJ) and LJB = LJ - LI + IHBP (where IHBP is the half band-with plus 1). Fourth, put C(NI,J) into CL(LI,LJB) if LJ is less than or equal to the total number of interior nodes of the K-th subregion. If LJ is greater than the total number of the interior nodes of the K-th subregion (recall that we number the intra-boundary nodes only after we have numbered all interior nodes of the subregion), then intra-boundary conditions are implemented by setting RL(LI) = RL(LI) - C(NI,J)*HW(NJ) (where RL(LI) is the LI-th component of the local load vector RL and HW(NJ) is the variable at the NJ-the global node from previous iteration). Finally, global boundary conditions are incorporated if the K-th subregion includes any global boundary node. The solution of the sub-matrix equation, CL(LI,LJB)*H(LI) = RL(LI), with a band matrix solver is straightforward (Yeh[3]). The construction of sub-matrix equations and their solution are performed for all subregions in each iteration until a convergent solution is obtained.

VERIFICATION AND APPLICATION

Verification with a Horizontal Column Diffusion Problem.
This example was selected to represent the simulation of a nonlinear one-dimensional moisture-diffusion problem with the SBI 3DFEMWATER. The total length of the column was 100 cm. The medium in the column was assumed to have a saturated diffusion coefficient of K_s = 10.0 cm^2/day, a saturated moisture content of θ_s = 0.4, a field capacity of θ_r = 0.1, and diffusion coefficient function given elsewhere (Philip[4]). The initial conditions were assumed to be θ = 0.1 throughout the entire column. The boundary conditions were: moisture content is held at 0.4 on the left side of the column and is given as a function of time as $\theta = \theta_r + (\theta_s - \theta_r)/(1 + 100/\sqrt{K_s t})^2$ on the right side of the column.

The column was discretized with 1 x 1 x 100 = 100 elements with element size = 1 x 1 x 1 cm, resulting in 2 x 2 x 101 = 404 node points. For model simulation, each of the four horizontal lines was considered a subregion. Thus, a total of four subregions, each with 101 node points, were used for the subregion block iteration simulation. A variable time step size was used. The initial time step size was 0.01 day and each subsequent time step size was increased by a factor of 0.25 with a maximum time step size of 2.0 days. A 40 day simulation was made with total of 40 time steps. The moisture content tolerance for nonlinear iteration was set to $1.0 \cdot 10^{-5}$, the nonlinear relaxation factor was set equal to 0.5, the tolerance for block iteration was set to $0.5 \cdot 10^{-5}$, and relaxation parameter of block iteration was set 1.0. The simulated moisture content is plotted in Fig. 2. It shows very close agreement between the analytical (Philip[4]) and the numerical solutions. Computer output indicates that the agreement is up to the fourth digit for long-time simulation.

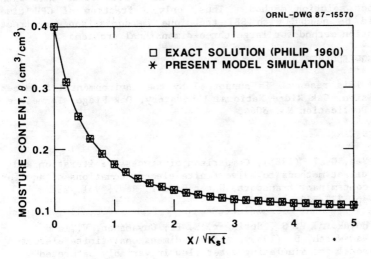

Fig. 2 Comparison of moisture content as simulated by exact and SBI numerical solutions.

Application to a Burial Trench Problem

In order to assess the perching and/or mounding of groundwater in burial trenches, the SBI 3DFEMWATER (Yeh[3]) was applied to a typical single trench (Solomon and Yeh[5]). A finite element mesh consisting of 25 x 17 x 11 = 4,675 nodes was used for simulating a solution region of 44 m long by 25 m wide and 15 m deep. For SBI simulations, the region is divided into 17 subregions, each subregion has 25 x 11 = 285 nodes. For the steady-state simulation, about 15 nonlinear iterations are required to reach a convergent solution because of very high nonlinearity of the problem. Within each nonlinear iteration, an average of 27 block iterations are required to achieve a convergent solution with maximum error tolerance of 0.5 cm. For most of the case studies, approximately 20 minutes of CPU time were required to reach a steady state solution on an IBM 370/3033. Had a direct band matrix solver been used, the band-width of the global coefficient matrix would be 399. This would have required a CPU storage of approximately 15 mega-bytes to store the global coefficient matrix. Since an IBM 370/3033 has only about 10 mega-bytes CPU storage, we would not be able to use a direct solution method to tackle this problem. On the other hand, the SBI technique requires about 1.0 mega-byte for the compressed global matrix, 1.1 mega-bytes for global stencil array, local stencil array, and node-mapping array, and 0.06 mega-bytes for the sub-matrix. Thus, a saving of about 85% of CPU storage with the SBI technique enable the simulation of this problem. Operational count indicates that the CPU time is proportional to the square of matrix band-width. Thus, the CPU time with SBI would have been only approximately equal to $(25/399)^2 \times 27 = 0.11$ times that with direct solution method. Thus, only a fraction of CPU time would be needed with SBI technique in comparison to direct solution method for large three-dimensional problems.

ACKNOWLEDGEMENTS

This research is supported by the Environmental Sciences Division, Oak Ridge National Laboratory, Oak Ridge, Tennessee. ESD Publication No. 3064.

REFERENCES

1. Yeh, G. T. (1985), Comparison of successive iteration and direct methods to solve finite element equations of aquifer contaminant transport, Water Resour. Res., Vol. 21, pp. 272-280.

2. Huyakorn, P. S., Springer, E. P., Guvanasen, V., and Wadsworth, D. (1986), A three-dimensional finite element model for simulating water flow in variably saturated porous media, Water Resour. Res., Vol. 22, pp. 1790-1808.

3. Yeh, G. T. (1987), 3DFEMWATER: A Three-Dimensional Finite Element Model of WATER Flow through Saturated-Unsaturated Media, ORNL-6386, Oak Ridge National Laboratory, Oak Ridge, Tennessee 37831.

4. Philip, J. R. (1960), General method of exact solution of the concentration-dependent diffusion equation, Aust. J. Phys., Vol. 13, pp. 1-12.

5. Solomon, D. K. and Yeh, G. T. (1987), Application of 3DFEMWATER to the Study of Trench "Bathtubbing", ORNL/RAP/LTR-87/89, Oak Ridge National Laboratory, Oak Ridge, Tennessee 37831.

SECTION 2C - MULTIPHASE FLOW

Numerical Simulation of Diffusion Rate of Crude Oil Particles into Wave Passes Water Regime

M.F.N. Abowei

Chemical/Petrochemical Engineering Department, Rivers State University of Science and Technology, P.M.B. 5080, Port Harcourt, Nigeria

ABSTRACT

Numerical models were developed to simulate the diffusion rate of crude oil particles into wave passes water regime. The developed models were used to correlate the diffusion rate as a function of the physical properties of the crude oil sample, wave dimensions and diffusion coefficient. The models were simulated with the aid of Hewlett Packard HP-85 computer and found that the diffusion rate was principally influenced by the crude oil density and wave phase differences. The correlated models are useful in predicting extent of crude oil diffusion as a result of oil spillage.

INTRODUCTION

Crude oil particles are discharged to the marine environment through pipelines or by release from ships and tankers; the density difference usually cause the crude oil particles to form into a surface plume or patch from which it subsequently diffuses into water. Because of the danger to affected communities and the recreational interests in inshore waters, it is important to be able to simulate the diffusion rate of crude oil particles into wave passes water regime.

Mechanism of dispersion of organic compounds on water regimes of flow was investigated by various authors [1-6]. Huh et al[1] developed correlation describing one-dimensional spreading rate of oil slicks on gravity surface tension viscosity control regime and similar model was also developed by Takahashi and co-authors[1] on surface tension - viscosity control regime. These studies were limited to only calm water surfaces. Yoran et al[2] conducted an investigation on mass transfer rates between phenol and water and only established the effect of bulk velocity in the diffusion rate of phenol and water. But there were no realistic mathematical models that correlate the physical properties of the oil, wave dimensions and diffusion coefficient.

It is the purpose of this paper therefore to formulate mathematical equations correlating functional parameters to simulate diffusion rate of crude oil particles into wave passes water regime. Crude oil samples from four oil wells in Nigeria were used for this study.

DIFFUSION RATE MODEL

The wave passes water regime takes simple harmonic motion (S.H.M.) and variation of the displacement, with time and phase difference, ϕ from the origin is given by

$$Y = \alpha \sin(\omega t - \phi) \tag{1a}$$

where

$$\omega = 2\pi f = 2\pi/T \text{ and } T = 1/f = \lambda/v.$$

$$\phi = 2\pi\chi/\lambda$$

Therefore equation (1) can be written as

$$Y = \alpha \sin 2\pi(t/T - \chi/\lambda) \tag{1b}$$

Since Y represents the displacement of a particle as the wave travels, the velocity, $\bar{v}$ of the particle of crude oil flowing at any instance is given by dy/dt and from equation (1b):

$$\bar{u} = dy/dt = \frac{2\pi\alpha}{T} \cos 2\pi(t/T - \chi/\lambda) \tag{2a}$$

or

$$\bar{u} = dy/dt = \omega\alpha \cos(\omega t - 2\pi\chi/\lambda) \tag{2b}$$

wave passes water regime of flow is actual result in turbulent diffusion of crude oil particles. A commonly used mass balance equation which discribed[3] turbulent situation is the following;

$$\partial c/\partial t - \frac{\partial}{\partial x}(uc - \varepsilon x \frac{\partial c}{\partial x}) - \partial/\partial y (vc - \varepsilon y \frac{\partial c}{\partial y}) - \partial/\partial z(wc - \varepsilon z \frac{\partial c}{\partial z}) = 0 \tag{3}$$

In its 1-dimensional average form as described[4,5] result in the following

$$\partial/\partial t(A\bar{c}) = -\partial/\partial z(Au\bar{c}) + \frac{\partial}{\partial z}(Ak_z \partial c/\partial z) \tag{4}$$

where

$$\varepsilon_z = Ak_z$$

Integration of equation (4) by variable separable technique result in

$$\bar{C}(z,t) = (M_o/A(4\pi k_z t)^{\frac{1}{2}}) \exp[-(z + \bar{u}t)/4k_z t] \tag{5}$$

But $M_o = \ell_o V_o$ and on substitution in equation (5) gives

$$C(z,t) = \ell_o V_o/A(4\pi k_z t)^{\frac{1}{2}}) \exp[-(z + ut)/4k_z t] \tag{6}$$

Substitution of equation (2a) or (2b) in (6) resulting in a general model for describing the diffusion rate of crude oil particles into wave passes water regime, thus,

$$C(z,t) = (\ell_o V_o/A.(4\pi k_z t)^{\frac{1}{2}}) \exp[-(z + \frac{2\pi t\alpha}{T}\cos 2\pi(t/T - \chi/\lambda)/4k_z t] \tag{10}$$

or in terms of angular velocity ω, it could be expressed as,

$$C(z,t) = (\ell_o V_o/A(4\pi k_z t)^{\frac{1}{2}}) \exp[-(z+\omega t\alpha \cos(\omega t - 2\pi\chi/\lambda)/4k_z t] \qquad (11)$$

DIFFUSION COEFFICIENT MODEL

Taylor[6] concluded the effective diffusion coefficient for turbulent flow regime as,

$$k_z = 10.06\psi\bar{u} \qquad (12)$$

The general mathematical model describing the diffusion coefficient of crude oil particles into wave passes water regime can be obtained by substituting equation (2a) or (2b) in (12).

Thus

$$k_z = ((20.12\psi\pi\alpha)/T)\cos 2\pi(t/T - \chi/\lambda) \qquad (13)$$

or

$$k_z = 10.06\omega\alpha\psi \cos(\omega t - 2\pi\chi/\lambda) \qquad (14)$$

COMPUTATIONAL PROCEDURE

The study was conducted in a Laboratory wind-wave tank shown schematically in figure 1. The water tank was 3m Long, 10m deep and 2m wide over which a wind tunnel 4.5m in length was mounted such that the air flow joined the water surface tangentially. The air flow as generated by a KDK standing fan mounted at the downward end of the tunnel. The following wave dimensions were measured; angular velocity, $\omega = 10\pi$ rad, frequency, f = 5 sec., amplude, α = 0.1m, period, T = 0.2 sec^{-1}, water depth, z = 10m and crude oil volume v = 0.5m³.

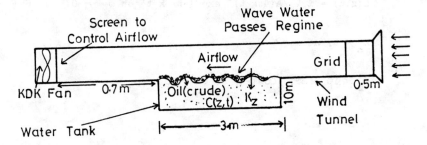

Fig.1: Wind-wave Tank used for experimental analysis

Physical properties of various crude oil samples were determined as listed in table 1 to ascertain their effect on the diffusion rate.

Table 1: Experimentally Determined Physical Properties of Crude Oil Samples

Crude Oil Sample	Surface Tension δ_o, N/M	Viscosity μ_o kg/m.h.	density ℓ_o, kg/m^3
Aran	0.0715	375.15	840
Adibawa	0.0400	599.76	880
Apara	0.0400	500.04	834
Obigbo	0.0360	119.88	915

RESULTS AND DISCUSSIONS

The data obtained were then fed into the computer using the developed mathematical models in equations (11) and (14) and the following results were obtained:

Phase Difference effect:

The phase difference which was attributed by the sinosoidal nature of the wave influences the diffusion rate. Complex and Real values were obtained as the time of diffusion increases at constant depth. The concentration content and the diffusion coefficients fluctuates, pending on the quadrant it falls upon as shown in table 2 and 3 respectively. The concentration content and the diffusion coefficients become independent of wave dimension at zero angular velocity which is a function of phase difference,

thus,
$$C(z,t) = \ell_o V_o/A(4\pi k_z t)^{\frac{1}{2}} \exp(-z/4k_z t) \text{ at } \omega \rightarrow 0$$

and
$$k_z = 0; \text{ at } \omega \rightarrow 0$$

but
$$k_z = 10.06\omega\alpha\psi \, \cos(-\phi) \text{ at } t \rightarrow 0$$

Table 2: Computed Results of Concentration Content of Crude Oil Particles into Wave Passes Water Regime

t, Sec.	C (z, t), kg/m³							
	Aran Crude		Adibawa Crude		Apara Crude		Obigbo Crude	
	$\phi=\pi$	$\phi=\pi/2$	$\phi=\pi$	$\phi=\pi/2$	$\phi=\pi$	$\phi=\pi/2$	$\phi=\pi$	$\phi=\pi/2$
0	0	0	0	0	0	0	0	0
5	8.4/i	8.4/i	8.8/i	8.8/i	8.3/i	8.3/i	9.2/i	9.2/i
10	12.6	8.4	13.2	8.8	12.5	8.3	13.7	9.15
15	4.2	5.5	4.4	5.7	4.2	5.4	4.6	6.00
20	3.3/i	3.4/i	3.5/i	3.6/i	3.3/i	3.38/i	3.6/i	3.7/i
25	2.94	2.98	3.08	3.12	2.92	2.96	3.20	3.25

Table 3: Computed Results of Diffusion Coefficient of Crude Oil Particles into Wave Passes Water Regime

t, sec.	K_z M²/Sec.	
	$\phi=\pi$	$\phi = \pi/2$
0	63.20	63.20
5	−47.4	−48.66
10	10.74	12.38
15	30.3	28.95
20	−58.0	−57.5
25	60.70	61.3

Density Effect

The density has a significant effect on overall diffusion rate but independent of diffusion coefficient as developed in equation (11) and (14) respectively. From comparative analysis of table 1 and 2, the concentration content as a result of diffusion is found to be higher with crude oil samples of high densities.

CONCLUSION

Mathematical correlations for the computation of diffusion rate and coefficient have been developed. Wave dimensions and crude oil physical properties influences the rate of diffusion. Density was found to have no effect on the diffusion coefficient.

NOMENCLATURES

A	=	Cross-sectional Area of water stream, m^2
ε_x, ε_y, ε_z	=	Diffusion coefficient of Longitudinal, transverse and vertical directions respectively, mol/m^3.Sec.
u, v, w,	=	Velocities of flow of Longitudinal, transverse and vertical direction respectively, m/sec.
ψ	=	Water stream diameter or radius, m
λ	=	Wave length, m

REFERENCES

1. Takahashi, T., Kitamura, Y., and Nakada, K. (1981) "Spreading of an Oil slick on calm water surfaces: Effect of the viscous drag of an oil slick on its spreading rate": International Chemical Engineering, vol.21, No.2, p.244-249.

2. Yoram, C., Donald, M., Wan, Y.S. (1980) "Mass Transfer Rates Between Oil slicks and water". Canadian Journal of Chemical Engineering Vol. 58, No.5, p.569-574.

3. Ippen, A.T. (1966) "Estuary and Coastline Hydrodynamics" Mac. Graw-Hill, Newyork pp.567.

4. Okubo, A (1964) "Equations Describing the Diffusion of an Introduced Pollutant in a one-dimensional estuary" In K. Yoshida (Ed), Studies in Oceanography, University of Washington Press P.216-226.

5. Abowei, M.F.N. and Wami, E.N. (1988) "Mathematical Modelling of Dissolution Rate of Crude Oil particles in water" ASME Periodicals, B, Vol.15 No.4, P.1-7 (France).

6. Taylor, G.T. (1954) "The Dispersion of Matter in Turbulent Flow through a pipe" Proc. R. Soc., Vol.A223, p.446-468.

A Decoupled Approach to the Simulation of Flow and Transport of Non-Aqueous Organic Phase Contaminants Through Porous Media

H.W. Reeves and L.M. Abriola
Department of Civil Engineering, The University of Michigan, Ann Arbor, MI 48109, USA

INTRODUCTION

The introduction and migration of slightly miscible non-aqueous organic fluids in the subsurface is an environmental problem that is receiving increasing attention. In the past several years, methods initially developed to model other types of multiphase flows in porous media (flow in petroleum reservoirs, for example) have been applied to this multiphase pollution problem. Several models have been presented in the literature which investigate immiscible fluid flow. Due to space constraints, these models will not be discussed herein. Mathematical models that consider both the flow of a non-aqueous phase liquid (NAPL) and the transport of organic contaminants in the other fluid phases include the compositional model of Abriola and Pinder[1] and the mathematical development in Corapcioglu and Baehr[2].

The only documented implementation of this "miscible" modeling approach which is capable of describing the flow and transport of several phases and species simultaneously is that presented in Abriola and Pinder[1] and Abriola[3]. In this model, species mass balance equations are solved simultaneously resulting in a formulation which can be highly inefficient and inflexible (Pinder and Abriola[4]). The addition of a new organic species for consideration in this model results in the addition of a new species mass balance equation, leading to matrix restructuring and a larger matrix problem for solution. For multi-species problems, such an approach becomes computationally prohibitive. In this paper, a decoupled modeling approach to the multiphase pollution problem is presented. This approach is similar in concept to that employed by Nghiem et al.[5] to model advanced oil recovery in the petroleum engineering literature. Two recent abstracts have made reference to a decoupled approach, but the details of the implementation are not yet available (Falta and Javandel[6]; Parker et al.[7])

EQUATION DEVELOPMENT

Flow Equations

For a multiphase mixture, the microscopic mass balance equation for a species i in a phase α may be written as

$$\frac{\partial}{\partial t}\left(\rho^\alpha \varepsilon_\alpha \omega_i^\alpha\right) + \nabla \cdot \left(\rho^\alpha \varepsilon_\alpha \omega_i^\alpha v^\alpha\right) - \nabla \cdot J_i^\alpha = E_i^\alpha \tag{1}$$

Where:

α = o, w, g; oil, water, gas and solid phases,
ρ^α = density of the α phase,
ε_α = volume fraction of the α phase,

ω_i^α = mass fraction of species i in the α phase,
$\mathbf{v}^\alpha$ = mass average velocity of the α phase,
$\mathbf{J}_i^\alpha$ = non-advective flux of species i in the α phase,
E_i^α = source/sink term arising from mass exchange between phases.

The flow equation for each fluid phase; o, w, g; is obtained by summing equation (1) over all species in each phase;

$$\frac{\partial}{\partial t}\left(\rho^\alpha \varepsilon_\alpha\right) + \nabla \cdot \left(\rho^\alpha \varepsilon_\alpha \mathbf{v}^\alpha\right) = E^\alpha \qquad (2)$$

Here the definition of mass fraction, requiring that the sum of all mass fractions in any phase be unity, has been employed and

$$E^\alpha = \sum_i E_i^\alpha$$

The following assumptions and ancillary equations are applied to equation (2) to develop flow equations for each phase:

(1) $\mathbf{v}^\alpha$ is described by the extended form of Darcy's Law to account for multiphase flow (Bear[8]),

$$\mathbf{v}^\alpha = -\frac{\mathbf{k} k_{r\alpha}}{\varepsilon_\alpha \mu^\alpha} \cdot \left(\nabla P^\alpha - \rho^\alpha \mathbf{g}\right) \qquad (3)$$

where:
$\mathbf{k}$ = intrinsic permeability tensor,
$k_{r\alpha}$ = relative permeability of α phase (a function of α phase saturation),
μ^α = kinematic viscosity of the α phase,
P^α = pressure of the α phase,
$\mathbf{g}$ = acceleration of gravity vector, positive down.

(2) Individual soil grains are incompressible, deformations are small and the soil behaves elastically.
(3) Contaminants are slightly soluble in water.
(4) Water is the wetting phase and phase saturations (s_α) may be described by a unique function of capillary pressure (hysteresis is neglected).

$$s_\alpha = s_\alpha(P_{ow}, P_{wg})$$

where

$$P_{\alpha\beta} = P^\alpha - P^\beta$$

is the capillary pressure and P^α and P^β are the pressures of two fluid phases.

(6) The gas phase is assumed to remain at a constant pressure. Contaminants are allowed to move through the gas phase via a concentration controlled diffusion mechanism.

The resultant equation describing the flow of the water or oil phase is of the form.

$$\varepsilon\left[\frac{\partial s_\alpha}{\partial P_{ow}}\frac{\partial P_{ow}}{\partial t} + \frac{\partial s_\alpha}{\partial P_{wg}}\frac{\partial P_{wg}}{\partial t}\right] + s_\alpha \alpha \frac{\partial P_{wg}}{\partial t} -$$

$$\nabla \cdot \left[\frac{\mathbf{k} k_{r\alpha}}{\mu^\alpha} \cdot \left(\nabla P^\alpha - \rho^\alpha \mathbf{g}\right)\right] = \frac{E^\alpha}{\rho^\alpha} - \frac{\varepsilon s_\alpha}{\rho^\alpha}\frac{\partial \rho^\alpha}{\partial t} \qquad \alpha = o, w \qquad (4)$$

where,
P^α = pressure of α phase, for the oil equation, P^o is replaced by $P_{ow} + P_{wg} + P^g$.
for the water equation, P^w is replaced by $P_{wg} + P^g$. Recall that P^g is constant.
ε = porosity,
α = compressibility of soil matrix.

The mass exchange terms, E^α, must be determined through the solution of species transport equations.

Transport Equations

Transport equations may also obtained from equation (1). A transport equation must be solved for each species of interest. Sorption of a species to the soil grains, chemical reactions within any phase, and biological degradation of the species are not considered herein. The transport equations are obtained by summing each species equation over the phases. Note that the exchange terms cancel out since mass conservation requires that the sum of E_i^α over α is zero. The transport equation for species i is,

$$\sum_\alpha \frac{\partial}{\partial t}\left(\rho^\alpha s_\alpha \varepsilon \omega_i^\alpha\right) + \nabla \cdot \left(\rho^\alpha s_\alpha \varepsilon \omega_i^\alpha \mathbf{v}^\alpha\right) - \nabla \cdot \left(D^\alpha \rho^\alpha s_\alpha \varepsilon \nabla \omega_i^\alpha\right) = 0 \quad (5)$$

Here, a Fickian form of J_i^α has been assumed.

In order to fully describe the transport problem, the mass fractions for a species i in each phase must be related. This model uses the familiar local equilibrium assumption to relate the mass fractions. Assuming that the mass fraction of a species i at any point is in equilibrium with all the phases present at that point allows the mass fractions to be related using a partition coefficient of the form,

$$K_i^{\alpha\beta} = \frac{\omega_i^\alpha}{\omega_i^\beta} \quad (6)$$

The mathematical model describing the flow and transport is the system of partial differential equations (4) and (5), coupled through the source and sink terms which are determined through the use of equation (1). These equations are, in general, non-linear and not amenable to analytical solution.

NUMERICAL IMPLEMENTATION

The partial differential equations represented by (4) and (5) are reduced to ordinary differential equations through the use of a Petrov-Galerkin finite element method. Let either set be represented by the function,

$$L(u) = 0. \quad (7)$$

A weighted residual equation may then be written as,

$$\int_\Omega L(u) \, M_i \, d\Omega = 0. \quad (8)$$

Unknowns and nonlinear coefficients may be expressed as:

$$u \approx \hat{u} = \sum_{i=o}^{N} u_i N_i \quad (9)$$

Where N_i are the familiar chapeau functions (Lapidus and Pinder[9]).

For equation (4) the variables approximated in this fashion are the capillary pressures, the phase saturations, the densities, the mobilities, the exchange terms and the saturation derivatives. Approximated variables in equation (5) are the mass fractions, the product of phase density and saturation and the exchange terms. Note that since the pressure has been approximated by a linear basis function in set (4), the velocity and dispersion coefficients appearing in (5) are calculated to be constant in each element. Since the divergence of the velocity term is not defined using this approximation, the derivatives in equation (5) are expanded by the chain rule and equation (2) is substituted yielding

$$\sum_\alpha \rho^\alpha \varepsilon s \frac{\partial \omega_i^\alpha}{\partial t} + \rho^\alpha \varepsilon s v^\alpha \cdot \nabla (\omega_i^\alpha) + \nabla \cdot \left(D^\alpha \rho^\alpha s_\alpha \varepsilon \nabla \omega_i^\alpha \right) = -\sum_\alpha \omega_i^\alpha E^\alpha \quad (10)$$

The Petrov-Galerkin method is characterized by the use of a different basis function to approximate the unknown variables than to weight the residual. (Hughes and Brooks[10]) An advantage of this method is that controlled upstream weighting techniques may be easily implemented. For this study the weighting functions are of the form,

$$M_i = N_i + p \frac{dN_i}{dx} \quad (11)$$

Note that the standard Galerkin technique is obtained by setting the coefficient p to zero.

Integration yields the matrix equation

$$[A]\left\{\frac{\partial \hat{u}}{\partial t}\right\} + [B]\{\hat{u}\} = \{f\} \quad (12)$$

where;
  [A], [B] = matrices arising from spatial discretization,
  $\{\hat{u}\}$ = unknown vector,
  $\{f\}$ = vector arising from boundary condition contributions.

The temporal derivatives in equation (12) are discretized using a variably weighted finite difference scheme. The solution procedure is as follows. The discretized forms of equation (4) are solved by a Picard iteration technique for the case of E^α equal to zero yielding the phase pressures. These pressures are used to calculate the velocity of each phase using equation (3). The discretized forms of (10) are then solved iteratively using this flow field. The E_i^α terms are lagged one iteration and evaluated through the use of equation (1). The source and sink terms are subsequently put into equation (4) and the pressures updated. This procedure is repeated until convergence is obtained in both sets of equations.

SAMPLE PROBLEM

To illustrate the use of the decoupled technique, results from a sample problem are now presented. The sample problem was described in detail in Abriola[3] and is similar to that in Abriola and Pinder[11]. The sample concerns the infiltration of a organic mixture consisting of a heavy oil and propane into a soil column with a residual water content. The partitioning of propane between the phases is described by a saturation dependent equation with constant molar partition coefficients. The boundary conditions are first-type conditions at the upstream node for the capillary pressures and mass fraction variables. Second-type conditions are used for the capillary pressures at the downstream node. The mass fraction is subject to a first-type condition at this node. The initial conditions consist of a water saturation at its residual level, 0.2, and a negligible organic saturation. The initial mass fraction of propane in the domain is 0.0. At the upstream node, propane has a mass fraction of 0.001. A fully implicit time approximation was used and the Petrov-Galerkin weighting factor, p, was set to zero for the simulation results presented below. Following are two figures which show the movement of the organic phase into the domain and the transport of the volatile propane in the gas phase. Notice that the contaminant plume in the gas phase extends beyond the organic saturation front due to the diffusional transport of the contaminant.

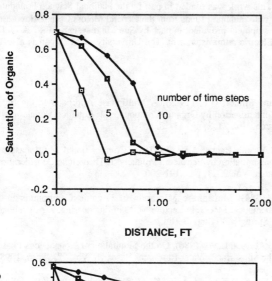

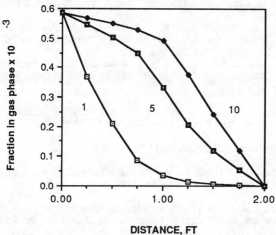

SUMMARY

In this paper a decoupled approach to the simulation of simultaneous flow and transport in porous media was presented. A sample problem illustrates application of this technique to the case of an infiltrating contaminant. Using this decoupled approach, additional species may be considered by increasing only the number of transport equations. The flow equations are not changed. The chemical expressions that define the relationships between mass fractions in each phase appear only in the transport equations. This formulation of the mass balance equations allows these expressions to be changed conveniently without major revision to the rest of the numerical model. The increased flexibility makes the decoupled technique desirable and the smaller matrix equations resulting from this formulation lead to a more efficient numerical model.

Acknowledgments This work was funded in part by the National Science Foundation under Grant ECE-8451469 with matching funds from the General Motors Corporation. Simulations were performed on equipment provided, in part, by Sun Microsystems, Inc. A Ph.D. student loan from General Electric administered by The University of Michigan is also gratefully acknowledged.

REFERENCES

1. Abriola, L.M. and Pinder, G.F. (1985), A Multiphase Approach to the Modeling of Porous Media Contamination by Organic Compounds, 1. Equation Development, Water Resources Research, Vol. 21, 1, pgs 11-18.

2. Corapcioglu, M.Y. and Baehr, A.L. (1987), A Compositional Multiphase Model for Groundwater Contamination by Petroleum Products, 1. Theoretical Considerations, Water Resources Research, Vol. 23, 1, pgs. 191-200.

3. Abriola, L.M. (1984). Multiphase Migration of Organic Compounds in a Porous Medium: A Mathematical Model. in Lecture Notes in Engineering, C.A. Brebbia and S.A. Orszag, editors. Springler-Verlag. Berlin.

4. Pinder, G.F. and Abriola, L.M. (1986), On the Simulation of Nonaqueous Phase Organic Compounds in the Subsurface, Water Resources Research, Vol. 22, 9, pgs. 109S-119S.

5. Nghiem, L.X.; Fong, D.K.; and Aziz, K. (1981), Compositional Modeling with an Equation of State, Society of Petroleum Engineers Journal, December, pgs. 687-698.

6. Falta, R.W., Jr. and Javandel, I. (1987), A Numerical Method for Multiphase Multicomponent Contaminant Transport in Groundwater Systems, EOS, Vol. 68, 44, pg. 1284.

7. Parker, J.C.; Kuppusamy, T.; and Lenhard, R.J. (1986), Modeling Organic Chemical Transport in Three Fluid Phase Porous Media, EOS, Vol. 67, 44, pg. 945.

8. Bear, J. (1979). Hydraulics of Groundwater. McGraw-Hill. New York.

9. Lapidus, L. and Pinder, G.F. (1982). Numerical Solution of Partial Differential Equations in Science and Engineering. John Wiley and Sons. New York.

10. Hughes, T.J.R. and Brooks, A.N. (1982). A Theoretical Framework for Petrov-Galerkin Methods with Discontinuous Weighting Functions: Applications to the Streamline-Upwind Procedure. in Finite Elements in Fluids. Gallagher, R.H. *et al.*, editors. Vol. 4. John Wiley and Sons. London.

11. Abriola, L.M. and Pinder, G.F. (1985), A Multiphase Approach to the Modeling of Porous Media Contamination by Organic Compounds, 2. Numerical Simulation, Water Resources Research, Vol. 21, 1, pgs. 19-26.

INVITED PAPER
The Transition Potentials Defining the Moving Boundaries in Multiphase Porous Media Flow
H.O. Schiegg
SIMULTEC AG, Burgrain 37, CH-8706 Meilen, Zürich, Switzerland

ABSTRACT

The transition potential is the fluid potential for the transition between fluid-dynamic existence of a fluid (funicular saturation) and its non-existence (residual saturation). The locus of such a transition defines a solution dependent boundary. This boundary condition is shown not only for the two-fluid-flow but also for the multiphase flow.

1. CRUCIAL POINTS FOR NUMERICAL SIMULATIONS

Multiphase flow in porous media means flow of non-mixing fluids. In a porous medium non-mixing fluids are subject to capillarity. Capillarity is specific for each pair of fluids. Consequently, the number of combinations of fluid pairs determines the number of different capillary influences: one for two fluids, three for three fluids and so on.

The locus of any transition between residual and funicular saturation represents a boundary. The numerical handling of such solution dependent boundaries as well as of the hysteresis of capillarity are known from the simulation of the groundwater surface.

For the computer simulation of multiphase flow there is another crucial point: The overlapping of the different capillary influences in general and specifically, the determination of the transition potentials, which is described in the following.

2. DEFINITIONS

The Transition Potential (Φ^T) is the potential of a fluid for transition from non-existence to existence and vice versa at a certain level z within a containment filled with one or more other non-mixing fluids. Such a containment may be a tube or any other container, a capillary or a pore space of a porous medium.

The Threshold Potential (Φ^{Tmin}) is the minimum transition potential. It is the potential required for fluid transition within a containment at all.

In a porous medium, transition from existence into non-existence and vice versa means transition from funicular saturation stage (existence) to residual saturation stage (non-existence) and vice versa. This fluid dynamic definition of existence is based on the concept of relative permeability, according to which funicular saturation implies a permeability greater than zero, whereas residual saturation means zero permeability.

3. TWO NON-MIXING FLUIDS

3.1 Level of interface

The level z of an interface, z^{Intf}, is a function of the potentials (Φ) of the two fluids. It can be determined by subtracting the two fluid potential equations, both for the level of the interface. In order to get a positive density (ρ) difference, the subtraction must be, if "h" stands for "heavy fluid", "l" stands for "light fluid", and "p" for pressure:

$$(\Phi_h - \Phi_l) = (p_h - p_l) + (\rho_h - \rho_l) g \, z^{Intf} \tag{1}$$

where (2) (3)

$$z^{Intf} = \frac{(\Phi_h - \Phi_l)}{(\rho_h - \rho_l)g} - \frac{(p_h - p_l)}{(\rho_h - \rho_l)g} \; ; \quad z^{Tab} = \frac{(\Phi_h - \Phi_l)}{(\rho_h - \rho_l)g}$$

3.1.1 <u>in a tube</u>: In a tube or any other containment with wide space between opposite walls the interface is a plane and as such called a table (Tab). The pressure difference between both sides of a table is zero. Consequently, the second term in Eq.2 vanishes.

3.1.2 <u>in a capillary</u>: Capillarity is due to anisotropy in the molecular force fields along an interface. They cause the interface tension (γ), the wetting angle (α) and the attraction of the wetting fluid, sucking it into the area of the non-wetting fluid. Consequences are a lower pressure in the wetting fluid and a curved interface, which is called a meniscus (Men). It is convex for the non-wetting (nw) fluid and concave for the wetting (w) fluid, since $p_{nw} > p_w$. The pressure difference due to capillarity is called capillary pressure (p_c), which, as a positive value, must be defined as follows

$$p_c = (p_{nw} - p_w) \tag{4}$$

or based on the Laplace-Eq. of capillarity as $p_c = \gamma \cdot \cos\alpha \cdot (2/R)$ where R represents the radius of the capillary or pore assuming $R = r \cdot \cos \alpha$, with r as the radius of curvature of the meniscus.

A capillary pressure (p_c) can be visualized by the capillary head (h_c), which is the capillary rise (a_c) under static conditions. For dynamic conditions the capillary rise changes compared to statics, whereas the capillary pressure (p_c), thus, also the capillary head (h_c) stay the same (e.g. Schiegg[1]).

Introduction of Eq.4 into Eq.2 yields

a) for h = w, l = nw: b) for h = nw, l = w: (5)

$$z^{Men} = \frac{(\Phi_h - \Phi_l)}{(\rho_h - \rho_l)g} + \frac{p_c}{(\rho_h - \rho_l)g} \ ; \qquad z^{Men} = \frac{(\Phi_h - \Phi_l)}{(\rho_h - \rho_l)g} - \frac{p_c}{(\rho_h - \rho_l)g}$$

As difference between the meniscus and the table the second terms in the equations above show the capillary head upwards (positive, above the table) and downwards (negative, below the table) depending on wettability, hence, $h_c = p_c / (\rho_h - \rho_l)g$.

3.1.3 <u>in a porous medium</u>: Microscopically an interface in a porous medium is composed of the outmost menisci and their inter-connections. Such an interface has an extremely wild topography with deep valleys and outranging high peaks. However, Darcy flow requires macroscopic view, i.e. smearing of the discrete microscopic conditions into a homogenized continuum. Macroscopically an interface in a porous medium is represented by the capillary curve which shows the fluid saturation (S) perpendicular to the table. The levels of such an interface vary from the level of the largest menisci with the smallest capillary head to the level of the smallest menisci with the biggest capillary head.

Transition means change between funicular and residual saturation (RS). Thus, the relevant saturation for the wetting fluid is its maximum pendular (pend) residual saturation (RSw) corresponding with the maximum capillary pressure (max.p_c), whereas for the non-wetting fluid it is its maximum insular (ins) residual saturation (RSnw) corresponding with the minimum capillary pressure (min.p_c).

3.2 The two Transition Potentials
The two Transition Potentials for any level z follow from Eq.1 with Eq.4 as:

for h = w, l = nw: (6) for h = nw, l = w: (7)

a): $\Phi_h^T = \Phi_l - p_c^{(*)} + (\rho_h - \rho_l)gz \qquad \Phi_h^T = \Phi_l + p_c^{(*)} + (\rho_h - \rho_l)gz$

b): $\Phi_1^T = \Phi_h + p_c^{(*)} - (\rho_h - \rho_1)gz \qquad \Phi_1^T = \Phi_h - p_c^{(*)} - (\rho_h - \rho_1)gz$

3.2.1 <u>Specification for a tube:</u> $\qquad p_c^{(*)} = 0$

3.2.2 <u>Specification for a capillary j:</u> $p_c^{(*)} = p_{cj}$

3.2.3 <u>Specification for a porous medium:</u>

- for a wetting fluid: $\qquad p_c^{(*)} = \max \cdot p_c = p_{c[RSw]}$
- for a non-wetting fluid: $\qquad p_c^{(*)} = \min \cdot p_c = p_{c[1-RSnw]}$

3.2.4 <u>Specification due to hysteresis</u>: Due to hysteresis of capillarity the capillary pressure must be specified according to imbibition (IM) or drainage (DR). Imbibition means the displacement of the non-wetting fluid by the wetting fluid. Drainage means the contrary.

	wetting fluid	non-wetting fluid:
Drainage:	$p_c^{(*)} = p_{c[RSw]}^{DR}$	$p_c^{(*)} = p_{c[1-RSnw]}^{DR}$
Imbibition:	$p_c^{(*)} = p_{c[RSw]}^{IM}$	$p_c^{(*)} = p_{c[1-RSnw]}^{IM}$

3.3 <u>The two Threshold Potentials:</u>
defined as minimum transition potentials, are determined for $h = w$, $1 = nw$ by the later Eq.15 with $W = h$ and $A = 1$.

4. THREE NON-MIXING FLUIDS

4.1 The three capillarities

If there are three non-mixing fluids, a heavy (h), a medium (m) and a light (l) one, there exist three pair combinations, thus, three capillary pressures and three capillary heads, which must be specified:

$$p_c\Big|_h^l, \ p_c\Big|_h^m, \ p_c\Big|_m^l \ ; \qquad h_c\Big|_h^l, \ h_c\Big|_h^m, \ h_c\Big|_m^l$$

4.2.1 <u>Wetting combinations</u>: For each of the three pairs (h/l, h/m, m/l with dividend = wetting, divisor = non-wetting fluid), there are two wetting possibilities (as h/l and l/h), hence, in total there are eight wetting combinations.

Examples are: i) water (W), oil (O), air (A) in quartz for h/l, h/m, m/l since $\rho_W > \rho_O > \rho_A$. In an organic porous medium, as humus, wettability between water and oil changes compared to quartz, thus, the wetting combination is: h/l, m/h, m/l.

4.2.2 Virgin and maculated capillarity

Once a virgin (virg) meniscus between the heavy and the light fluid is maculated by the medium fluid, the virgin meniscus will disappear. Instead, a double meniscus will be relevant. It is composed of the two coinciding menisci of the medium fluid, one against the heavy and one against the light fluid. The double meniscus is called maculated (mac) meniscus. Accordingly exists a maculated capillary curve. Experimental confirmation: see Schiegg[2](p.81).

4.2.3 Thickness of layer of medium fluid

$H_m^{Tube} = z^{Tab}|_{lm}^{l} - z^{Tab}|_h^m$ is the thickness of the medium fluid in a tube. With Eq.3 it can be determined as function of the three potentials. For Φ_m follows explicitly

$$\Phi_m = \left[H_m^{Tube} \cdot g \cdot (\rho_m - \rho_l)(\rho_h - \rho_m) + \Phi_h(\rho_m - \rho_l) + \Phi_l(\rho_h - \rho_m) \right] \frac{1}{(\rho_h - \rho_l)} \quad (8)$$

In a capillary or porous medium the thickness of a layer of the medium fluid is based on Eqs.5. It depends on the wetting combination. In a porous medium it is variable, as shown by the saturation picture.

4.2.4 Saturation picture

A saturation picture represents the influence of the three capillarities. In the following context only the wetting combination (h/l, h/m, m/l; see sect.4.2.1) is looked at by means of the fluids water (W), oil (O) and air (A) with $\rho_W > \rho_O > \rho_A$ in quartz, see Fig.1.

4.2.5 Relative Oil-Potential

A change of potential in oil is effective only as much as it is a change against the other two potentials. For convenience as well as for dynamic conditions the relative oil potential (Φ_r) is introduced. It is characterized by taking the table between water and air as reference (z_0). For the relative potential it follows with $z^{Tab} = z_0 = 0$ and $\Phi_A \sim 0$ because of $\rho_W > \rho_O \gg \rho_A \sim 0$ from Eq.3 $\Phi_W = \Phi_A \sim 0$ and furthermore from Eq.8 with $\rho_W \doteq 1$

$$\Phi_r = H_O^{Tube} \cdot g \cdot \rho_O \cdot (1 - \rho_O) \quad (9)$$

4.3 The Transition Potentials

4.3.1 Without any oil

The virgin capillary curve is the only relevant one. The Transition Potentials are determined according to Eqs.6, Sect.3.2.3 and Fig.1 (see c,d):

$$\Phi_W^T = \Phi_A - p_{c(RSW)}^{virg} + (\rho_W - \rho_A)gz; \quad \Phi_A^T = \Phi_W + p_{c(1-RSA)}^{virg} - (\rho_W - \rho_A)gz \quad (10)$$

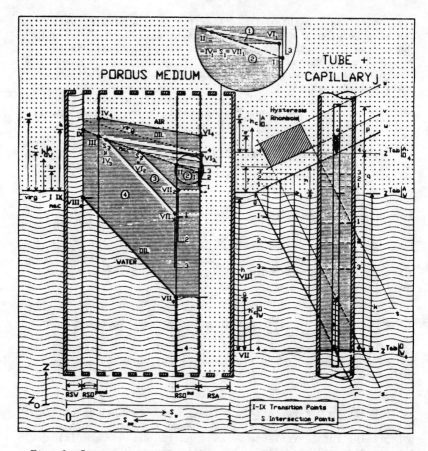

Fig. 1. Saturation picture for water, oil and air in quartz

$a = h_{c[RSW]}^{mac}$; $\quad b = h_{c[1-RSA]}^{mac}$; $\quad c = h_{c[1-RSA]}^{virg}$; $\quad d = h_{c[RSW]}^{virg}$

$e = h_c\big|_{O[1-RSA]}^{A}$; $\quad f = h_c\big|_{O[RSW+RSO^{pend}]}^{A}$; $\quad g = h_c\big|_{O[1-RSA-RSO^{ins}]}^{A}$

$h = h_c\big|_{W[RSW]}^{O}$; $\quad y = h_c^{S_{II}}{}_{[1-RSA-RSO^{ins}]}^{mac\ IMIM}$; $\quad i = h_c\big|_{W[1-RSA-RSO^{ins}]}^{O\ IM}$

$k = h_c\big|_{W[1-RSA-RSO^{ins}]}^{O\ DR}$; $\quad\quad l = \Phi_r$ (see sect. 4.2.5 Eq.9)

$m/n = (\rho_W-\rho_O)/\rho_O$; $\quad o = h_c\big|_{O[1-RSA-RSO^{ins}]}^{A\ IM}$; $\quad p = h_c\big|_{O[1-RSA-RSO^{ins}]}^{A\ DR}$

$q = H_0^{Tube}$ (see sect. 4.2.3); $\quad r = z^{Tab}\big|_{W(\Phi_r)}^{O}$; $\quad s = z^{Men}\big|_{W(\Phi_r)}^{O\ IM}$

$t = z^{Men}\big|_{W(\Phi_r)}^{O\ DR}$; $\quad u = z^{Tab}\big|_{O(\Phi_r)}^{A}$; $\quad v = z^{Men}\big|_{O(\Phi_r)}^{A\ IM}$; $\quad w = z^{Men}\big|_{O(\Phi_r)}^{A\ DR}$

4.3.2 With funicular oil only:

With funicular oil only the transition for oil must be due to a primary entry which causes the irreversible change from virgin to maculated conditions.

When the relative oil potential reaches the Threshold Potential, see table positions "1" in Fig.1, the oil shows up for the first time in the porous medium and this in point II, as explained in more detail in the next section 4.4. With increasing relative oil potential the oil enters the porous medium, as indicated in Fig.1, along the maculated capillary curve. After the intersection point S has started in point II with the table positions "1" and has followed the maculated capillary curve, as indicated by S_2 for the table positions "2" and S_3 for the table positions "3", S finally reaches point III. Subsequently, the point of transition follows along the vertical upwards as shown by point IV for table positions "4".

The transition potentials are determined according to Eq.6. Above the maculated capillary curve, the capillary curve between oil and air is relevant, below it is the capillary curve between water and oil. Depending on whether the level z under consideration is below z_{II}, above z_{III} or between z_{II} and z_{III}, different capillary pressures are relevant as can be seen from Fig.1 and the Eqs. 12–14. z_{II} and z_{III} are determined according to Eq.5a.

$$z_{II} = z^{Tab}\Big|_W^A + p_c^{mac}[1-RSA-RSO^{ins}]; \quad z_{III} = z^{Tab}\Big|_W^A + p_c^{mac}[RSW+RSO^{pend}]$$
(11)

The dependence on the potentials of water and air enters by Eq.3 determining the level of the table in Eq.11.

for $z > z_{II}$:
$$\Phi_O^T = \Phi_A - p_c\Big|_O^A[RSW+RSOpend] + (\rho_O - \rho_A)gz \quad (a)$$
$$\Phi_A^T = \Phi_O + p_c\Big|_O^A[1-RSA] - (\rho_O - \rho_A)gz \quad (b)$$
(12)

for $z < z_I$:
$$\Phi_W^T = \Phi_O - p_c\Big|_W^O[RSW] + (\rho_W - \rho_O)gz \quad (a)$$
$$\Phi_O^T = \Phi_W + p_c\Big|_W^O[1-RSA-RSO^{ins}] - (\rho_W - \rho_O)gz \quad (b)$$
(13)

for $z_I < z < z_{II}$:
$$\Phi_O^T = \Phi_A - p_{c(z)}^{(*)} + (\rho_O - \rho_A)gz \quad (a)$$
$$\Phi_A^T = \Phi_O + p_c\Big|_O^A[1-RSA] - (\rho_O - \rho_A)gz \quad (b)$$
(14)

$$p_{c(z)}^{(*)} = p_c^{min} + (p_c^{max} - p_c^{min})(z - z_I)/(z_{II} - z_I)$$

with $\quad p_c^{max} = p_c\Big|_O^A[RSW+RSO^{pend}]; \quad p_c^{min} = p_c\Big|_O^A[1-RSA-RSO^{ins}]$

4.3.3 <u>With residual oil</u>: Existence of residual oil indicates that the relative oil potential, at least once, must have been greater. In this case in the saturation picture the areas for insular residual oil (RSO^{ins}) and pendular residual oil (RSO^{pend}) are occupied with residual oil instead of water and air as in Fig.1. However, the determination of the Transition Potentials still follows Eqs.12-14.

4.4 The Threshold Potentials

As defined in sect.2 the Threshold Potential is the minimum Transition Potential and is the potential required for a fluid to be existent within a containment at all. According to Eqs.6 and 7 the following statements hold:

1. With a constant potential in the lighter fluid the transition potential for the heavier fluid is the lower, the smaller z, since the gravity difference is positive by definition.

2. With a constant potential in the heavier fluid the transition potential for the lighter fluid is the lower, the bigger z.

These two statements determine the Threshold Potentials by inserting the following z into Eqs.6 and 7:

- For the heavy fluid the smallest z, i.e. the lower boundary of the containment is relevant. The potential for the heavy fluid is reduced by the Entry Presssure ($p_c(*)$) in case the heavy fluid is the wetting one, see Eq.6a and correspondingly increased, if the heavy fluid is the non-wetting one, see Eq.7a.

- For the light fluid the biggest z, the upper boundary of the containment is relevant. Again, the potential for the light fluid is reduced by the Entry Pressure ($p_c(*)$) in case the light fluid is the wetting one, see Eq.7b and correspondingly increased, if it is the non-wetting one, see Eq.6b.

- For the medium fluid the relevant z is the z of the interface between the heavy and light fluid due to the contrary course of the two statements above.

Applied to water, oil and air in quartz this last recognition says, that the interface between water and air forms the threshold for the oil. In order to become existent in the containment the oil must overcome this threshold with its interface against air. This fact gave the Threshold Potential its name.

The z-levels of the interfaces which must be equal, are those of oil and air and of water and air, specifically of the maculated interface (see sect. 4.2.2), both determined according to Eq.5a. In a porous medium (quartz) the threshold for the oil is, at microscopic view, formed by the deepest valleys of the interface between water and air, i.e. at macroscopic view that the threshold is determined by z_I and that for Eq.5a the relevant capillary pressures are those for [1-RSA]. However, these capillary pressures are theoretical. Before oil physically can enter a porous medium, it must reach funicular saturation, thus the insular saturation must be built up first. Therefore, the effective threshold is equal to z_{II} and in order to determine the Threshold Potential for oil the relevant capillary pressures are those for [1-RSA-RSO^{ins}] as can be seen from the following Eqs.16c and 17. As shown in Fig.1 at Threshold Potential, which corresponds with the table positions "1", the oil first shows up in point II with its minimum distribution in the porous medium in form of a thin layer crossing the range of insular oil.

Summarizing, the Threshold Potentials for water, oil and air in quartz are according to Eqs.10 for Eqs.15, Eq.13a for Eq.16a. Eq.12b for Eq.16b and Eq.14a for Eq.16c.

<u>without any oil</u>:

$$\Phi_W^{Tmin} = \Phi_A - P_{c[RSW]}^{virg} + (\rho_W - \rho_A) \cdot g \cdot z_{(\text{lower boundary})} \quad (15)$$

$$\Phi_A^{Tmin} = \Phi_W + P_{c[1-RSA]}^{virg} - (\rho_W - \rho_A) \cdot g \cdot z_{(\text{upper boundary})}$$

<u>with oil</u>:

$$\Phi_W^{Tmin} = \Phi_O - P_c \Big|_W^O [RSW] + (\rho_W - \rho_O) \cdot g \cdot z_{(\text{lower boundary})} \quad (a$$

$$\Phi_A^{Tmin} = \Phi_O + P_c \Big|_{O[1-RSA]}^A - (\rho_O - \rho_A) \cdot g \cdot z_{(\text{upper boundary})} \quad (b \quad (16)$$

$$\Phi_O^{Tmin} = \Phi_A - P_c \Big|_{O[1-RSA-RSO^{ins}]}^A + (\rho_O - \rho_A) \cdot g \cdot z_{[1-RSA-RSO^{ins}]}^{mac} \quad (c$$

where $z_{[1-RSA-RSO^{ins}]}^{mac} = \dfrac{\Phi_W - \Phi_A}{(\rho_W - \rho_A)g} + \dfrac{P_c^{mac}[1-RSA-RSO^{ins}]}{(\rho_W - \rho_A)}$

which reduces for the Relative Treshold Potential according to sect. 4.2.5 to

$$\Phi_r^{Tmin} = \rho_o \cdot g \cdot P_{c[1-RSA-RSO^{ins}]}^{mac} - P_c \Big|_{O[1-RSA-RSO^{ins}]}^A \quad (17)$$

This relationship says that the Relative Threshold Potential must produce an oil column in air of equal height as the threshold represented by the minimum maculated capillarity between water and air but may be reduced by the capillarity between oil and air.

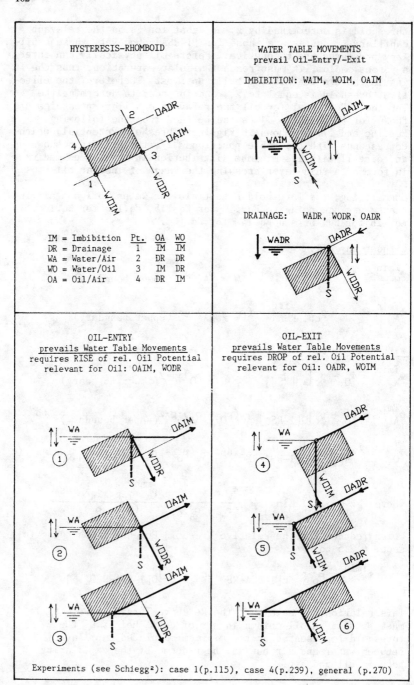

Fig. 2 Hysteresis of transition

4.5 Hysteresis Rhomboid

By taking the hysteresis of capillarity into consideration each of the three capillary heads is subject to hysteresis and may vary between its two extreme values, the one for drainage (DR) and the other one for imbibition (IM). As a consequence and since a maculated capillary head assumes a double meniscus composed of two coinciding oil menisci, it results for the maculated capillary head not only a one dimensional hysteresis range but a two dimensional one, the so called Hysteresis Rhomboid, as shown in Fig.1. Its lower extreme is composed of the two imbibition menisci for oil; its upper extreme consists of the two drainage menisci.

Within the range $z_{II} < z < z_{III}$ each position of the point of intersection (S) is provided with a specific Hysteresis Rhomboid for its maculated capillarity and, thus, also for the relevant transition potential, according to Eq. 14a as for the threshold potential determined by Eq.16c.

The position for a point S within its Hysteresis Rhomboid or primarily along its borders depends on the prevailing dynamics and under static conditions on the dynamics they have resulted from. These superposing dynamic influences concern oil entry (rise of relative oil potential) and oil exit (drop of relative oil potential) on the one hand and the movement (rise or drop) of the table between water and air on the other hand.

Fig.2 shows for the various prevalences of the mentioned dynamic influences the positions of S depending on the level of the interface between water and air. The position of S along the horizontal indicates the relative potential. Accordingly, the Transient Potential for oil can be lowest for case 6 and highest for case 2 in Fig.2. However, until both oil menisci begin to move, the required relative potential may even be higher, see cases 1 and 3 in Fig.2.

References

1. Schiegg, H.O. (1977). Experimental Contribution to the Dynamic Capillary Fringe. IAHR-Symposium on Hydrodynamic Diffusion and Dispersion in Porous Media, (Ed. Istituto di Idraulica dell'Università di Pavia), Pavia, Italy, 307-314.

2. Schiegg, H.O. (1979). Verdrängungs-Simulation dreier nicht mischbarer Fluide in poröser Matrix. (Mitteilung Nr. 40 der Versuchsanstalt für Wasserbau, Hydrologie und Glaziologie, ETH Zürich). (English Translation in press).

An Enhanced Percolation Model for the Capillary Pressure-Saturation Relation

W.E. Soll, L.A. Ferrand and M.A. Celia

Ralph M. Parsons Laboratory, Department of Civil Engineering, Massachusetts Institute of Technology, Cambridge, MA 02139, USA

INTRODUCTION

Analysis of immiscible flow in porous media requires constitutive relationships between fluid content(s), fluid pressure(s) and relative permeabilities. Laboratory measurement of these relationships, especially for the case of more than two fluid phases, is an expensive and time consuming undertaking. An enhanced percolation model offers a theoretical framework in which these relationships can be predicted based on measurement of simple fluid and matrix properties. An application of this technique to the hysteretic relationship between capillary pressure and saturation for a two-fluid system is presented.

THEORY

The inherent complexity of fluid movement in porous media is due, in large part, to the tortuosity of the solid matrix. A fundamental property of pore space topology is the 'connectedness' or 'branchiness' of flow paths. When two or more immiscible fluids coexist in a medium, the flow physics are further complicated by phase interactions at the pore scale. Both interconnected topology and interfacial dynamics are incorporated into a consistent theory in the enhanced percolation model.

Fundamental to the present work is construction of a topologically complex yet mathematically tractable model of pore space. This takes the form of a regular geometric pattern of spherical nodes connected by cylindrical bonds. Specification of bond and node radii, bond length and the pattern of interconnectedness serves to completely define the geometry of the pore space. The sizes of nodes and bonds are typically chosen from probability distributions that best characterize the solid matrix of interest. The connectedness of the network is quantified by the coordination number, which is defined as the average number of bonds that emanate from a node. The present model derives from the classical work of Fatt (1956) and subsequent contributions of Chandler et al. (1982) and Chatzis and Dullien (1985), among others.

Two-phase flow in porous solids at low capillary number has been recognized as a percolation process (Broadbent and Hammersley, 1957). The branch of mathematical physics known as percolation theory provides substantial theoretical underpinnings for the two-phase flow problem. In the present model, drainage and imbibition are modeled as bond and site percolation processes, respectively. However, classical percolation theory is unable to accommodate several important physical phenomena, including residual saturations. The present model employs an enhanced theory that incorporates mechanisms for trapping fluids in the network, thereby leading to the equivalent of residual saturations for both wetting and nonwetting fluids.

The computational algorithms used herein are based on the fundamental pore scale physics of fluid-fluid-solid interactions in porous media. Immiscible fluid pressures may be expressed in terms of the pressure difference across fluid-fluid interfaces or capillary pressure

$$P_{c\alpha\beta} \equiv P_\alpha - P_\beta, \tag{1}$$

where P is pressure and subscript α indicates the wetting fluid, β the nonwetting fluid. The content of each fluid in some sample volume of the medium may be defined in terms of volumes of fluid per volume of pore space or relative saturations, S_α and S_β, with

$$S_\alpha + S_\beta = 1. \tag{2}$$

Analogous variables may be defined for systems containing more than two fluids.

In the pore scale network model, it is assumed that only one fluid resides at a given location at a given time. Once the geometry of the network is fixed, the redistribution of fluids in response to pressure changes is calculated using the Young-Laplace equation (Dullien, 1979)

$$P_{c\alpha\beta} = (2\sigma_{\alpha\beta}\cos\theta)/R \tag{3}$$

where $\sigma_{\alpha\beta}$ is surface tension, θ is contact angle and R is the effective radius of curvature of the fluid-fluid interface. If a cylindrical tube of radius R_T is initially filled with wetting fluid β and is in contact with nonwetting fluid α at one end, α will not displace β until

$$P_{c\alpha\beta} > (2\sigma_{\alpha\beta}\cos\theta)/R_T. \qquad (4)$$

Repeated application of (4) allows fluid-fluid interfaces to be tracked throughout the network as capillary pressure is varied. When a volume of fluid within the network loses hydraulic connection with its external reservoir it is considered to be trapped. Computation of occupied volumes for each fluid at each step gives relative saturation values which can be used to generate capillary pressure - saturation relations for the network.

APPLICATION

The ultimate goal of the current work is the development of a three-dimensional model of pore space in which the movement of three immiscible fluid phases can be simulated. It is clear that a three-dimensional network will best predict multifluid behavior since it is impossible to construct bi- or tri-continua in two-dimensional space. However, our initial efforts have been aimed at developing a two-dimensional network model for two fluids which qualitatively simulates both drainage-imbibition hysteresis and residual saturations.

The two-dimensional network used in the example described in this section is made up of triangular elements whose vertices are nodes and whose sides are bonds (see Figure 1). Node and bond volumes are determined by assigning radii from specified random distributions in a pre-processing step. This example, and most of the work completed to date, assumes log-normal pore size distributions, although a number of other distributions can be generated by the simulation code. In the current model, a node radius is always larger than that of connecting bonds.

Boundary and initial conditions generally correspond to those imposed in a laboratory experiment to determine the capillary pressure - saturation relation for a porous medium. In the example simulation, one end of the network is assumed to be in contact with a reservoir of wetting fluid, the opposite end with a reservoir of nonwetting fluid. At the start of a simulation, the network is assumed to be filled with wetting fluid. Initial fluid reservoir pressures are identical, i.e., initial capillary pressure is zero.

In each simulation, capillary pressure is increased in a stepwise fashion. The magnitude of the first pressure change is chosen such that the largest bonds in the network are drained, as determined by (4). Subsequent changes are chosen to drain successively smaller bonds. Nodes are non-limiting in this case because they are always larger than adjacent bonds. At each capillary pressure level the grid is searched for bonds which are both large enough to conduct nonwetting fluid and connected to nodes which contain wetting fluid. The search procedure is repeated until a scan of the network, from a specified corner to the most recently drained pore, is completed without a change in fluid distribution. Relative saturation is then computed for this capillary pressure. The next pressure change is then imposed, and the procedure repeated.

If the capillary pressure is decreased, wetting fluid imbibes into the network. The imbibition process is controlled by nodes rather than bonds; only nodes of sufficiently small radius can conduct fluid at a given pressure difference. Previously trapped wetting fluid may be reconnected to its reservoir. Complete hysteretic capillary pressure – saturation curves can be generated, as can intermediate scanning curves.

Figure 1 illustrates the redistribution of fluids within a 10x12 node network in response to pressure changes. At the start of the simulation (not shown), all nodes and bonds are filled with wetting fluid. Side boundaries are assumed to be impermeable. Examples of intermediate drainage steps are shown in Figures 1a and 1b. Solid black areas represent wetting fluid, hatched areas nonwetting fluid. Solid pores completely surrounded by hatched pores and/or impermeable boundaries represent trapped wetting fluid. The fully drained condition of the network is shown in Figure 1c. Calculation of the volume of trapped fluid at this step gives the residual wetting fluid saturation ($S_{\beta r}$). Capillary pressure is then decreased in a stepwise fashion (not shown) to simulate imbibition. The fully imbibed state is shown in Figure 1d. Note that previously trapped wetting fluid has been reconnected to the continuous phase. Calculation of the volume of trapped fluid (hatched areas) gives residual nonwetting fluid saturation ($S_{\alpha r}$) for the network.

Figure 2 shows the relationship between capillary pressure and wetting fluid saturation for primary drainage (drainage from $S_\beta=1$), secondary imbibition (imbibition from $S_\alpha=1-S_{\beta r}$) and secondary drainage (drainage from $S_\beta=1-S_{\alpha r}$) for a 60x60 node network. The curves capture the essential behavior of equivalent relationships found for laboratory samples of porous media. Unlike previously reported techniques, the enhanced percolation model simulates both drainage-imbibition hysteresis and residual wetting and nonwetting fluid saturations. Efficient search and tracking algorithms allow relatively complete inclusion of pore scale physics while minimizing computational effort.

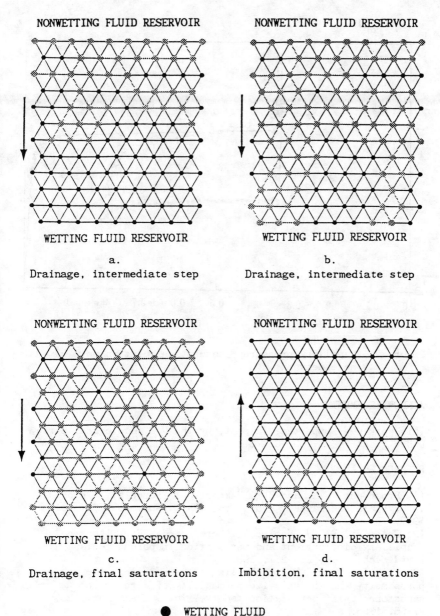

Figure 1. Four stages of a capillary pressure – saturation simulation for a 10x12 node network

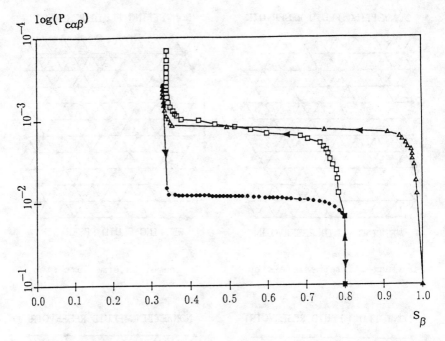

Figure 2. Capillary pressure - saturation relation on a 60x60 node network

ACKNOWLEDGEMENTS

This work was supported, in part, by the U.S. Geological Survey under grant 14-08-0001-G1473.

REFERENCES

(1) Broadbent, S.K. and J.M. Hammersley, "Percolation processes I. Crystals and mazes," Proc. Camb. Phil. Soc., 53, 629-641, 1957.
(2) Chandler, R., J. Koplic, K. Lerman and J.F. Willemsen, "Capillary displacement and percolation in porous media," J. Fluid Mech., 119, 249-267, 1982.
(3) Chatzis, I. and F.A.L. Dullien, "The modeling of mercury porosimetry and the relative permability of mercury in sandstones using percolation theory," Intl. Chem. Eng., 25, 47-66, 1985.
(4) Dullien, F.A.L., Porous media: Fluid transport and pore structure, Academic Press, New York, 1979.
(5) Fatt, I., "The network model of porous media, I. Capillary pressure characteristics," Trans. AIME Petr. Div., 207, 144-159, 1956.

SECTION 2D - STOCHASTIC MODELS

A High-Resolution Finite Difference Simulator for 3D Unsaturated Flow in Heterogeneous Media

R. Ababou
Princeton University, Department of Civil Engineering and Operations Research, Princeton, NJ 08544, USA

L.W. Gelhar
Massachusetts Institute of Technology, Department of Civil Engineering, Parsons Laboratory, Cambridge, MA 02139, USA

ABSTRACT:

The nonlinear equation of three-dimensional unsaturated flow is solved by a finite difference method for large single realizations of random field coefficients, based on a stochastic approach of field heterogeneity. The most difficult case considered for solution is a transient strip-source infiltration in a large domain discretized into 300,000 grid points, with a spatially random unsaturated conductivity curve. Numerical issues are briefly discussed, along with simulation results.

INTRODUCTION:

The flow of water in naturally heterogeneous, unsaturated porous formations is not easily accessible to detailed observation at the large scale outside the laboratory. On the other hand, the highly nonlinear equation of unsaturated flow cannot be solved exactly for the non-trivial type of spatial heterogeneity encountered in the field. A numerical approach might be feasible, but requires some additional assumption to compensate for the lack of detailed input data. In order to reduce the complexity of the problem, we adopt here the view that a stochastic representation of three-dimensional spatial variability will be adequate for investigating the global behaviour of large scale unsaturated flow systems. The stochastic approach is justified in more detail in the comprehensive work by Ababou [1], which includes a review of field data.

The case of stochastic unsaturated flow was previously studied by Mantoglou and Gelhar [2], using a linearized spectral perturbation method to obtain a statistical description of the flow in closed form. Their solutions revealed complex interactions between heterogeneity and nonlinearity, resulting in global effects like effective anisotropy and hysteresis, both nonlinear functions of the mean ambient water pressure. However, the accuracy of their results remains essentially unknown, due to the linearization and perturbation approximations that were made.

In contrast, we submit that carefully designed numerical experiments based on the stochastic approach may lead to a more realistic and detailed representation of heterogeneous unsaturated flow processes, under less restrictive assumptions than those needed for obtaining analytical results. Given current supercomputer capabilities, it appears now feasible to simulate complex flow phenomena over reasonably large scales, with a much higher resolution than achievable in the field. The ultimate goal of such a program is to obtain a global characterization of flow behaviour in the presence of heterogeneities spanning many scales of length. In the sequel, we give a brief overview of the numerical/stochastic method developed in Ababou [1], and present results for some large simulations of transient strip-source infiltration in three-dimensional, randomly heterogeneous soils.

SINGLE REALIZATION APPROACH:

In the stochastic approach to subsurface flow, the hydrodynamic coefficients of the governing flow equation are assumed to be random fields. This basic assumption can be interpreted in various ways. The single-realization approach adopted here considers that field conditions are best represented by a single large replicate of the random porous medium. This is in contrast with the Monte-Carlo simulation method, where one solves the flow problem for many independent replicates of the coefficients, generated across a hypothetical ensemble space. The latter approach may be useful in the context of risk analysis. However, the single realization approach appears more suitable for analyzing the physics of heterogeneous flow systems at the large scale.

The specific type of flow problem to be discussed here is transient strip-source infiltration in a dry porous medium whose unsaturated conductivity curve varies randomly in 3D space. Both the saturated conductivity (Ksat) and the slope (α) of the log-conductivity/pressure curve are assumed random. Accordingly, one large realization of each random field parameter is generated over the grid, by using the 3D turning band method of Tompson et al. [3]. Thus, there is a different conductivity curve for each node of the grid. The nonlinear (Van Genuchten) water retention curve is spatially constant.

NUMERICAL METHOD:

A special-purpose flow simulator was developed in order to resolve the fine details of subsurface flow in saturated as well as unsaturated stochastic porous formations (Ababou [1]). Here, we focus strictly on the case of unsaturated flow in random soils. According to the single realization approach outlined just above, the flow simulator is required to solve the Richards equation with nonlinear (pressure dependent) and highly variable (random) coefficients in a large 3D domain.

The numerical code is based on a finite difference approximation of Richards equation, with a seven point centered scheme in three-dimensional space and a fully implicit backward difference in time. This discretization produces a sparse algebraic system, seven diagonal symmetric, and furthermore positive-definite if the coefficients are frozen. However, for the problem at hand, the system is nonlinear, spatially random, and may be quite large.

Briefly, the strategy to solve the large sparse nonlinear system is as follows. At each time step, the system is linearized iteratively by using a modified Picard scheme, and the resulting matrix system is then solved by the "strongly implicit procedure" (SIP), a fast sparse iterative matrix solver based on an approximate LU factorization.

COMPUTATIONAL REQUIREMENTS:

A heuristic stability analysis of the nonlinear finite difference system suggested a grid Peclet number constraint on the vertical mesh size (Pe = $\alpha \Delta x \leq 2$). Moreover, a stringent constraint on the time step could be required if the Peclet number condition is not everywhere satisfied. More details can be found in Ababou [1]. In practice, the limitations on the time step appear quite severe for transient infiltration in dry soils. However, when the time step is small, only a few SIP iterations are required for matrix solution of the linearized system. The advantage of the method lies in the fact that the CPU time for each SIP iteration is only proportional to N, the 3D size of the grid, compared to a power 7/3 of N for direct solution methods like band-Gauss substitution. This is important in view of the large discrete systems implied by the single-realization approach.

Indeed, the computational grid must be particularly large in order to resolve the detailed fluctuations of the flow field over a reasonably large 3D domain. The mesh size must be smaller than the typical fluctuation scale of random field inputs (say a fraction of correlation length) and is also bounded by the Peclet number constraint. On the other hand, the size of the domain of interest increases with the time scale of simulation, and should be presumably much larger than the correlation lengths for a meaningful analysis of global flow behaviour. As an example, the largest random flow problem discussed below was discretized on a finite difference grid comprising 300,000 nodes, and required several hours of supercomputer time for a 20 day real time simulation.

SIMULATION RESULTS:

Figure (1) illustrates the results obtained for a relatively modest size problem (30,000 nodes) of strip-source infiltration in a random, statistically isotropic soil. The random field parameters Ksat and α were log-normal, independent, with isotropic exponential covariance functions. The figure shows two perspective views of a pressure surface contour, which corresponds to the location of a fairly sharp wetting front. The simulation required only a few hours of Microvax-2 time for a few days of real time infiltration.

Figure (2) illustrates the results for a much larger simulation (300,000 nodes) on a statistically anisotropic soil, with correlation lengths 1 m horizontally and 0.2 m vertically. The 3D domain size was 15 m horizontally and 5 m vertically. The pressure contour lines are shown for three different vertical planes along the strip (10 m length). The larger horizontal correlation causes the moisture plume to spread laterally, as predicted by Mantoglou and Gelhar [2]. The simulation of 10 days of infiltration and 10 subsequent days of drainage consumed 5 CPU hours of Cray 2 time, and required about 48 Mbytes (6 Mwords) of central memory.

REFERENCES:

1. Ababou R. (1988): Three-Dimensional Flow in Random Porous Media, Ph.D. thesis, Massachusetts Institute of Technology, Department of Civil Engineering, Cambridge, MA 02139, U.S.A., pp. 833.

2. Mantoglou A. and Gelhar L. W. (1987):
   - Stochastic Modeling of Large-Scale Transient Unsaturated Flow Systems, Water Resour. Res., 23(1), 37-46;
   - Capillary Tension Head Variance, Mean Soil Moisture Content, and Effective Specific Soil Moisture Capacity of Transient Unsaturated Flow in Stratified Soils, Water Resour. Res., 23(1), 47-56;
   - Effective Hydraulic Conductivities of Transient Unsaturated Flow in Stratified Soils, Water Resour. Res., 23(1), 57-67.

3. Tompson A. F. B., Ababou R., and Gelhar L. W. (1987): Applications and Use of the Three-Dimensional Turning Band Random Field Generator in Hydrology: Single Realization Problems, Tech. Report No. 313, March 1987, Parsons Laboratory, M.I.T., Cambridge, MA 02139, U.S.A.

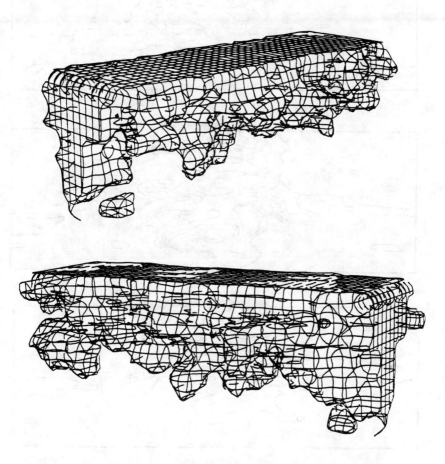

Figure 1 Two perspective views of the pressure contour surface h = - 90 cm at t = 2 days for strip-source infiltration in a statistically isotropic soil with initial pressure h_{in} = - 150 cm (K_s and α random, perfectly independent).

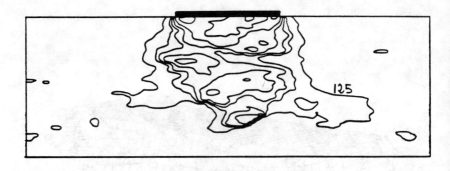

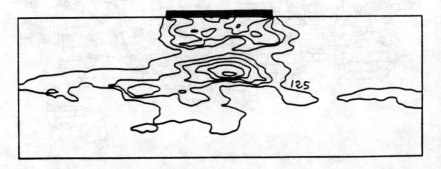

Figure 2 Contour lines of pressure head in three vertical-transverse slices during the simulated strip-source experiment after 10 days of infiltration (t = 10 days). From top to bottom: slices Y = 2m, Y = 4.8m, Y = 9.8m.

Solving Stochastic Groundwater Problems using Sensitivity Theory and Hermite Interpolating Polynomials
D.P. Ahlfeld and G.F. Pinder
Department of Civil Engineering and Operations Research, Princeton University, Princeton, NJ 08544, USA

INTRODUCTION

The stochastic groundwater problem consists of solving the classical equations for groundwater flow and contaminant transport with uncertainty associated with the parameters[1]. This uncertainty is quantified by representing the parameters as random variables with some distribution and covariance structure. If this structure is known then the solution of this problem will consist of a probability distribution of the dependent variables. In this paper we are concerned with solving the equations for groundwater flow and convective-dispersive transport for the distribution of concentration when a distribution of hydraulic conductivity is given. We shall describe and demonstrate a methodology for determining this distribution which is computationally efficient.

A number of approaches have been proposed for addressing the stochastic contaminant transport problem. A common approach is the use of random sampling methods such as the Monte Carlo method or stratified sampling[2]. For our case, these methods would require repeated drawing of realizations from the conductivity distribution, simulation of concentration using each realization of conductivity and accumulation of these concentration realizations to form a distribution. While in theory these methods will produce the entire concentration distribution as the number of realizations approaches infinity, the computational cost of repeated simulation limits their use.

First order methods have been used by many researchers (for example, Wagner and Gorelick)[3] and are capable of determining the first and second moments of the distribution of concentration. However, these methods can not be used to determine the distribution itself. While these methods are computationally efficient relative to sampling methods, they are limited in their applicability to cases where the coefficient of variation of the uncertain parameter is much less than one[4], a condition not satisfied for many problems of interest.

In this paper we propose a method of solving the stochastic groundwater problem which is based on a limited form of stratified sampling of the parameter distribution and interpolation of those sampling results using first order sensitivity information and Hermite polynomial interpolators.

PROCEDURE FOR SOLUTION BY LIMITED SAMPLING

The stratified sampling methods described above generate the distribution of concentrations based on the distribution of hydraulic conductivities in a three step procedure. For simplicity of exposition we shall consider the distribution of concentration at only one node in space at a fixed time and further consider uniform conductivity over the domain as the single parameter.

Consider the cumulative distribution of the random variable conductivity, K,

$$F(k) = Prob\{K \leq k\} \tag{1}$$

In the first step of the stratified sampling a series of realizations of conductivity, k_i, are generated from the inverse of the cumulative distribution function, F^{-1}. Each realization comes from its own stratum where each stratum has equal probability of occurrence. The second step consists of finding the concentration realization, c_i, associated with each conductivity realization. This is represented as a functional relationship between the realization k_i and the random variable c.

$$c_i = c(k_i) \tag{2}$$

Realizations of c_i based on all k_i are collected and in the third step are used to construct the distribution of c by histogram analysis or by assuming some functional form for the distribution.

In most practical cases, determining c_i as described in (2) requires the solution of a numerical groundwater simulation model for each realization i. The number of realizations needed may be on the order of 100 or 1000, thus the computational burden of this method lies largely in repeated solution of the simulation model. This concern leads us to consider alternate ways of representing the functional relationship between concentration and conductivity.

APPROXIMATING THE $c(K)$ RELATIONSHIP

We propose representing the functional relationship between conductivity and concentration (2) by piecewise interpolation by Hermite polynomials[5], of a limited number of computed values of (2). Hermite polynomials are first order continuous interpolators which represent a function by its value and derivative. Here the independent variable is the conductivity rather than the conventional variables of space. Thus for a single conductivity parameter (one dimension) we represent (2) as

$$c(k_i) \approx \sum_{j=1}^{N} c_j \phi_{0j}(k_i) + \frac{\partial c_j}{\partial k_j} \phi_{1j}(k_i) \tag{3}$$

where

ϕ_{0j} = Hermite basis function for the function at node j,
ϕ_{1j} = Hermite basis function for the derivative at node j,
c_j = the function value at node j,
$\frac{\partial c_j}{\partial k_j}$ = the function derivative at node j,
N = number of nodes in the interpolation scheme.

Note that these basis functions have the same characteristics of limited support as those used in conventional interpolation over space so that interpolation at any point requires querying only those nodes which bound the interval containing the point.

The function value and its derivative are determined by direct evaluation of the simulation model at the points j. The derivative of the function can be found by the application of sensitivity theory.

Sensitivity theory provides a computationally efficient means of computing the derivatives of simulation model outputs with respect to simulation model input. To compute these derivatives the simulation model is solved for the concentrations (the "forward problem") and then a related "backward problem" is solved for the adjoint sensitivity vector[6]. With this vector the sensitivity of the solution with respect to any number of different parameters can be computed with a single inner product. The "backward problem" requires approximately the same computational effort as the forward problem for confined flow and convective-dispersive transport[7], thus many derivatives can be computed with the computational effort of about 2 forward simulations.

AN EXAMPLE

To demonstrate this approach to solving stochastic groundwater problems we present a simple example on a hypothetical aquifer were we consider the calculation of the distribution of concentration at single point based on the assumed known distribution of a single uniform conductivity parameter.

The hypothetical aquifer domain is shown in Figure 1.

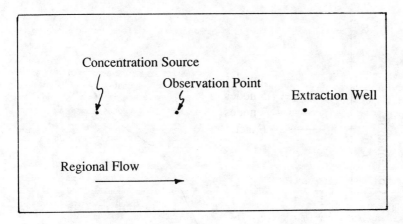

Figure 1 : **Hypothetical Aquifer Domain and Features**

The transport dynamics of the system are driven by a constant contaminant source (modeled as a Dirichlet condition) at the indicated location and an extraction well at the indicated location. The flow stress is imposed on a system which has a regional flow from left to right. The concentration is observed at a single point indicated in Figure 1. A homogeneous aquifer is used with deterministic parameters given in Table 1.

Aquifer Thickness	$50\,ft$
Longitudinal Dispersivity	$50\,ft$
Transverse Dispersivity	$10\,ft$
Porosity	0.2
Diffusion Coefficient	$0.0001\,ft^2/day$
Number of Nodes	180
Space Discretization	$100\text{-}200\,ft$
Time Step for Mass Transport	10.5 days

Table 1: **Hydraulic and Numerical Parameters Used in Hypothetical Aquifer**

The conductivity is lognormally distributed. This distribution has a mean of 154 ft/day and standard deviation of 253 ft/day, statistics that make first order analysis inapplicable.

APPROXIMATING THE SIMULATION MODEL

The central part of this work is the ability to approximate the function (2) by the interpolating function (3). Clearly, the more function values used in the interpolation scheme (i.e. nodes) the better the interpolation will be. To analyze the ability of (3) to represent (2) we have generated this $c(K)$ based on 1000 values of k_i which are drawn from the probability distribution using a stratified sampling approach so that each value comes from a strata with equal probability of occurrence. The curve so generated is the solid line in Figure 2.

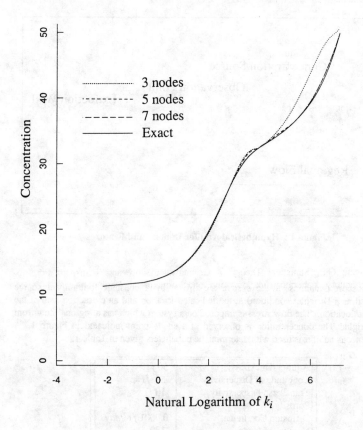

Figure 2 : **Approximate and Exact Relationships of the Log Hydraulic Conductivity vs Concentration**

Our Hermite interpolator is tested by computing the function and derivative values at the ends of the function and at evenly spaced points across the function. Although the use of sensitivity theory for computation of the derivatives is part of our proposed formulation (3), for this example we use finite differences to compute these derivatives in order to test the interpolation concept. The Hermite interpolator (3) is then used to approximate the function at the same conductivity values used to generate the solid curve in Figure 2. The results of using different numbers of nodes in (3) are superimposed on the solid curve in Figure 2. With as few as five nodes (including nodes at the endpoints of the distribution) the simulation function can be well represented. Beyond seven nodes the difference between the curves is graphically indistinguishable.

COMPARISON OF GENERATED DISTRIBUTIONS

Ultimately we are interested in the concentration distribution generated by our approach. The various distributions, generated by histogram analysis using eleven intervals are shown in Figure 3. The solid curve is the exact distribution as computed by 1000 simulations. The distributions generated by (3) are represented by the other lines as indicated on the figure and quickly converge to the exact distribution so that with eleven or more nodes an exact match is obtained.

While, for our example, an exact duplication of the concentration distribution required only 11 nodes (at a cost of roughly 22 simulations), over 500 simulations were needed to converge to the distribution using the conventional Monte Carlo approach. Thus the proposed method presents the opportunity for significant computational savings, in this case, about a 20 fold improvement in the cost of computing the distribution.

CONCLUSIONS

A methodology for determining the distribution of concentration based on uncertainty in hydraulic conductivity has been presented. The methodology approximates the function used to relate the conductivity and the concentration using a limited number of function evaluations, sensitivity theory to compute the function derivatives and Hermite interpolating polynomials.

References

1. Dagan, G., "Statistical Theory of Groundwater Flow and Transport: Pore to Laboratory, Laboratory to Formation, and Formation to Regional Scale," *Water Resources Research*, vol. 22, pp. 120S-134S, August 1986.
2. McKay, M. D., Beckman, R. J., and Conover, W. J., "A Comparison of Three Methods for Selecting Values of Input Variables in the Analysis of Output from a Computer Code," *Technometrics*, vol. 21(2), pp. 239-245, May 1979.
3. Wagner, B. J. and Gorelick, S. M., "Optimal Groundwater Quality Management Under Parameter Uncertainty," *Water Resources Research*, vol. 23(7), pp. 1162-1174, July 1987.
4. Dettinger, M. D. and Wilson, J. L., "First Order Analysis of Uncertainty in Numerical Models of Groundwater Flow Part 1. Mathematical Development," *Water Resources Research*, vol. 17(1), February 1981.
5. Lapidus, Leon and Pinder, George F., *Numerical Solution of Partial Differential Equations in Science and Engineering*, Wiley, New York, 1982.
6. Sykes, J.F., Wilson, J.L., and Andrews, R.W., "Sensitivity Analysis for Steady State Groundwater Flow Using Adjoint Operators," *Water Resources Research*, vol. 21, no. 3, pp. 359-371, March, 1985.
7. Ahlfeld, D. P., "Designing Contaminated Groundwater Remediation Systems Using Numerical Simulation and Nonlinear Optimization," *Ph.D. dissertation*, Dept. of Civil Engineering, Princeton University, 1986.

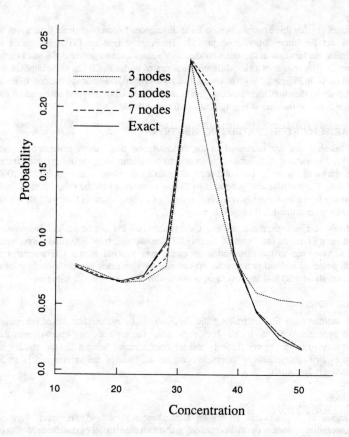

Figure 3: **Approximate and Exact Probability Distributions of Concentration**

Supercomputer Simulations of Heterogeneous Hillslopes
A. Binley and K. Beven
Department of Environmental Science, University of Lancaster, UK
J. Elgy
Department of Civil Engineering, Aston University, UK

ABSTRACT

A fully three dimensional model of variably saturated flow has been developed to investigate the hydrological effects of spatial variability of saturated hydraulic conductivity on a 150 m by 100 m hillslope and assess the validity of the concept of effective hydraulic conductivity. The model is based on the Galerkin approximation to the finite element method. Accessibility to the vector processor of a CDC Cyber 205 supercomputer permitted numerical solutions on grids containing several thousand node points and two thousand time steps. The results suggest that under conditions of predominant subsurface flow the concept of equivalent uniform properties are valid.

INTRODUCTION

Over the past decade much attention has been devoted to the development of distributed physically based models of catchment hydrology. A number of these models, such as the Système Hydrologique Européen (SHE) (Abbot et al.[1]) and the Institute of Hydrology Distributed Model (IHDM) (Beven et al.[4]), have now reached the testing stage and will soon be available for use by engineering consultants. Although these models are based on theoretically acceptable equations, there are still limitations in the present modeling strategy. Of interest to this study are the current modeling assumptions made concerning the effects of soil spatial variability.

Ample evidence of soil spatial variability is presented in the literature, the incorporation of which is permissible in physically based models, although immense data requirements have restricted the use of this information. It is generally assumed that areas of the flow domain can be represented by some equivalent soil property (or effective model parameter). Recently Binley et al.[5,6], carried out a fully three dimensional analysis of transient variably saturated flow on a heterogeneous hillslope. Their results suggest that under conditions of dominating subsurface flow the concept of an equivalent hydraulic conductivity may be valid. Due to the immense computer requirements of three dimensional analysis, Binley *et al.* restricted their investigation to the response of two storm events. This paper reports on continuing work on heterogeneous hillslope runoff generation a and presents the results of three dimensional simulations of multiple storm events.

EFFECTIVE HYDRAULIC CONDUCTIVITIES FOR HILLSLOPE RUNOFF GENERATION

A numerical approximation to Richards equation (Richards[10]) was derived using the Galerkin approximation to the finite element method. Three dimensional brick type elements were used together with Gauss quadrature integration of the finite element integral equations. The literature contains several examples of finite element solutions to three dimensional variably saturated flow, for example Frind and Verge[8] and Babu and Pinder[2], although all have restricted their analysis to grids containing several hundred elements. In the case of hillslope runoff simulation, the high nonlinearity of large regions of unsaturated flow and recognition of soil heterogeneity necessitate the use of grids containing several thousand elements, even for small hillslopes.
The computing requirements of such analysis prohibit the use of normal scalar computers. In order to carry out the numerical simulations the CDC Cyber 205 at the University of Manchester Regional Computer Centre, UK was used. Restructuring of the code permitted considerable use of the vector architecture of the Cyber 205. Particular decrease in processing time was achieved using the Jacobi conjugate gradient sparse equation solver of Kincaid *et al.*[9].

A hypothetical straight hillslope 150 m wide and 100 m long was adopted for this study. The soil is 1 m deep and forms a 1 in 6 slope. The hillslope was discretized into 6300 node points (5280 elements). Maximum vertical and horizontal discretization were 0.1 m and 7.5 m respectively. Due to the restrictive computer requirements only one case of soil spatial variability was considered for the hillslope. The unsaturated soil water properties were described using the Brooks-Corey relationships (Brooks and Corey[7]), that is:

$$K_r = (\psi/\psi_0)^{-2+3\alpha}$$

$$\theta_r = (\psi/\psi_0)^{\alpha}$$

Where K_r is the relative hydraulic conductivity, θ_r is the relative moisture content, ψ is the pressure head, ψ_0 is the air entry potential and α is a constant. The following parameters were adopted: $\alpha = -0.129$, $\psi_0 = -0.356$ m.

Using the turning bands method, a spatially correlated field of log normally distributed hydraulic conductivity was generated at ten soil depths, the log variance at each depth being specified as 1.5. The ten random fields also contained a specified correlation structure with depth.

The mean hydraulic conductivity was allowed to vary with depth according to:

$$K_s = K_0 \exp\{fz\}$$

Where z is the soil depth, K is the mean saturated hydraulic conductivity at the soil surface and f is a parameter. Using data for real soils (Beven[3]) values of K_0 and f were selected to be 0.15 m hr^{-1} and -3.33 m^{-1} respectively.

To remove the effects of initial conditions, the pressure head was set to -2 m at all nodes and a series of four events totalling 96 mm rainfall over a duration of 33 days was applied to the hillslope. The pressure head field at the end of this period was then adopted as initial conditions for the series of events under investigation. This 'setting up' period required 1140 time steps and approximately 3 hours CPU time on the Cyber 205. The series of events following the 'setting up' period totalled 60.2 mm rainfall over 9 days, which was discretized into 780 time steps and required 2.5 hours CPU time.

In order to assess the validity of an equivalent homogeneous hillslope, the following equation was used to define a number of uniform slopes based on the conductivity distribution of the heterogeneous slope,

$$K_{eff} = 0.5 K_G \exp\{p\sigma^2\} \qquad (1)$$

Where K_{eff} is the effective hydraulic conductivity, K_G is the geometric mean of the log normally distributed conductivity values, σ^2 is the log variance and p is a parameter. Values of p equal to -1, 0 and 1 result in equation (1) being equal to the harmonic, geometric and arithmetic means of the distribution respectively.

A total of nine values of p ranging from -3 to 3 were selected and the corresponding uniform slopes generated. Note that the slopes still retained the exponential decline of conductivity as in the original heterogeneous system. Adopting the same initial pressure head field and 'setting up' period, the 9 day series of events were simulated for each uniform slope. Analysis of error variance over the 9 day period, for each value of p, revealed an optimum value of approximately 0.8 for both subsurface and total (subsurface plus surface) flow (figure 1). Identical trends were shown by similar analysis of the individual events of the 9 day period.

A comparison of subsurface and total flow hydrographs for the heterogeneous slope and the 'best' uniform slope (p=1) is shown in figure 2. The subsurface response of the homogeneous slope, in general, overestimates that of the heterogeneous slope but fails to reproduce the magnitude of the main hydrograph peak during the third day. A comparison of the total flow responses reveals similar behaviour. A noticeable feature of the total flow hydrograph for the uniform slope is the earlier peak and steeper falling limb. This is due to the build up of surface saturation being limited to the base of the uniform slope whereas in the case of the heterogeneous slope large areas of low conductivity up slope may force surface saturation away from the base thus producing surface runoff with a greater travel time.

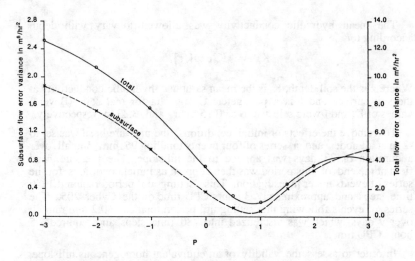

Figure 1. Variation of error variance with effective hydraulic conductivity, expressed in terms of parameter p of equation (1), for subsurface and total flow.

CONCLUSIONS

The results presented suggest that under conditions of hydrograph domination by subsurface flow (either by seepage or return flow) the concept of an equivalent uniform hillslope may be valid. Although a perfect match is unlikely, close agreement between heterogeneous and homogeneous responses appears plausible under a range of external conditions. The definition of the uniform property in terms of the underlying distribution parameters still requires further investigation. A value greater than the geometric mean and close to the arithmetic mean was found to be the optimum value for the single realization of variability considered here. Such a value compliments the results of previous investigations of simpler flow systems.

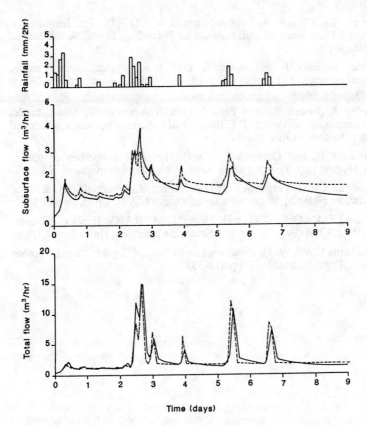

Figure 2. Subsurface and total flow hydrographs for heterogeneous (solid line) and 'best' homogeneous slope (p=1) (dashed line).

REFERENCES

1. Abbot M.B. Bathurst J.C. Cunge J.A. O'Connel P.E. and Rasmussen J. (1986), An introduction to the European Hydrological System - Système Hydrologique Européen, "SHE", 2. Structure of a physically-based, distributed modelling system, J.Hydrology, Vol.87,pp.61-77.

2. Babu D.K. and Pinder G.F. (1984), A finite element - finite difference alternating direction algorithm for three dimensional groundwater transport, Adv. in Water Resources, Vol.7, pp.116-119.

3. Beven K. (1983), Introducing spatial variability into TOPMODEL: Theory and preliminary results, Unpublished Rept., Dept. Environmental Sci., Univ. Virginia, USA.

4. Beven K. Calver A. and Morros E.M. (1987), The Institute of Hydrology Distributed Model, Institute of Hydrology Rept. 98, Institute of Hydrology, UK.

5. Binley A. Elgy J. and Beven K. (1988), A physically-based model of heterogeneous hillslopes. I Runoff production, submitted to Water Resources Research.

6. Binley A. Beven K. and Elgy J. (1988), A physically-based model of heterogeneous hillslopes. II Effective hydraulic conductivities, submitted to Water Resources Research.

7. Brooks R.H. and Corey A.T. (1964), Hydraulic properties of porous media, Hydrol. pap. 3., Agric. Eng. Dept., Colorado State Univ., USA.

8. Frind E.O. and Verge M.J. (1978), Three dimensional modeling of groundwater systems, Water Resources Research, Vol. 14(4), pp.844-856.

9. Kincaid D.R. Oppe T.C. and Young D.M. (1984), Itpackv 2C users guide, Rept. CNA-191, Center for Numerical Analysis, Univ. Texas, USA.

10. Richards L.A. (1931), Capillary conduction of liquids through porous mediums, Physics, Vol. 1(5), pp.318-333.

A Comparison of Numerical Solution Techniques for the Stochastic Analysis of Nonstationary, Transient, Subsurface Mass Transport

W. Graham and D. McLaughlin

Department of Civil Engineering, Massachusetts Institute of Technology, Cambridge, MA 02139, USA

ABSTRACT

Stochastic methods are applied to the analysis and prediction of large-scale solute transport in saturated heterogeneous porous media. A set of coupled partial differential equations describing the propagation of the concentration mean and covariance is derived from the classical advection-dispersion equation. In this derivation it is assumed that the major factor contributing to mass transport prediction uncertainty is the unknown spatial distribution of the underlying steady-state groundwater velocity field.

The coupled set of concentration moment equations maintains the general form of a transport equation, and must be solved numerically. Both an iterative principal direction finite element algorithm, and an iterative conventional Galerkin finite element algorithm are developed to solve the equations. The performance of these solution algorithms is verified by Monte Carlo simulation, and the accuracy, efficiency, and computer storage requirements of the three solution techniques are contrasted.

INTRODUCTION

Groundwater contamination is an increasing problem which is affecting aquifers and threatening water supplies across the United States. Unfortunately, groundwater contamination is less visible, and therefore more difficult to predict and detect than other forms of environmental pollution. Complicating the assessment of groundwater contamination is the fact that the hydrogeologic properties which govern the fate and transport of subsurface pollutants vary widely in space, over distances that are small in relation to the extent of a typical contaminant plume. In addition, gathering information about the spatial variability of these properties is an expensive and time consuming process.

In the absence of perfect knowledge of the spatial distribution of the hydrogeologic properties at a contamination site, stochastic analysis provides a logical and convenient method for predicting the movement of a contaminant plume, and summarizing the accuracy of this prediction. Using this method, groundwater velocity and contaminant concentration are treated as random functions of space. Known spatial moments of the input random variables (in this case groundwater velocity) are used to derive moments of the output random variables (in this case groundwater contaminant concentration) based on knowledge of the underlying physical processes of mass transport. The first concentration moment, or mean, provides a best estimate of the extent of the contaminant plume at any particular time, while the second moment, or variance, provides a measure of the accuracy of this estimate.

METHODS

The basic assumption underlying this research is that the largest factor contributing to mass transport prediction uncertainty is the unknown spatial distribution of the groundwater velocity field. It is assumed that the classical advection-dispersion equation perfectly describes the movement of mass in the subsurface environment, given the true spatial velocity distribution. Thus mass transport is described as large-scale spreading due to spatial variations in velocity, and smoothing due to pore-scale dispersion. Molecular diffusion, chemical reactions, and biological degradation are neglected in this analysis. A steady-state velocity field is assumed.

In this study two techniques of stochastic modelling, Monte Carlo simulation and perturbation analysis, are used to derive concentration moments based on a random velocity field with a known mean and covariance. Monte Carlo simulation is essentially a numerical technique for calculating sample moments based on a collection of random replicates generated from the random field in question. Perturbation techniques use analytical methods to derive equations for the ensemble moments of the random field. Further details on each of these methods are summarized below.

Perturbation Methods

The stochastic mass transport equation can be written:

$$\frac{\partial c(x)}{\partial t} + \frac{\partial [v_i(x)c(x)]}{\partial x_i} - \frac{\partial}{\partial x_i}D_{ij}\frac{\partial c(x)}{\partial x_j} = 0 \tag{1}$$

where the Einstein summation convention is used and

- c random concentration (M/L^3)
- v_i random velocity in the ith direction (L/T)
- D_{ij} deterministic pore scale dispersion coefficient (L^2/T)

Each of the random variables in Equation 1 (c and v_i) can be expanded into the sum of a spatially variable mean and a small perturbation around this mean. Taking the expected value of the expanded equation produces an equation describing the mean subsurface mass transport, $\bar{c}$:

$$\frac{\partial \bar{c}(x)}{\partial t} + \bar{v}_i(x)\frac{\partial \bar{c}(x)}{\partial x_i} - \frac{\partial}{\partial x_i}D_{ij}\frac{\partial \bar{c}(x)}{\partial x_j} + \frac{\partial}{\partial x_i}Pv_ic(x,x) = 0 \tag{2}$$

The mean mass transport equation has the same form as the traditional advection-dispersion equation with one additional term—the divergence of the expected value of the product of the velocity and concentration perturbations, $Pv_ic(x,x)$. This term is the ensemble macrodispersive flux which accounts for the increased spreading of the mean contaminant plume due to velocity perturbations not captured by the mean advective process. It should be emphasized that this term accounts for differences among the random replicates of the ensemble, and does not describe the macroscopic spreading of a single replicate.

A first order approximation to the equation for the macrodispersive flux can be obtained from Equation 1 by first subtracting the mean equation, then multiplying each term of the resulting concentration perturbation equation by a velocity perturbation, and finally taking expected values throughout. Products of small perturbations are neglected to obtain closure. The resulting partial differential equation governing the propagation of the ensemble macrodispersive flux through space and time is:

$$\frac{\partial Pv_kc(x',x)}{\partial t} + \bar{v}_i(x)\frac{\partial Pv_kc(x',x)}{\partial x_i} - \frac{\partial}{\partial x_i}D_{ij}\frac{\partial Pv_kc(x',x)}{\partial x_j} + \frac{\partial \bar{c}(x)}{\partial x_i}Pv_kv_i(x',x) = 0 \quad k=1,3 \tag{3}$$

The macroscopic flux equation also retains the basic form of a mass transport equation, however it is dependent on two spatial vectors x and x', and a new forcing term is introduced. This forcing term is the product of the mean concentration gradient at a point x, and the covariance of the velocity perturbation at that point and all other points in the domain, $Pv_k v_i(x',x)$.

In a similar manner, a first-order equation for the concentration covariance around the mean plume described by Equation 2 can be derived:

$$\frac{\partial Pcc(x,x')}{\partial t} + \bar{v}_i(x)\frac{\partial Pcc(x,x')}{\partial x_i} - \frac{\partial}{\partial x_i}D_{ij}\frac{\partial Pcc(x,x')}{\partial x_j} + \frac{\partial \bar{c}(x)}{\partial x_i}Pv_i c(x,x')$$

$$+ \bar{v}_i(x')\frac{\partial Pcc(x',x)}{\partial x_i'} - \frac{\partial}{\partial x_i'}D_{ij}\frac{\partial Pcc(x',x)}{\partial x_j'} + \frac{\partial \bar{c}(x')}{\partial x_i'}Pv_i c(x',x) = 0 \qquad (4)$$

Note that the concentration covariance equation depends on both the mean concentration field and the macrodispersive flux. The equation contains derivatives with respect to both the x and the x' vectors, which complicates its solution.

Equations 2 through 4 form a system of five coupled partial differential equations which describe the propagation of the mean concentration plume, the macrodispersive flux, and the concentration covariance through a random velocity field with mean $\bar{v}(x)$ and covariance $Pv_i v_k(x,x')$. These equations must in general be solved numerically, using an iterative scheme to account for their interdependence. In this work both a conventional Galerkin and a principal direction finite element algorithm were developed to solve these equations.

<u>Finite Element Algorithm</u> A two-dimensional conventional Galerkin finite element algorithm using bilinear basis functions was developed to solve Equations 2 through 4. Since each of the equations has the same basic form, the same solver was used for the entire system. For each equation the forcing terms were evaluated at the half time step, in an iterative fashion.

From Equation 3, it is apparent that the macrodispersive flux depends on two space vectors x and x'. However since only derivatives with respect to x appear in the equation, x' can be treated as a parameter. Thus Equation 3 can be solved by repeatly solving an equation only in x, for each x' in the domain. The matrix $Pv_i c(x,x')$ is then compiled from the individual solution vectors.

Equation 4 for $Pcc(x,x')$ contains derivatives with respect to both x and x', and therefore cannot be separated in the same manner as Equation 3. To solve this equation, the portion containing derivatives with respect to x' was treated as an extended forcing term and the equation was solved iteratively.

Since Equations 2 through 4 all maintain the form of a transport equation, the spatial and temporal discretization requirements depend on the grid Peclet number and Courant number.

<u>Principal Direction Algorithm</u> A prinicpal direction algorithm, as described by Frind[1], was also set up to solve these equations. Using this method an orthogonal curvilinear coordinate system is set up where x follows the streamlines. In this system advection only occurs in the x direction, and therefore the equation can be split into two steps -- an equation implicit in x, but explicit in y, which involves both advection and dispersion terms, and an equation implicit in y, but explicit in x, which involves only dispersive terms. These equations can be solved sequentially using Galerkin finite elements with linear basis functions.

Use of the principal direction technique can save both computer run time and storage since only a tridiagonal matrix is solved at each step. In addition discretization requirements can be theoretically be relaxed in the y direction since there is no advection in this direction, and therefore no possibility for numerical dispersion. However the split solution technique introduces other numerical errors, and the discretization in the y direction must be fine enough to capture the fluctuations of the forcing terms.

The same iterative techniques detailed above for the Galerkin finite element algorithm were also used in the principal direction technique.

Monte Carlo Methods

Using Monte Carlo methods, the moments of the concentration field are derived by repeatedly solving the subsurface mass transport equation using different random velocity replicates. In this study, a multivariate, multidimensional turning bands algorithm (Shinozuka and Jan[2]) was developed to generate the random velocity replicates. A conventional Galerkin finite element code was then used to solve for the concentration plume resulting from each random replicate. The mean and covariance of the concentration field were determined using sample statistics calculated from the ensemble of random replicates. The number of replicates required to give accurate sample statistics was determined based on the concentration moments obtained from the perturbation method.

RESULTS

A simple two-dimensional problem was set up to illustrate and compare the above methods for obtaining concentration moments for an existing contaminant plume. In this problem a continuous source releases contaminant, at the solubility limit, into a confined aquifer at a known location. Values assumed for the required input parameters are summarized in Table 1. The analytical velocity covariance function used was derived from the steady-state specific discharge spectrum presented by Gelhar[3]. The concentration moments are calculated at 25 days after the start of contaminant release. The same grid, consisting of 26 nodes in the x direction and 11 nodes in the y direction, was used for each method to facilitate comparison of the results.

Table 1   Inputs for the Sample Problem

Mean Velocity	0.1 m/day
Longitudinal Dispersion Coeffiecient	0.0125 m^2/day
Transverse Dispersion Coefficient	0.005 m^2/day
Log Hydraulic Conductivity Standard Deviation	0.5
Log Hydraulic Conductivity Correlation Length	1.0 m
Spatial Discetization	0.25 m
Temporal Discretization	1.0 day

Figures 1 through 4 show the comparison between the concentration moments calculated by the three stochastic modelling methods. In general the finite element covariance algorithm matches the Monte Carlo simulation more closely than the principal direction covariance algorithm. This is probably due to the fact that the Monte Carlo simulation was run using a Galerkin finite element solver. The differences between the finite element and principal direction covariance results are due to numerical differences between the two solution techniques.

Figure 2 shows the comparison of the concentration standard deviations. Maximum concentration uncertainty is found along the centerline of the plume in areas of high concentration gradient. The nonsymmetries exhibited by the Monte Carlo method are due to the fact that this method computes sample statistics from a finite number of replicates.

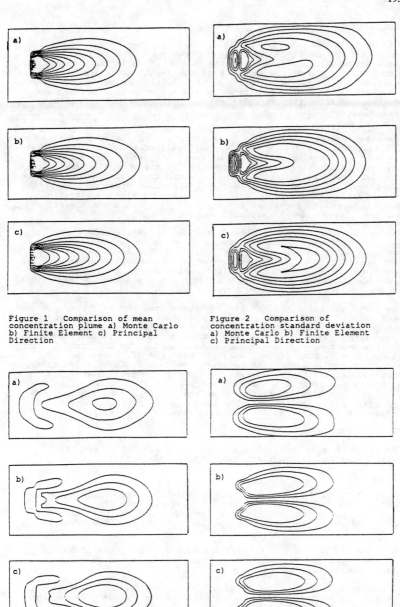

Figure 1 Comparison of mean concentration plume a) Monte Carlo b) Finite Element c) Principal Direction

Figure 2 Comparison of concentration standard deviation a) Monte Carlo b) Finite Element c) Principal Direction

Figure 3 Comparison of longitudinal macrodispersive flux a) Monte Carlo b) Finite Element c) Principal Direction

Figure 4 Comparison of transverse macrodispersive flux a) Monte Carlo b) Finite Element c) Principal Direction

CONCLUSIONS

Both Monte Carlo and perturbation methods provide only approximations of the actual ensemble moments. The accuracy of these approximations depends on the degree of variability in the actual random field, the accuracy of the input statistics, and the accuracy of the numerical and analytic solution methods, among other things. Each of the demonstrated stochastic modelling techniques possesses characteristics which makes it suitable for a particular type of problem.

The Monte Carlo method has the advantage that it does not rely on a small perturbation assumption, and therefore may be more applicable for aquifers with highly variable soil characteristics. However as the soil variability increases, so does the number of replicates required to produce reliable concentration moments. Therefore, the computer costs associated with using Monte Carlo simulation escalate rapidly with increasing soil variability. Other disadvantages of the Monte Carlo method include numerical errors in the concentration moments caused by numerical errors in the turning bands algorithm and the finite element transport solver.

The perturbation method possesses the obvious disadvantage that it is only applicable to random fields in which perturbations around the mean solution are relatively small. Preliminary results indicate that the first-order concentration moments are adequate for log hydraulic conductivity standard deviations up to approximately 1.0. In addition, errors are introduced by the iterative algorithms used to evaluate the forcing terms.

The perturbation method has the advantage that the solution grid need only be designed to capture the details of the moments of the random fields, rather than the variation of the random fields themselves. Thus a covariance solution grid will typically require fewer nodes than a Monte Carlo solution grid. The principal direction algorithm is less accurate than the finite element algorithm for problems with uniform mean flow fields and relatively large dispersivities. However the principal direction covariance algorithm requires less computer storage and runs faster given the same solution grid, particularly as the number of nodes increases. The principal direction solution technique is thus best suited for large-scale problems with small transverse dispersivity.

ACKNOWLEDGMENTS

The research described in this paper was supported in part by National Science Foundation Grant Number ECE-8514987.

REFERENCES

1. Frind E. O. (1982). The Principal Direction Technique: A New Approach to Groundwater Contaminant Transport Modelling, Proceedings, Fourth International Conference on Finite Elements in Water Resources, Hanover, Germany. Springer Verlag, New York, pp.13-25 to 13-42.

2. Shinozuka M. and Jan C.M.(1972). Digital Simulation of Random Processes and its Applications, Journal of Sound and Vibration, 25(1), pp.111 to 128.

3. Gelhar L.W. and Axness C. L.(1983). Three-Dimensional Stochastic Analysis of Macrodispersion in Aquifers, Water Resources Research, 19(1), pp. 161 to 180.

4. Burnett R.D. and Frind E.O. (1987). Simulation of Contaminant Transport in Three Dimensions, 1, The Alternating Direction Galerkin Technique, Water Resources Research, 23(4), pp.683 to 694.

Modelling Flow in Heterogeneous Aquifers: Identification of the Important Scales of Variability

L.R. Townley
CSIRO Division of Water Resources, Private Bag, Wembley, W.A., 6014, Australia

ABSTRACT

Although the spatial distribution of transmissivity in real aquifers is highly variable, adequate simulation is often achieved using spatially averaged transmissivity for model input. A fundamental question concerns the choice of the largest suitable averaging scale at which adequate simulation is achieved, but above which, prediction accuracy deteriorates. Prediction accuracy for a hypothetical aquifer is evaluated numerically using first-order calculations of the covariance of predicted heads. Uncertainty in transmissivity is represented by spatial covariance functions which take into account the length scales of averaging. The prediction accuracy is found to deteriorate when the averaging scale exceeds the correlation length of the underlying log transmissivity field.

INTRODUCTION

Steady flow in regional aquifers is commonly described by the aquifer flow equation (Bear[1]) :

$$\frac{\partial}{\partial x}\left(T\frac{\partial \phi}{\partial x}\right) + \frac{\partial}{\partial y}\left(T\frac{\partial \phi}{\partial y}\right) + Q = 0 \qquad (1)$$

where $\phi(x,y)$ represents piezometric head, $T(x,y)$ is transmissivity and Q is recharge. A corresponding finite element model can be written in the form :

$$Kh = f \qquad (2)$$

where h ($n \times 1$) contains heads at all nodes, K ($n \times n$) depends on transmissivities and the geometry of the finite element grid, and f ($n \times 1$) depends on boundary conditions and recharge.

Both recharge and transmissivity can be highly variable in space, thus considerable effort is required to prepare data for field applications. In the limit, individual values of each parameter could be prescribed for each node or element of a finite element grid, but in practice, constant values are assigned to a small number of zones which fill the aquifer domain. These constant values can be conceptualised as spatial averages, but there is no theoretical basis for selection of the size of these zones.

Previous work by Townley[2] considered the case of variable recharge which is spatially averaged before input to a model. This paper addresses the analogous problem of variable transmissivity.

ILLUSTRATION

Suppose that $Y = \log_{10} T$, measured at the finest possible scale, is represented by the stationary random field $Y_0(x, y)$ and that the corresponding distribution of heads is $\phi_0(x, y)$. Now suppose that Y_0 is smoothed with a "moving average filter", which calculates the average of Y_0 over a circle of radius L centred at (x, y). The resulting distribution, $Y_L(x, y)$, is an approximation to Y_0, and the predicted distribution of heads $\phi_L(x, y)$, calculated using $Y_L(x, y)$, is also an approximation to ϕ_0.

As L increases towards or beyond the length of the domain D, perhaps, in the limit, to ∞, $Y_L(x, y)$ approaches a constant spatial average, Y_∞, over the whole domain. In many circumstances, the resulting prediction ϕ_∞ is not a good approximation to ϕ_0. An obvious question to ask is "How large can L be, before ϕ_L is no longer a reasonable approximation to ϕ_0?". It seems reasonable to hypothesise that there exists some intermediate scale, between $L = 0$ and $L \approx D$, where adequate predictions can be obtained, without the need for excessive detail in the distribution of transmissivity.

THEORY

The effects of spatial averaging could be assessed by deterministic simulation, or by stochastic techniques such as the Monte Carlo method. Another stochastic technique, known as the first-order second-moment method (Sagar[3]; Dettinger and Wilson[4]; Townley and Wilson[5]), is based on a Taylor series expansion of the finite element model in all the uncertain parameters, and has substantial computational advantages.

Suppose that Y_0 has constant mean m_Y, variance σ_Y^2 and exponential correlation structure $\rho_0(r)$ with length scale λ, i.e.

$$Cov(Y_0, Y_0) = Cov_0(r) = \sigma_Y^2 \, \rho_0(r) = \sigma_Y^2 \exp(-r/\lambda) \qquad (3)$$

where r is the separation between points (x_i, y_i) and (x_j, y_j). The spatially averaged field of log transmissivity can be defined as :

$$Y_L(x_i, y_i) = \frac{1}{A_i} \int_{A_i} Y_0(x, y) \, dA_i \qquad (4)$$

where the domain of integration A_i is a circle of radius L centred at (x_i, y_i), dA_i is an element of area within the circle, and A_i is also used to denote the area of the circle. It follows that the expected value of Y_L is m_Y and that the covariance between Y_L at points i and j is:

$$Cov(Y_L, Y_L) = Cov_L(r) = \sigma_Y^2 \frac{1}{A_i} \frac{1}{A_j} \int_{A_j} \int_{A_i} \rho_0(s)\, dA_i\, dA_j \qquad (5)$$

where s takes the values of distances between all pairs of points in A_i and A_j. The cross-covariance $Cov(Y_0, Y_L)$ is similarly defined.

Suppose y (not to be confused with the cartesian coordinate) is a $(p \times 1)$ vector of parameters which defines the spatial distribution of Y. If y is random with expected value m_y and covariance P_{yy}, it follows that h is random, with mean $m_h = h(m_y)$ and covariance:

$$P_{hh} = \left(\frac{dh}{dy'}\right) P_{yy} \left(\frac{dh}{dy'}\right)' \qquad (6)$$

where P_{hh} $(n \times n)$ is symmetric and the $(n \times p)$ sensitivity matrix dh/dy' is evaluated at $y = m_y$. This result is equivalent to Townley and Wilson's[5] Equation 65, although their Equation 69 is computationally faster.

The contents of P_{yy} depend, of course, on the choice of parameterisation of transmissivity. If transmissivity is defined separately at each node of a finite element grid, for example, then p is equal to n and the (i,j)'th term of P_{yy} contains a covariance based on the distance between the i'th and j'th nodes. In the context of this paper, suitable covariances are given by Equations 3 and 5, for point and spatially averaged processes, respectively.

The covariance of the difference between Y_0 and Y_L can be written as:

$$Cov[Y_0 - Y_L, Y_0 - Y_L] = Cov[Y_0, Y_0] + Cov[Y_L, Y_L] - 2Cov[Y_0, Y_L] \qquad (7)$$

which depends on r. It follows that the covariance of the difference between h_0 and h_L can be obtained by using Equation 7 to fill P_{yy} in Equation 6.

RESULTS

Townley[6] (Figure 6) showed good agreement between Monte Carlo and first-order second-moment results for a two-dimensional flow system originally studied by Smith and Freeze[7]. The same geometry is studied here.

Consider a domain of lengths 200 and 100 in the x- and y-directions, respectively. A fairly coarse finite element grid is used with 99 nodes and 160 elements (Figure 1a). Fixed head boundaries are assumed along the top of the domain ($\phi = 1.0$) and along the lower three-eighths of the right hand side ($\phi = 0.0$). The underlying transmissivity distribution is defined by $m_Y = -2$, $\sigma_Y = 0.43$ and $\lambda = 17$ (i.e. nearly equal to the correlation scales used by Smith and Freeze). Using Equation 3 to fill P_{yy}, the standard deviation of heads (from the diagonal of P_{hh} in Equation 6) due to uncertainty in Y is as shown in Figure 1b.

Evaluating Equation 5 by numerical integration results in a range of covariance functions for six averaging scales, L, as shown in Figure 3a. As the averaging scale increases, the corresponding standard deviations of predicted heads (Figure 2) diverge further and further from Figure 1b, thus indicating that averaging the log transmissivity field tends to reduce the variability seen in the corresponding model predictions.

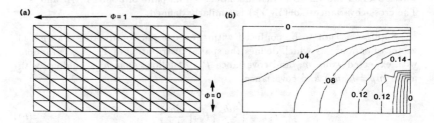

Figure 1. (a) Finite element grid, and (b) Standard deviations of predicted heads using Equation 3 to fill P_{yy}.

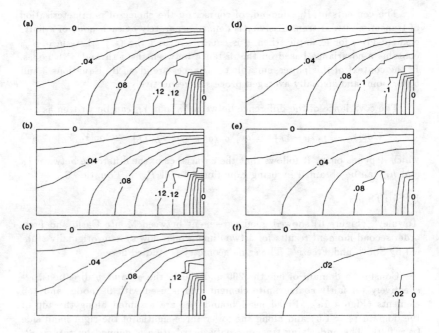

Figure 2. Standard deviations of predicted heads obtained using Equation 5 to fill P_{yy}, with (a) $L = 0.1\lambda$, (b) $L = 0.2\lambda$, (c) $L = 0.5\lambda$, (d) $L = \lambda$, (e) $L = 2\lambda$, and (f) $L = 5\lambda$.

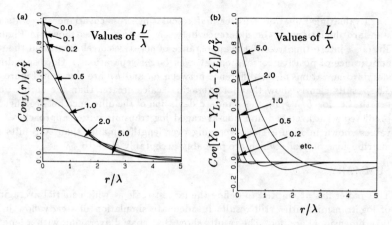

Figure 3. (a) Covariance of Y_L (Equation 5), and (b) Covariance of the difference between Y_0 and Y_L (Equation 7), for various L/λ.

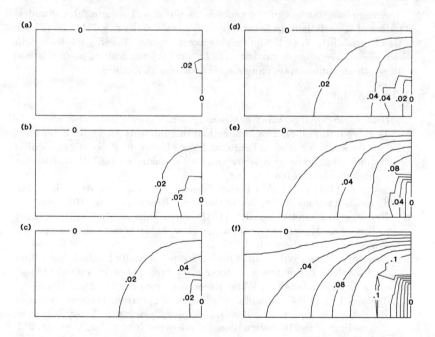

Figure 4. Standard deviations of the difference between heads predicted with and without spatial averaging, using Equation 7 to fill P_{yy}, with (a) $L = 0.1\lambda$, (b) $L = 0.2\lambda$, (c) $L = 0.5\lambda$, (d) $L = \lambda$, (e) $L = 2\lambda$, and (f) $L = 5\lambda$.

Evaluating Equation 7 (after first calculating cross-covariances) results in standard deviations of the difference between Y_0 and Y_L as shown in Figure 3b. (It is interesting to note the appearance of negative correlations, and the reappearance of positive correlations at even larger separations.) Corresponding standard deviations of the difference between h_0 and h_L are shown in Figure 4. The results clearly show that for averaging scales greater than the correlation length (i.e. for $L/\lambda \geq 1$), the standard deviation of the difference between heads predicted with averaged and non-averaged log transmissivities approaches the values shown in Figure 1b. Averaging over lengths smaller than λ results in distributions of heads which are probably acceptably close to ϕ_0.

SUMMARY

This paper has attempted to define the largest scale at which spatial averaging of log transmissivities still results in adequate simulation of steady flow in a two-dimensional aquifer. The results show that spatial averaging with a length scale less than that of the underlying point process has a negligible effect on the variability of corresponding predicted heads. This result is intuitively pleasing, as it gives further meaning to the length scale of the underlying process.

A complete solution to the problem as posed will require further understanding of the relationship between : (i) the scale of the flow field, D; (ii) the scale of variability, λ, of the underlying point process, Y_0; (iii) the scale with which spatial averaging is carried out, L; and (iv) the scale of discretisation, with which an approximate finite element solution is obtained.

REFERENCES

1. Bear J. (1979). Hydraulics of Groundwater, McGraw-Hill. New York.
2. Townley L.R. (1988), The Implications of Spatial Averaging in Estimating Net Recharge for Regional Aquifer Flow Models, in Proceedings, CSIRO Groundwater Recharge Symposium, Perth, Australia, 1987. A.A. Balkema Publishers, Amsterdam.
3. Sagar B. (1978), Galerkin Finite Element Procedure for Analyzing Flow Through Random Media, Water Resources Research, 14(6), 1035-1044.
4. Dettinger M.D. and Wilson J.L. (1981), First-Order Analysis of Uncertainty in Numerical Models of Groundwater Flow, 1, Mathematical Development, Water Resources Research, 17(1), 149-161.
5. Townley L.R. and Wilson J.L. (1985), Computationally Efficient Algorithms for Parameter Estimation and Uncertainty Propagation in Numerical Models of Groundwater Flow, Water Resources Research, 21(12), 1851-1860.
6. Townley L.R. (1984), Second Order Effects of Uncertain Transmissivities on Predictions of Piezometric Heads, in Finite Elements in Water Resources, Proceedings of the 5th International Conference, Burlington, Vermont, 251-264. Springer-Verlag.
7. Smith L. and Freeze R.A. (1979), Stochastic Analysis of Steady State Groundwater Flow in a Bounded Domain, 2. Two-Dimensional Simulations, Water Resources Research, 15(6), 1543-1559.

SECTION 2E - SALTWATER INTRUSION

Modelling of Sea Water Intrusion of Layered Coastal Aquifer
A. Das Gupta and N. Sivanathan
Division of Water Resources Engineering, Asian Institute of Technology, P.O. Box 2754, Bangkok 10501, Thailand

ABSTRACT

The effect of concentration on solute transport is considered to analyze sea water intrusion problem in two-layered aquifer system. An iterative solution scheme is adopted to solve the flow and transport equation simultaneously by using implicit finite difference approximation of equation. The model has been applied to a hypothetical but typical representation of a field situation to assess the effect of pumpage on sea water intrusion.

INTRODUCTION

Exploitation of ground water supplies for domestic, agricultural and industrial purposes is often hampered in coastal area due to encroachment of sea water. With the development of ground water supplies and subsequent lowering of the water table or piezometric surface, the dynamic balance between the fresh and sea water is disturbed, thus permitting the sea water to intrude further into the aquifer. An assessment of the phenomenon of sea water intrusion in specific coastal area is needed before initiating a major ground water development program. This could be achieved with the help of mathematical model analysis of the phenomenon considering the geological and hydrologic conditions of the specific area.

The complexity of the phenomenon of flow of fresh water underlain by sea water has led many investigators to make numerous idealizations in attempts to reduce mathematical description of the phenomenon to a tractable form. The first attempt to determine quantitatively the pattern of movement of the transition zone was made by Henry in 1964. Following this, several researchers developed mathematical models to analyze the sea water intrusion phenomenon using miscible flow condition (for example by Pinder and Cooper, 1970; Huyakorn and Taylor, 1976; Frind, 1980; and Volker and Rushton, 1982). It has been indicated that the behavior of the transition zone can be best described by the variable density hydrodynamic dispersion model.

This paper deals with a quasi-three-dimensional solute transport model for simulating area transport in aquifer considering the interlinkage of flow and transport of solute between the aquifers in vertical plane for a two-layered aquifer system with an unconfined aquifer overlying a confined aquifer. The model has been applied to a hypothetical but typical representation of a field situation to assess the effect of pumpage on sea water intrusion.

SOLUTE TRANSPORT MODEL

The mathematical formulation of the phenomenon of sea water intrusion can be defined in terms of two partial differential equations.

Flow equation

The flow equation is obtained by combining Darcy equation expressed for variable density fluid with continuity equation

(a) for unconfined aquifer

$$\frac{\partial}{\partial x_i} \{h[A_{ij}(\frac{\partial h}{\partial x_j} - \varepsilon \frac{C}{C_s} \frac{\partial z}{\partial x_j})]\} = S_y \frac{\partial h}{\partial t} + W \quad (1)$$

(b) for confined aquifer

$$\frac{\partial}{\partial x_i}[A_{ij}(\frac{\partial h}{\partial x_j} - \varepsilon \frac{C}{C_s} \frac{\partial z}{\partial x_j})] = S_s \frac{\partial h}{\partial t} + \frac{W}{b} \quad (2)$$

with $\quad A_{ij} = K_{ij}\frac{\mu_f}{\mu} = (\frac{k_{ij}\rho g}{\mu_f})\frac{\mu_f}{\mu}; \quad \varepsilon = \frac{\rho_s - \rho_f}{\rho_f} \quad (3)$

$$\rho = \rho_f + (\rho_s - \rho_f)\frac{C}{C_s} \quad (4)$$

$$\frac{\mu}{\mu_f} = 1 + 1.46 \times 10^{-3} C + 3.1 \times 10^{-6} C^2 \quad (5)$$

In Equations (1) through (5), h is the fresh water hydraulic head; K_{ij} and k_{ij} are respectively hydraulic conductivity and intrinsic permeability; ρ_f and μ_f are respectively fresh water density and dynamic viscosity; ρ_s is the density of sea water; ρ and μ are density and dynamic viscosity of fluid mixture, respectively; C is the concentration of the dissolved salt; C_s is the maximum salt concentration; z is the elevation head measured positive downward from a horizontal reference plane; g is the gravitational acceleration; S_y is the specific yield of unconfined aquifer; S_s is the specific storage of confined aquifer; b is the saturated thickness of confined aquifer; and W stands for source, sink and leakage flux through aquitard.

Transport equation

The general equation used to describe the two-dimensional area transport and dispersion of nonreactive dissolved salt in flowing ground water considering the leakage of salt between aquifers is (Konikow and Grove, 1977)

$$\frac{\partial}{\partial x_i}\left(D_{ij}\frac{\partial C}{\partial x_j}\right) - \frac{\partial}{\partial x_i}(CV_i) = \frac{\partial C}{\partial t} + \frac{C'W}{bn} \qquad (6)$$

where V_i is the fluid velocity in i-direction; and C' is the concentration of source or sink fluid or concentration associated with leakage flux between aquifers. The components of dispersion tensor D_{ij} in Equation (6) for two-dimensional flow are expressed as (Bear, 1972)

$$D_{xx} = \alpha_L \frac{V_x^2}{|V|} + \alpha_T \frac{V_y^2}{|V|}; \qquad D_{yy} = \alpha_T \frac{V_x^2}{|V|} + \alpha_L \frac{V_y^2}{|V|}$$

$$D_{xy} = D_{yx} = (\alpha_L - \alpha_T)\frac{V_x V_y}{|V|}; \qquad |V| = \sqrt{V_x^2 + V_y^2} \qquad (7)$$

where V_x and V_y are fluid velocity components in x and y direction respectively; and α_L and α_T are respectively the longitudinal and transverse dispersivities. The flow and solute transport equations are reduced to a set of simultaneous equations using implicit finite difference approximation of the derivatives. Because of flow being dependent on concentration distribution, the solution is being iterated between the flow and transport equation. The stability and accuracy of the results are governed by Courant and Neumann criteria. The accuracy of the numerical solution is checked with analytical solution for one-dimensional advection-dispersion of a solute through a semi-infinite isotropic porous medium given by Ogata and Banks (1961). The numerical solution is found to be accurate for grid Peclet number upto 1.0. However, for higher values of Peclet number, the numerical solution displays appreciable numerical dispersion ahead of the concentration front. But the solution is still stable without any overshooting or undershooting.

MODEL APPLICATION

Hypothetical but typical field problems are considered to analyze the sea water intrusion problem under pumping condition as well as to assess the effect of concentration term in flow model for area transport and to analyze the effect of leakage flux through aquitard on concentration distribution in the system when two-layer aquifer systems are considered. An area extent of a two-layer aquifer system (unconfined above confined) of 3.25 km square with three pumping wells parallel to the shoreline, as shown in Figure 1 is considered. Aquifer parameter values are indicated in the figure. Initially, a fresh water flow rate of 0.022 m/day is maintained towards sea by having a suitable head distribution over the domain. Initial head values along shoreline and along the landward boundary are maintained as time-independent boundary conditions for transient analysis.

In the first case, analysis is done for horizontal confined aquifer only under two levels of source concentration. The purpose is to assess the extent of effect of variation of viscosity on concentration distribution. Figure 2 indicates that when the source concentration C_o = 30 g/ℓ, the relative concentration distribution pattern with time at a specific location is identical from solutions with viscosity independent case (μ/μ_f = 1) and with viscosity dependent case [μ/μ_f defined by Equation (5)]. However with higher source concentration, a lower level of concentration distribution is attained with viscosity dependent solution. With higher μ/μ_f ratio at higher concentration level, head value decreases. This results in lower flow velocity and subsequently lower convective and dispersive transport. It is to be noted that the density dependent term (θ) and the concentration term associated with $\partial z/\partial x$ are neglected in this analysis because of aquifer bed being horizontal. The concentration of salt in sea water is in the range of 25-30 g/ℓ and for this level of concentration, the effect of variation of viscosity on concentration distribution is negligible.

When aquifer bed is sloping, a substantial change in flow magnitude is expected because of the significance of density dependent term (θ) and concentration term associated with $\partial z/\partial x$ in flow equation, as indicated in Figure 3. With adverse (i.e. negative) slope, $\partial z/\partial x$ is positive which indicates a reduction in flow velocity and as such, contaminant from source moves relatively faster, thereby indicating higher level of relative concentration at the location compared to that for sustaining (i.e. positive) slope. This analysis is for confined aquifer only. The effects of leakage flux between aquifers on concentration distribution are indicated in Figure 4. In this case, initial and boundary conditions are same as before and the pumpage is from confined aquifer only. Leakage from unconfined aquifer to confined aquifer results in reduction in level of concentration in unconfined aquifer. Whereas with influx to confined aquifer, the piezometric level drop due to pumping is less compared to the case when there is no leakage. As a result, the level of concentration with time at a particular point is low compared to that for fully confined situation.

With increase in K' value, further reduction is expected, as indicated in Figure 4.

CONCLUSIONS

Analysis with a coupled flow and transport model considering the effect of salt concentration on fluid properties indicates that for sea water intrusion problem with source concentration normally in the range of 25 - 30 g/ℓ, the effect of concentration change in viscosity is not significant. For area transport analysis, the slope of the aquifer bed introduces appreciable effect on concentration distribution as the flow velocity becomes significantly concentration dependent. Analysis with hypothetical but typical field problems indicates the effect of vertical leakage for a two-layered aquifer system on concentration distribution.

REFERENCES

1. Bear, J. (1972) Dynamics of Fluids in Porous Media, American Elsevier, New York.

2. Frind, E.O. (1986), Sea Water Intrusion in Continuous Coastal Aquifer-Aquitard Systems, Proc., 3rd. Int. Conf. Finite Element on Water Resources, University of Mississippi, Mississippi, pp. 2.177 - 2.198.

3. Huyakorn, P. and Taylor, C. (1976), Finite Element Models for Coupled Ground Water Flow and Convective Dispersion, Proc. 1st Int. Conf. Finite Element in Water Resources, Princeton University, Princeton: pp. 1.131 - 1.151

4. Konikow, L.F. and Grove, D.B. (1977), Derivation of Equations Describing Solute Transport in Ground Water, U.S. Geol. Survey Water Resour. Inves. 77-19.

5. Ogata, A. and Banks, R.B. (1961), A Solution of the Differential Equation of Longitudinal Dispersion in Porous Media, U.S. Geol. Survey Prof. Paper 411-A.

6. Pinder, G.F. and Cooper, H.H. Jr. (1970), A Numerical Technique of Calculating the Transient Position of the Saltwater Front, Water Resour. Res., Vol. 6, No. 3, pp. 875-880.

7. Volker, R.E. and Rushton, K.R. (1982), An Assessment of the Importance of some Important Parameters for Seawater Intrusion in Aquifers and a Comparison of Dispersive and Sharp-Interfall Modelling Approaches, Jour. Hydrology, Vol. 56, pp. 239-250.

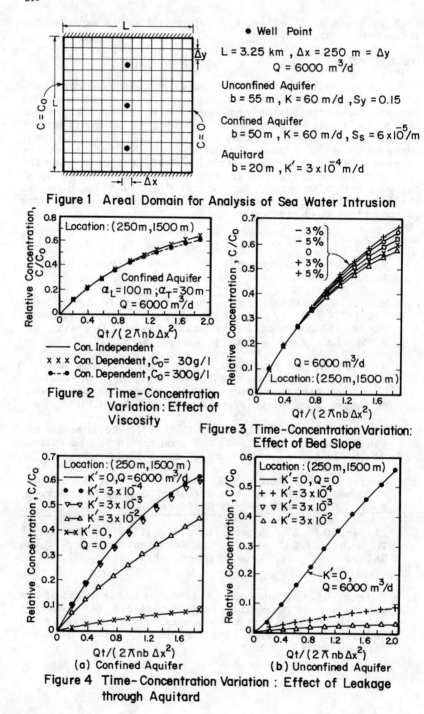

Figure 1 Areal Domain for Analysis of Sea Water Intrusion

Figure 2 Time-Concentration Variation: Effect of Viscosity

Figure 3 Time-Concentration Variation: Effect of Bed Slope

Figure 4 Time-Concentration Variation: Effect of Leakage through Aquitard

A Comparison of Coupled Freshwater-Saltwater Sharp-Interface and Convective-Dispersive Models of Saltwater Intrusion in a Layered Aquifer System

M.C. Hill

U.S. Geological Survey, Denver Federal Center, P.O. Box 25046, M.S. 413, Lakewood, CO 80225, USA

ABSTRACT

Simulated results of the coupled freshwater-saltwater sharp-interface and convective-dispersive numerical models are compared by using steady-state cross-sectional simulations. The results indicate that in some aquifers the calculated sharp interface is located further landward than would be expected.

INTRODUCTION

A quasi-three-dimensional, sharp-interface numerical model of coupled salt and fresh ground-water flow was recently developed by Essaid[1]. Unlike many sharp-interface models in which the saltwater is assumed to be static, this model includes the storage and flow dynamics of the saltwater.

A comparison of Essaid's model and the two-dimensional convective-dispersive model developed by Voss[5] was required for the joint use of the two models in a study of a coastal aquifer system in Cape May County, New Jersey. Previous comparisons of convective-dispersive and sharp-interface models (for example, Henry[3], and Volcker and Rushton[4]) were insufficient because the sharp-interface models used in the comparisons were developed by using different assumptions, and(or) the simulated physical systems were not as complex as the field site considered. In the present work, an idealized cross section that generally represents the aquifer system in coastal areas of Cape May County, New Jersey is simulated. Only steady-state conditions are considered. The purpose of this report is to present the results of the simulations. A discussion of the hydrogeology of the study area is beyond the scope of this paper, but geohydrology was discussed in detail by Gill[2].

Inch-pound units are used in this paper. The unit of weight is pound force (lbf); the unit of mass is the pound force divided by the acceleration of gravity (lbf-s^2/ft).

MODELS USED IN THIS STUDY

The coupled freshwater-saltwater, sharp-interface model solves continuity equations for the fresh and saltwater regions of a simulated aquifer system. The sharp interface that separates the two regions is located such that the hydraulic pressures on each side of the interface are equivalent, and the continuity of pressure in the system is maintained. The model is quasi-three-dimensional in that only horizontal flow is simulated within aquifer layers and only vertical flow is simulated between aquifer layers. The finite-difference method is used to discretize the simulated system.

The two-dimensional, convective-dispersive model used in this study is commonly known as SUTRA (Saturated-Unsaturated TRAnsport). The model solves continuity and transport equations in which pressure and solute mass fraction are the dependent variables, and fluid density depends on solute mass fraction. The model can simulate flow and transport in two dimensions. In the present work, simulated solute mass fractions are presented as isochlors, which are lines of equal chloride concentration relative to seawater. Thus, the 1.0-isochlor indicates seawater; the 0.0-isochlor indicates freshwater. The finite-element method is used to discretize the simulated area or cross section.

SIMULATED CROSS SECTION

The cross section of the aquifer system simulated in this work depicts a water-table aquifer, two confined aquifers, and two confining beds (fig. 1a). The finite-element mesh of the convective-dispersive model that is superimposed on the modeled region is rectangular, extending vertically from 10 to 190 ft below sea level, and horizontally across the 84,000-ft length of the section (fig. 1b). The finite-difference grid used for the sharp-interface model consists of three numerical layers, each representing one of the three aquifers, and 22 columns of finite-difference cells, each 4,000 feet wide. The finite-difference cells are block-centered, so the furthest landward column of cells was centered at the landward-vertical boundary of the simulated region. The two grids are different because of the different stability and convergence criteria of the models.

The simulated boundary conditions are as follows: the bottom and landward-vertical boundaries are impermeable, the hydraulic head or pressure at the seaward-vertical boundary and the top-offshore boundary are specified to represent static saltwater, and the hydraulic head or pressure along the top-onshore boundary is specified to represent water-table altitudes that vary parabolically between sea level at the shore and 12 ft above sea level at the landward-vertical boundary. In the sharp-interface model, the head along the top of the region is fixed in an inactive overlying layer connected to the uppermost model layer, forming a head-dependent boundary condition.

The hydraulic-conductivity values used in the sharp-interface model are shown in figure 1a; the other parameter values used in

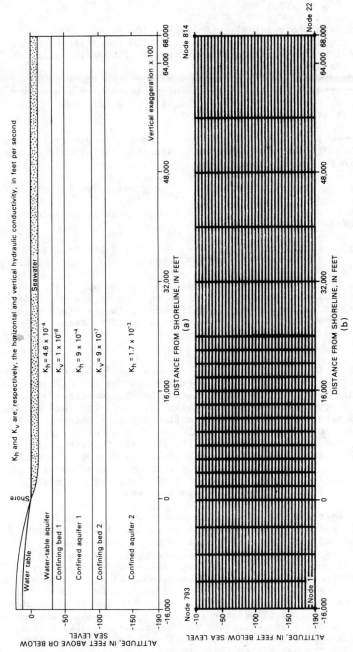

Figure 1 : (a) Modeled cross section and (b) finite-element mesh for the convective-dispersive model.

the model are as follows:

Horizontal to vertical anisotropy : 100 in aquifers,
10 in confining beds,
Porosity = 0.1,   Dynamic viscosity = 2.09 × 10⁻⁵ lbf-s/ft²,
$(\rho_s - \rho_o)/\rho_o = 0.025$,

where ρ_s and ρ_o are the densities of saltwater and freshwater, respectively, and the dynamic viscosity is consistent with a ground-water temperature of 68.4°F.

The permeability (k) required in the convective-dispersive model was calculated from the hydraulic conductivity (K) used in the sharp-interface model as follows:

k = K × (Dynamic viscosity) / (ρ_o × gravity).

Parameters required exclusively by the convective-dispersive model (Voss[5]) are:

$\partial\rho/\partial C = 1.457$ lbf-s²/ft⁴, $\rho_o = 1.94$ lbf-s²/ft⁴,
$\alpha_{Lmax} = 25$ feet, $\alpha_{Lmin} = 2.5$ feet, $\alpha_T = 2.5$ feet,

where C is the solute mass fraction, α_{Lmax} and α_{Lmin} are the longitudinal dispersivities in the direction of maximum and minimum permeability, respectively, and α_T is the transverse dispersivity. Molecular diffusivity is equal to zero.

DISCUSSION

Results of the simulations (fig. 2) show the following: 1) In confined aquifer 1 the sharp interface crosses the 0.2- and 0.4-isochlors, but in the water-table aquifer and confined aquifer 2, the sharp interface is as much as 14,000 ft landward of the 0.2 isochlor; 2) the freshwater inflow and most internal flow rates were nearly equivalent in landward areas of the two models, but are substantially different near the interfaces; and, 3) hydraulic heads in the confined aquifers of the sharp-interface model are as much as 0.4 ft higher than heads calculated from hydraulic pressures simulated at the center of each layer in the convective-dispersive model.

The proximity of the sharp interface to the isochlors is strongly influenced by the simulated values of dispersivity (Volcker and Rushton[4]). A closer match between the sharp interface and the isochlors would be expected by reducing the transverse dispersivity in the convective-dispersive model (C.I. Voss, oral commun., 1988). Results simulated using transverse dispersivity values equal to 0.0, 0.1 and 0.5 ft indicated that: in the water-table aquifer, results were similar to those shown in figure 2a; in confined aquifer 1, the 0.5 isochlor approaches the calculated sharp interface as the transverse dispersivity becomes small; and, in confined aquifer 2, the 0.2-isochlor approaches the calculated sharp interface at the toe, but remains as much as 4,000 ft seaward of the sharp interface in the upper part of the aquifer.

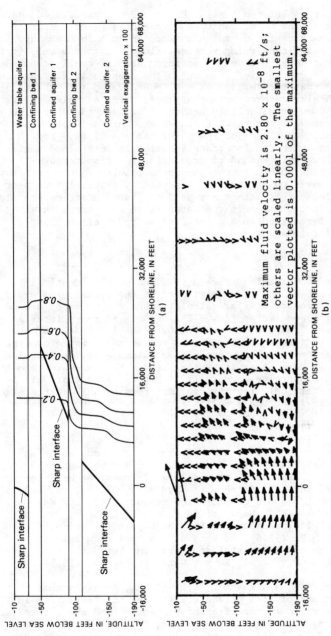

Figure 2 : Results of steady-state simulations : (a) isochlors and the sharp interface; (b) fluid velocity vectors from the convective-dispersive model.

The differences between the isochlors and the calculated sharp interfaces of the water-table aquifer and confined aquifer 2 are, to some extent, due to an assumption made in the sharp-interface model concerning vertical flow. In that model, it is assumed that leakage through confining beds is small compared to the water in the aquifer into which the leakage is flowing, and, therefore, that the leakage can be incorporated into the water type which already occurs in that layer. For example, the water flowing upward from confined aquifer 1 between the shoreline and 18,000 ft seaward starts as fresh-water, but is incorporated into the saltwater zone of the water-table aquifer. For aquifers in which the interface location is determined mostly by lateral flow, this method of accounting for flow through confining beds should produce accurate results. However, if the interface position is strongly affected by vertical flow near the interface, this method may produce sharp-interface positions which are inaccurate. In the water-table aquifer of this problem, it resulted in a calculated sharp interface that is much too far inland.

CONCLUSIONS

The two models described herein produced nearly identical flow rates in much of the freshwater part of the system, and would, therefore, produce similar calculated travel times for use in studies of ground-water flow. The calculated sharp interface commonly was landward of even the 0.2 isochlor in two of the aquifers, indicating that, for the conditions considered, the sharp-interface model may produce an estimate of the location of saltwater in coastal aquifers that is further landward than the convective-dispersive model. It is hypothesized that this is due to how vertical flow through confining beds is represented in the sharp-interface model.

REFERENCES CITED

1. Essaid, H.I. (1986), A comparison of the coupled fresh water-salt water flow and the Ghyben-Herzberg sharp interface approaches to modeling of transient behavior in coastal aquifer systems. J. Hydrol., Vol. 86, pp. 169-193.
2. Gill, H.E., 1962, Ground-water resources of Cape May County, New Jersey, N.J. Dept. of Cons. and Econ. Dev., Div. of Water Policy and Supply, Special Report 18, 171 p.
3. Henry, H.R. (1964), Effects of dispersion on salt encroachment in coastal aquifers, In: Sea Water in Coastal Aquifers, U.S. Geol. Surv., Water-Supply Pap. 1613-C, pp. 70-84.
4. Volker, R.E. and Rushton, K.R. (1982), An assessment of the importance of some parameters for seawater intrusion in aquifers and a comparison of dispersive and sharp-interface modeling approaches. J. Hydrol., Vol. 56, pp. 239-250.
5. Voss, C.I. (1984), A finite-element simulation model for saturated-unsaturated fluid-density-dependent ground-water flow with energy transport or chemically-reactive single-species solute transport, U.S. Geol. Surv., Water Resour. Invest. Rep. 84-4369, 409 pp.

Can the Sharp Interface Salt-Water Model Capture Transient Behavior?

G. Pinder and S. Stothoff

Department of Civil Engineering, Princeton University, Princeton, NJ 08854, USA

INTRODUCTION

While the simulation of groundwater flow in coastal aquifers encountering salt water should formally employ the three dimensional flow and transport equations, the reduction of this problem to two spatial dimensions in the areal plane, using a sharp interface assumption, has often been employed to render field problems tractable. The assumptions inherent in this simplification have been found appropriate for steady state conditions, but there appears to be little evidence supporting this approach for transient problems. In this paper, the transient sharp-interface areal two-dimensional formulation and the three dimensional solute transport formulation are compared, and an example of divergence between the two approaches is presented.

GOVERNING EQUATIONS

Solute Transport

The equations governing three dimensional density-dependent transport are

$$\varepsilon \frac{\partial c}{\partial t} + \mathbf{q} \bullet \nabla c - \nabla \bullet D \bullet \nabla c = 0 \tag{1}$$

$$\nabla \bullet \rho \mathbf{q} = 0 \tag{2}$$

$$\nabla \bullet \mathbf{q} = 0 \tag{3}$$

$$\mathbf{q} + \frac{\mathbf{k}}{\mu} \bullet (\nabla P - \rho \mathbf{q}) = 0 \tag{4}$$

In the above expressions $c(\mathbf{x}, t)$ is the solute concentration, ε is the porosity, $\mathbf{q}(\mathbf{x}, t)$ is the phase average fluid velocity, $\rho(\mathbf{x}, t)$ is the fluid density, $\mathbf{k}(\mathbf{x}, t)$ is the intrinsic permeability, $\mathbf{D}(\mathbf{x}, t)$ is the dispersion coefficient, $\mu(\mathbf{x}, t)$ is the kinematic viscosity, $P(\mathbf{x}, t)$ is the fluid pressure, and g is the gravitational acceleration. This system of equations is nonlinear and must be solved iteratively.

Sharp Interface Equations

The sharp interface equations are developed by writing the fluid flow equations in the fresh- and salt-water zones, integrating these equations vertically to eliminate that dimension, and coupling the resulting equations through conditions imposed at the salt-water/fresh-water interface. The vertical integration is justified when vertical head gradients are negligible. The vertically averaged equations are written (Huyakorn and Pinder[2])

$$\bar{\nabla} \cdot l_f \mathbf{K}_f \cdot \bar{\nabla} \bar{h}_f + q_{fc} - (l_f S_{fs} + S_y + \varepsilon \rho_f{}^*)\frac{\partial \bar{h}_f}{\partial t} + \varepsilon \rho_s{}^* \frac{\partial \bar{h}_s}{\partial t} = 0 \qquad (5)$$

$$\bar{\nabla} \cdot l_s \mathbf{K}_s \cdot \bar{\nabla} \bar{h}_s - q_{sa} + q_{sb} - (l_s S_{ss} + \varepsilon \rho_s{}^*)\frac{\partial \bar{h}_s}{\partial t} + \varepsilon \rho_f{}^* \frac{\partial \bar{h}_f}{\partial t} = 0 \qquad (6)$$

where the variables l_s, l_f, $\bar{h}_f$, $\bar{h}_s$, a, b, and c are presented in Figure 1. In addition

$$\bar{h}_f(x_1, x_2, t) \equiv \frac{1}{l_f}\int_b^c h_f(\mathbf{x}, t)dx_3 \quad ; \quad \bar{h}_s(x_1, x_2, t) \equiv \frac{1}{l_s}\int_a^b h_s(\mathbf{x}, t)dx_3 \quad (7)$$

$$\mathbf{K}_s \equiv \frac{\rho_s g \mathbf{k}}{\mu_s} \quad ; \quad \mathbf{K}_f \equiv \frac{\rho_f g \mathbf{k}}{\mu_f} \tag{8}$$

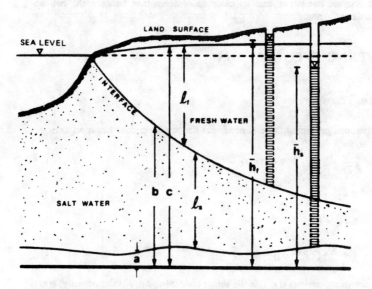

Figure 1: Schematic Description of Flow System (After Huyakorn and Pinder[2].)

$$\rho_s{}^* \equiv \frac{\rho_s}{(\rho_s - \rho_f)} \quad ; \quad \rho_f{}^* \equiv \frac{\rho_f}{(\rho_s - \rho_f)} \tag{9}$$

$$l_f(x_1, x_2, t) = c(x_1, x_2, t) - b(x_1, x_2, t) \tag{10a}$$

$$l_s(x_1, x_2, t) = b(x_1, x_2, t) - a(x_1, x_2, t) \tag{10b}$$

where $\overline{\nabla}(\bullet)$ is the two dimensional cartesian gradient operator defined in the areal plane, $h_f(\mathbf{x}, t)$ is the fresh-water head, $h_s(\mathbf{x}, t)$ is the salt-water head, q_{fc} is the flux through the water-air interface, q_{sa} is the flux of saltwater through the boundary $a(x_1, x_2)$, S_{fs} and S_{ss} are the fresh- and salt-water specific storages respectively, and S_y is the specific yield. The term q_{sa} includes the well discharge when appropriate.

Equations (5)-(10) constitute a set of coupled non-linear equations wherein the nonlinearity is due to the thickness terms l_f and l_s. Note that the interface dynamics depend upon the variables ε, S_{fs}, S_{ss}, S_y, $\rho_f{}^*$, and $\rho_s{}^*$, and vertical hydraulic conductivity does not appear, either explicitly or implicitly, in any equation. This parameter is lost in the process of vertical integration.

COMPARING THE MODELS

Problem Definition

The question posed in this paper focuses on the transient behavior of the sharp interface as described by equations (5)-(10). It is hypothesized that the interface does not respond to a stress causing vertical fluid gradients in a manner consistent with the dynamic movement of the salt-water/fresh-water front as described by equations (1)-(4).

To test this hypothesis, salt-water upconing beneath a pumping well was simulated using both the transport equations (in cylindrical coordinates) (Hsieh[1]) and the sharp interface equations defined in the areal plane. In both cases, the system is initially in hydrostatic equilibrium, and the outside boundary remains so through first-type conditions. The remaining boundaries are no-flux boundaries, except at the well. The finite element mesh used to define this system as a transport problem is presented in Figure 2. The finite element mesh used to define this system as as a sharp-interface formulation in the areal plane is presented in Figure 3. Auxiliary information selected for this investigation is presented in Table 1.

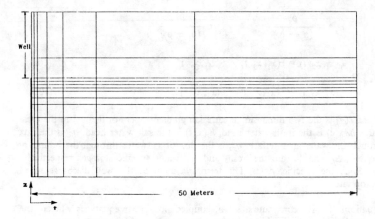

Figure 2: Discretization for the Transport Simulation

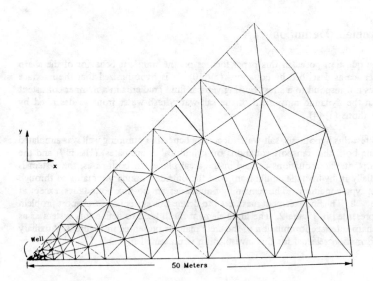

Figure 3: Discretization for the Sharp Interface Simulation

Table 1 - Simulation Dimensions and Parameters		
Parameter	*Transport*	*Interface*
Simulation Radius (m)	50.	50.
Aquifer Thickness (m)	25.	25.
Initial Seawater Thickness (m)	10.	10.
Pumping Rate (m³/sec)	0.015	0.015
Intrinsic Permeability(m²)	2.e-11	2.e-11
Porosity	0.25	0.25
Seawater Density (kg/m³)		1025.
Seawater Salt Concentration (kg/m³)	35.	
Radial Dispersivity (m)	0.5	
Vertical Dispersivity (m)	0.1	

Simulation Results

In response to the pumping stress, the interface moves vertically upward. Assuming, somewhat arbitrarily, that the interface is represented by a chloride concentration of 10 kg/m³, the interface movement is illustrated in Figure 4. The movement of the interface, as simulated by the sharp interface model, is presented in Figure 5. A comparison of these two figures indicates that the two simulations yield entirely different solutions.

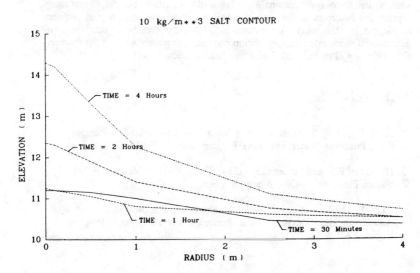

Figure 4: Time History of Interface Movement for Transport Simulation

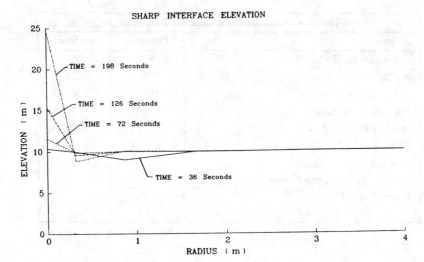

Figure 5: Time History of Interface Movement for Sharp Interface Simulation

CONCLUSIONS

The assumptions inherent in the vertically integrated sharp interface formulation, while at first glance quite reasonable, in fact lead to equations that fail to accurately capture the dynamics of the salt-water/fresh-water interface for the test problem considered in this paper. By extension, it may be concluded that the two-dimensional sharp interface formulation does not adequately account for the transient behavior of regimes exhibiting significant vertical flow.

REFERENCES

1. Hsieh P. (1977), Simulation of Salt-water Upconing Beneath a Pumping Well, Princeton Water Resources Program Report 77-WR-10.

2. Huyakorn P.S. and Pinder G.F. (1983), Computational Methods in Subsurface Flow, pp. 99-109, Academic Press. New York.

3. Page, R. (1979), An Areal Model for Sea-Water Intrusion in a Coastal Aquifer: Program Documentation. Princeton Water Resources Program Report 79-WR-11.

SECTION 3 - MODELING SURFACE WATER FLOWS

SECTION 3A - TIDAL MODELS

A Consistency Analysis of the FEM: Application to Primitive and Wave Equations

J. Drolet
Department of Civil Engineering, Princeton Univesity, Princeton, NJ 08544, USA

W.G. Gray
Department of Civil Engineering, University of Notre Dame, IN 46615, USA

Introduction

The order of consistency is a direct measure of the speed at which a model converges to the exact solution as the elements become finer. As such, consistency unveils the behavior of the numerical solution, in the limit, where the grid spacing is reduced to zero. Here, the analysis is restricted to the spatial semi-discretization.

Often, one may establish the consistency of an approximation by proving that the local truncation error goes to zero in the limit as the grid size is reduced to zero. When the solution is well behaved, the natural tool for such an analysis is the Taylor series expansion[1]. Then, the approximation is expanded around a point. This leads to an expression containing the exact differential equation plus some error terms. If those leading error terms can be written as the power of the grid spacing, Δx, and if this order is greater or equal to one, then the approximation is consistent. The order of this lowest order term is defined as the order of consistency. Because the order of consistency represents the rate at which the approximate solution approaches the exact solution as the mesh is refined, it is desirable to build models with a high order of consistency.

For FEM, a definition of consistency in a global sense is more appropriate. The following definition, similar to the one of Cakmak et al.[2] is proposed here. Consider the error, ε, between a differential equation, f, and its FE approximation, $\hat{f}$.

$$\frac{\iint_\Omega f \phi_i \, d\Omega - \iint_\Omega \hat{f} \phi_i \, d\Omega}{\iint_\Omega \phi_i \, d\Omega} = \varepsilon[O(\Delta x^p)]$$

The approximation, $\hat{f}$, is consistent if the error, ε, is of the order of Δx^p. Where p is at least one.

The Taylor series expansion has proven to be a reliable tool for proving the consistency of approximations over regular meshes. However, approximation over irregular or scattered grids may require a slightly more sophisticated tool in order to show consistency. Such a tool is presented, in detail. It will prove to be useful in instances where the classical method fails to show consistency.

The Equations

The following equations (1-3) are representative of the vertically integrated shallow water equations in their primitive form[3] and in their wave form[4,5]. For the sake of simplicity, the coriolis term, the atmospheric forcing term and the convective term have been left out the analysis. Because the friction term is spatially discretized in the same way as the time derivative term, there is no need to consider this term since it will not affect the result of the analysis.

Primitive momentum $\quad \dfrac{\partial \mathbf{Q}}{\partial t} + c^2 \nabla \zeta = 0 \quad$ (1)

Wave continuity $\quad \dfrac{\partial^2 \zeta}{\partial t^2} - c^2 \nabla \cdot \nabla \zeta = 0 \quad$ (2)

Wave momentum $\quad \dfrac{\partial^2 \mathbf{Q}}{\partial t^2} - c^2 \nabla (\nabla \cdot \mathbf{Q}) = 0 \quad$ (3)

where $c^2 = gh$.

The Galerkin method is then applied to each of these equations. The primitive momentum equation (1) becomes,

$$\iint_\Omega \left[\dfrac{\partial \hat{\mathbf{Q}}}{\partial t} + c^2 \nabla \hat{\zeta} \right] \phi_i \, d\Omega = 0 \quad (4)$$

After application of Green's theorem, equation (4) becomes,

$$\iint_\Omega \dfrac{\partial \hat{\mathbf{Q}}}{\partial t} \phi_i \, d\Omega - c^2 \iint_\Omega \hat{\zeta} \nabla \phi_i \, d\Omega + c^2 \int_{\partial \Omega} \hat{\zeta} \mathbf{n} \, \phi_i \, d\Gamma = 0 \quad (5)$$

or,

x - direction $\quad \iint_\Omega \dfrac{\partial \hat{Q}_x}{\partial t} \phi_i \, d\Omega - c^2 \iint_\Omega \hat{\zeta} \dfrac{\partial \phi_i}{\partial x} \, d\Omega + c^2 \int_{\partial \Omega} \hat{\zeta} \mathbf{n} \, \phi_i \, d\Gamma = 0 \quad$ (6a)

y - direction $\quad \iint_\Omega \dfrac{\partial \hat{Q}_y}{\partial t} \phi_i \, d\Omega - c^2 \iint_\Omega \hat{\zeta} \dfrac{\partial \phi_i}{\partial y} \, d\Omega + c^2 \int_{\partial \Omega} \hat{\zeta} \mathbf{n} \, \phi_i \, d\Gamma = 0 \quad$ (6b)

Equations (6a and 6b) reduce to equations (7a and 7b) in the interior and to equations (8a and 8b) at a boundary parallel to the y axis such as the boundaries shown in figure 1.

interior x - direction $\quad \iint_\Omega \dfrac{\partial \hat{Q}_x}{\partial t} \phi_i \, d\Omega - c^2 \iint_\Omega \hat{\zeta} \dfrac{\partial \phi_i}{\partial x} \, d\Omega = 0 \quad$ (7a)

y - direction $\quad \iint_\Omega \dfrac{\partial \hat{Q}_y}{\partial t} \phi_i \, d\Omega - c^2 \iint_\Omega \hat{\zeta} \dfrac{\partial \phi_i}{\partial y} \, d\Omega = 0 \quad$ (7b)

boundary x - direction $\quad \iint_\Omega \dfrac{\partial \hat{Q}_x}{\partial t} \phi_i \, d\Omega - c^2 \iint_\Omega \hat{\zeta} \dfrac{\partial \phi_i}{\partial x} \, d\Omega + c^2 \int_{\partial \Omega} \hat{\zeta} \, \phi_i \, dy \quad$ (8a)

y - direction $\quad \iint_\Omega \dfrac{\partial \hat{Q}_y}{\partial t} \phi_i \, d\Omega - c^2 \iint_\Omega \hat{\zeta} \dfrac{\partial \phi_i}{\partial y} \, d\Omega = 0 \quad$ (8b)

Similarly, the wave continuity equation (2) becomes,

$$\iint_\Omega \left[\dfrac{\partial^2 \hat{\zeta}}{\partial t^2} - c^2 \nabla \cdot \nabla \hat{\zeta} \right] \phi_i \, d\Omega = 0 \quad (9)$$

applying Green's theorem leads to (10) at the boundary. The boundary integral vanishes in the interior.

$$\iint_\Omega \dfrac{\partial^2 \hat{\zeta}}{\partial t^2} \phi_i \, d\Omega + c^2 \iint_\Omega \nabla \hat{\zeta} \cdot \nabla \phi_i \, d\Omega - c^2 \int_{\partial \Omega} \nabla \hat{\zeta} \cdot \mathbf{n} \, \phi_i \, d\Gamma = 0 \quad (10)$$

Finally, application of the Galerkin method to the wave momentum equation (3) leads to,

$$\iint_\Omega \left[\frac{\partial^2 \hat{Q}}{\partial t^2} - c^2 \nabla \cdot (\nabla \cdot \hat{Q}) \right] \phi_i \, d\Omega = 0 \qquad (11)$$

With the help of tensor notation, it is easily shown that,

$$\iint_\Omega \nabla (\nabla \cdot f) \phi_i \, d\Omega = -\iint_\Omega (\nabla \cdot f) \nabla \phi_i \, d\Omega + \int_{\partial\Omega} \nabla \cdot f \, \phi_i \, n \, d\Gamma \qquad (12)$$

This identity allows equation (11) to be rewritten as,

$$\iint_\Omega \frac{\partial^2 \hat{Q}}{\partial t^2} \phi_i \, d\Omega + c^2 \iint_\Omega (\nabla \cdot \hat{Q}) \nabla \phi_i \, d\Omega - c^2 \int_{\partial\Omega} (\nabla \cdot \hat{Q}) \, n \, \phi_i \, d\Gamma = 0 \qquad (13)$$

or,

x - direction
$$\iint_\Omega \frac{\partial^2 \hat{Q}_x}{\partial t^2} \phi_i \, d\Omega + c^2 \iint_\Omega \left(\frac{\partial \hat{Q}_x}{\partial x} + \frac{\partial \hat{Q}_y}{\partial y} \right) \frac{\partial \phi_i}{\partial x} \, d\Omega - c^2 \int_{\partial\Omega} \left(\frac{\partial \hat{Q}_x}{\partial x} + \frac{\partial \hat{Q}_y}{\partial y} \right) \phi_i \, dy = 0 \qquad (14a)$$

y - direction
$$\iint_\Omega \frac{\partial^2 \hat{Q}_y}{\partial t^2} \phi_i \, d\Omega + c^2 \iint_\Omega \left(\frac{\partial \hat{Q}_x}{\partial x} + \frac{\partial \hat{Q}_y}{\partial y} \right) \frac{\partial \phi_i}{\partial y} \, d\Omega = 0 \qquad (14b)$$

As before, in the interior, the boundary integral vanishes.

Spatial Discretizations

Four spatial discretizations are considered. The first one is an hexagonal arrangement. It contains six identical isosceles triangles. Then, two arrangements, B and C, which often arise together when deriving a triangular grid from a rectangular grid, are analyzed individually. This arrangement is often preferred to other square triangle arrangements because its diagonals do not all go in the same direction. Finally, the interaction of B and C is studied with arrangement D.

Several boundary arrangements can be derived from the arrangements in the interior. The boundary arrangements shown in figure 1 were arbitrarily chosen from the several possibilities offered to us.

Order of Consistency

The classical Taylor series expansion approach to the problem of consistency was applied to configurations A, B and C. Equations (7,8,10 and 14) were integrated exactly using linear basis functions. A finite difference equivalent was obtained for each equation[6]. The resulting FD approximations were expanded in Taylor series. Those results are presented in table I.

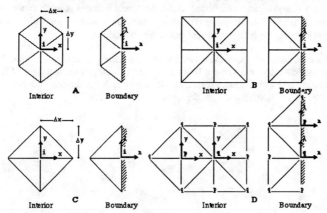

Figure 2.4 Spacial discretizations for the consistency analysis

Table I Order of Consistency (classical approach)

Arrangement	Equation		Interior	Boundary
A	Primitive Momentum	(x direction)	2	2
		(y direction)	2	2
	Wave Continuity		2	2
	Wave Momentum	(x direction)	2	2
		(y direction)	2	2
B	Primitive Momentum	(x direction)	2	2
		(y direction)	2	1
	Wave Continuity		0	0
	Wave Momentum	(x direction)	0	0
		(y direction)	0	0
C	Primitive Momentum	(x direction)	2	2
		(y direction)	2	1
	Wave Continuity		0	0
	Wave Momentum	(x direction)	0	0
		(y direction)	0	0

The same type of analysis was conducted for equation (1-3) in one dimension. Again, the Galerkin method with linear basis function was employed. Exact integration was also performed and, their FD equivalents were expanded in Taylor series. All three equations were found to be second order accurate in the interior. At a boundary node, both wave equations also exhibited second order accuracy while the primitive momentum equation only showed first order accuracy.

From the one dimensional analysis, one would expect the wave equations, in 2-D, to be more accurate than the primitive equations. On the contrary, the Taylor series expansion of the wave equation approximations over the meshes B and C fails to prove consistency although experience has shown these equations to be convergent[7]. However, arrangement B and C usually co-exist in a mesh. It is possible that their errors cancel. In order to verify this hypothesis, it is necessary to analyze arrangements A and B simultaneously.

Consistency Analysis involving one node or more

The consistency analysis can be expanded to involve approximations at more than one node. This method makes use of the Fourier transform and leads to a modified equation[8,9]. Following an example is probably the best way to understand this method.

Consider the wave continuity equation (2) over a mesh containing arrangements B and C in the interior, more specifically, arrangement D in the interior. It is understood that figure 1 D is only part of a larger grid with the same regularity. Obviously such a grid along with equation (2) lends itself to two solutions, p and q, of the type,

$$\hat{p} = \sum_k p_k\, e^{i(\sigma_x x + \sigma_y y + \beta t)} \tag{15a}$$

$$\hat{q} = \sum_k q_k\, e^{i(\sigma_x x + \sigma_y y + \beta t)} \tag{15b}$$

It is possible to evaluate the Galerkin expression (10) in term of space shift operators. Clearly, two equations are then obtained:

at node p,
$$-\beta^2 \frac{2\Delta}{6} [4p + (E_x + E_x^{-1} + E_y + E_y^{-1})q]$$

$$+ c^2 \left\{ \frac{(\Delta y)^2}{2\Delta} [2p - (E_x + E_x^{-1})q] + \frac{(\Delta x)^2}{2\Delta} [2p - (E_y + E_y^{-1})q] \right\} = 0 \quad (16a)$$

at node q,
$$-\beta^2 \frac{2\Delta}{6} [(8 + (E_x + E_x^{-1})(E_y + E_y^{-1}))q + (E_x + E_x^{-1} + E_y + E_y^{-1})p]$$

$$+ c^2 \left\{ \frac{(\Delta y)^2}{2\Delta} [2q - (E_x + E_x^{-1})p] + \frac{(\Delta x)^2}{2\Delta} [2q - (E_y + E_y^{-1})p] \right\} = 0 \quad (16b)$$

where Δ is the area of an element,

$$\Delta = \frac{\Delta x \Delta y}{2} \quad (17)$$

The system of equation (16) can be written as the matrix equation (18) in spectral space by substituting p and q (15a and b) into (16). This is equivalent to taking a discrete Fourier transform of (16).

$$\begin{bmatrix} -4\beta^2 \frac{\Delta}{6} + \frac{c^2}{\Delta}[(\Delta x)^2 + (\Delta y)^2], & -\beta^2 \frac{\Delta}{6}(\hat{E}_x + \hat{E}_x^{-1} + \hat{E}_y + \hat{E}_y^{-1}) \\ & -\frac{c^2}{2\Delta}[(\Delta y)^2(\hat{E}_x + \hat{E}_x^{-1}) + (\Delta x)^2(\hat{E}_y + \hat{E}_y^{-1})] \\ -\beta^2 \frac{\Delta}{6}(\hat{E}_x + \hat{E}_x^{-1} + \hat{E}_y + \hat{E}_y^{-1}) & -\beta^2 \frac{\Delta}{6}[8 + (\hat{E}_x + \hat{E}_x^{-1})(\hat{E}_y + \hat{E}_y^{-1})] \\ -\frac{c^2}{2\Delta}[(\Delta y)^2(\hat{E}_x + \hat{E}_x^{-1}) + (\Delta x)^2(\hat{E}_y + \hat{E}_y^{-1})], & +\frac{c^2}{\Delta}[(\Delta x)^2 + (\Delta y)^2] \end{bmatrix} \begin{pmatrix} \hat{p} \\ \hat{q} \end{pmatrix} = [0] \quad (18)$$

The Fourier transform of the space shift operators is given by[10],

$$\hat{A} = \frac{A \cdot e^{i(\beta t + \sigma_x x + \sigma_y y)}}{e^{i(\beta t + \sigma_x x + \sigma_y y)}}$$

This system of equation has a rank of 1. Therefore, its non-trivial solution may be written in terms of one parameter. If $\hat{p}$ is eliminated from this system, one obtains one equation in terms of the parameter $\hat{q}$. This equation is also a polynomial in β^2. An approximate value of β^2 can be found using perturbation techniques[11]. β^2 and, the space shift operators, $\hat{E}_x, \hat{E}_{x-1}, \hat{E}_y$ and $\hat{E}_{y-1}$ are expanded in powers of Δx and Δy as follow.

$$\beta^2 = \beta_0 + \beta_x \Delta x + \beta_y \Delta y + \beta_{xx} \Delta x^2 + \beta_{xy} \Delta x \Delta y + \beta_{yy} \Delta y^2 + \ldots \quad (20)$$

$$\hat{E}_x = e^{i\sigma_x \Delta x} = 1 + i\sigma_x \Delta x + (i\sigma_x \Delta x)^2/2 + (i\sigma_x \Delta x)^3/6 + \ldots$$

$$\hat{E}_y = e^{i\sigma_y \Delta y} = 1 + i\sigma_y \Delta y + (i\sigma_y \Delta y)^2/2 + (i\sigma_y \Delta y)^3/6 + \ldots$$

$$\hat{E}_x^{-1} = e^{-i\sigma_x \Delta x} = 1 - i\sigma_x \Delta x + (i\sigma_x \Delta x)^2/2 - (i\sigma_x \Delta x)^3/6 + \ldots$$

$$\hat{E}_y^{-1} = e^{-i\sigma_y \Delta y} = 1 - i\sigma_y \Delta y + (i\sigma_y \Delta y)^2/2 - (i\sigma_y \Delta y)^3/6 + \ldots \quad (21)$$

Equating to zero, equal powers of Δx and Δy in the reduced system of equations, one obtains that,
$$\beta_0 = c^2(\sigma_x^2 + \sigma_y^2)$$
$$\beta_x = 0$$
$$\beta_y = 0 \quad (22)$$

Thus β_2 is,

$$\beta^2 = c^2[\sigma_x^2 + \sigma_y^2 + O(\Delta x^2, \Delta x \Delta y, \Delta y^2)] \qquad (23)$$

Taking an inverse Fourier transform leads to a modified equation[8,9].

$$\frac{\partial^2 \zeta}{\partial t^2} = c^2 \left(\frac{\partial^2 \zeta}{\partial x^2} + \frac{\partial^2 \zeta}{\partial y^2} + O(\Delta x^2, \Delta x \Delta y, \Delta y^2) \right) \qquad (24)$$

This differential equation is equivalent to the numerical approximation. It is found, by inspection, that this equation is the wave continuity equation (2) plus an error term of the order of Δx^2, $\Delta x \Delta y$ and Δy^2. Thus this approximation is second order consistent.

This method was also applied to the wave momentum equation. The results are summarized in table II.

Table II Order of Consistency (simultaneous analysis of p and q nodes)

Arrangement	Equation		Interior	Boundary
D	Primitive Momentum	(x direction)	2	2
		(y direction)	2	2
	Wave Continuity		2	1
	Wave Momentum	(x direction)	2	1
		(y direction)	2	2

It is to be noted that for the first order consistent approximations, the leading error term is multiplied by a cross derivative. Thus, if there is no variation in either the x direction or the y direction, those approximations also become second order accurate.

Conclusion

The local truncation error analysis of a finite element approximation may not be sufficient to determine if the approximation is consistent.

An alternative method for studying the order of consistency of a FD or FE approximation has been presented. This method allows for the analysis of an approximation at several nodes simultaneously. Thus considering consistency in a global sense.

The method was successfully applied to some well known approximations of primitive and wave equations. Although some of these approximation have local truncation errors of an order less than one, the alternative consistency analysis method showed them to be consistent in a global sense. These results are in agreement with numerical experiments[12].

References

1 - Richmeyer R. D. and Morton K. W. (1967). Difference Methods for Initial-Value Problems, Wiley Intersciences. New York.
2 - Cakmak R. S., Botha J. F. and Gray W. G. (1987). Computational and applied mathematics for engineering analysis, Computational Mechanics. Southampton.
3 - Leenndertse J. J. (1967). Aspects of a Computational Model for Long - Period Water Wave Propagation, Rand Memorandum. (RM - 5294 - PR)
4 - Lynch D. R. and Gray W. G. (1979). A Wave Equation Model for Finite Element Tidal Computations, Computers and Fluids.
5 - Gray W. G. (unpublished). FLEET.
6 - Gray W. G. and Pinder G. F. (1976). On the relationship between the finite element and finite difference methods, International Journal for Numerical Methods in Engineering, Vol. 10 ,pp. 893-923.
7 - Lynch D. R. (1985). Mass Balance in Shallow Water Simulations, Communication in Applied Numerical Methods, Vol. 1.
8 - Warming R. F. and Hyett B. J. (1974). The Modified Equation Approch to the Stability and Accuracy Analysis of Finite - Difference Methods, Journal of Computational Physics.
9 - Trefethen L. N. (1982). Wave Propagation and Stability for Finite Difference Schemes, University Microfilms International.
10 - Vichnevetsky R. and Bowles J. B. (1982). Fourier Analysis of Numerical approximations of Hyperbolic Equations, SIAM. Philadelphia.
11 - Bender C. M. and Orszag S. A. (1978). Advanced Mathematical Methods for Scientists and Engineers, McGraw Hill. New York.
12 - Gray W. G., Drolet J. and Kinnmark I. P. E. (1987). A Simulation of Tidal Flow in the Southern Part of the North Sea and the English Channel, Advances in Water Resources.

A Comparison of Tidal Models for the Southwest Coast of Vancouver Island

M.G.G. Foreman
Institute of Ocean Sciences, P.O. Box 6000, Sidney, B.C. V8L4B2, Canada

INTRODUCTION

In the spirit of the Tidal Forum that is part of this conference, this paper presents the preliminary results of a comparison between one finite difference and two finite element models. These models were applied to the southwest coast of Vancouver Island with the eventual aim of predicting barotropic tidal currents for an upcoming field program. The finite difference model was developed by Flather[1] while the finite element models were based on early work by Lynch and Gray[2] and developed further by Werner and Lynch[3] and Walters[4]. Accuracy is evaluated by comparing model results against observations from twelve tide gauges and twelve current meters.

The roles of advection and bottom friction in the correct representation of the M_2 and K_1 tidal constitutents are studied. As the K_1 constituent contains a substantial contribution in the form of a continental shelf wave, it is of particular interest.

THE MODELS

Flather's[1] finite difference model was developed to study K_1 shelf waves along the Vancouver Island coast. It is a barotropic, explicit time-stepping, shallow water model with approximately 1900 grid rectangles, a resolution of approximately 6 km., and a domain that is approximately twice the area of the finite element models. Although the advective terms were not included in Flather's[1] original simulation, they were included in some of these tests.

The two finite element models are based on the Lynch and Gray[2] 'wave equation' formulation of the shallow water equations. Both models can be fully nonlinear. Whereas Werner and Lynch's[3] model employs time-stepping, Walter's[3] approach uses a harmonic method and calculates the nonlinear interactions iteratively. Both models have been applied successfully to the English Channel/southern North Sea data set.

The triangular grid for the finite element models (shown in Figure 1) was generated with the package of computer programs developed by Henry[5]. It has 1548 nodes, 2723 elements, and the triangle sides vary in length from 42 to 2.2 km.

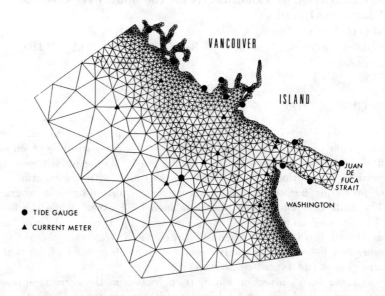

Figure 1: The finite element grid and observation sites.

Elevation-specified boundary conditions, with values interpolated from the results of the Flather model and Crean's[7] model of Juan de Fuca and Georgia Straits, were used for both finite element models. Radiation/specified boundary conditions, with values interpolated from Flather's[6] coarser resolution (18 km.) Northeast Pacific tidal model and Crean's[7] model, were used for the Flather model.

Both time stepping models were started from rest and run for 12 days. Δt was 24 and 144 seconds for the Flather and Lynch/Werner models respectively. Amplitudes and phases were calculated with a harmonic analysis of the hourly elevations and velocities over the last 5 days. The 12 day runs required approximately 403 (Flather) and 392 (Lynch/Werner) minutes on a VAX785. Each iteration of the Walters model required approximately 38 minutes. Both finite element models were run with a minimum depth of 10 meters.

RESULTS

Results from the three numerical models were compared with observations from the twelve tide gauge and twelve current meter sites shown in Figure 1. In all cases, the observation was compared to the model value at the nearest node or grid point.

For the runs testing the effects of advection, it was intended that the Chezy parameter, C_f, be set to 62.63. However, in order to stabilize the Lynch/Werner model with advection, it was necessary to decrease C_f to 44.286. (Interestingly, when Flather's model was run with advection, $C_f = 62.63$, and elevation specified boundary conditions, it was also unstable.) Subsequent tests with the Lynch/Werner showed that the instability arose either adjacent to an elevation specified boundary or in a region where the triangle size changed rapidly.

Tables 1 and 2 show average errors without and with advection. The average elevation error was calculated by first expressing the tide for each constituent (with frequency σ, amplitude a, and Greenwich phase lag g) as

$$a\cos(\sigma t - g) = b\cos(\sigma t) + c\sin(\sigma t).$$

If (b_ℓ, c_ℓ) and (B_ℓ, C_ℓ) respectively denote the (b, c) parameters for the model and observations at site ℓ, the tidal elevation error is defined as

$$E_t = \sum_{\ell=1}^{12} \frac{((b_\ell - B_\ell)^2 + (c_\ell - C_\ell)^2)^{1/2}}{12(B_\ell^2 + C_\ell^2)^{1/2}}.$$

Constituent	Error	Flather	Lynch/Werner	Walters
K_1	E_t	0.077	.051	.055
K_1	E_c	0.734	.569	.696
M_2	E_t	0.114	.083	.082
M_2	E_c	0.365	.404	.413

Table 1: E_t and E_c with no advection.

For currents, the error is defined analogously. If $(b_{u\ell}, c_{u\ell})$ and $(b_{v\ell}, c_{v\ell})$ denote the (b, c) parameters for the u and v model velocities at site ℓ and $(B_{u\ell}, C_{u\ell})$, $(B_{v\ell}, C_{v\ell})$ denote their observed counterparts, the current error is defined as

$$E_c = \sum_{\ell=1}^{12} \frac{((b_{u\ell} - B_{u\ell})^2 + (c_{u\ell} - C_{u\ell})^2 + (b_{v\ell} - B_{v\ell})^2 + (c_{v\ell} - C_{v\ell})^2)^{1/2}}{12(B_{u\ell}^2 + C_{u\ell}^2 + B_{v\ell}^2 + C_{v\ell}^2)^{1/2}}.$$

Constituent	Error	Flather	Lynch/Werner	Walters
K_1	E_t	.075	.039	.055
K_1	E_c	.773	.541	.694
M_2	E_t	.113	.087	.082
M_2	E_c	.389	.388	.413

Table 2: E_t and E_c with advection.

Tables 1 and 2 show that the Lynch/Werner model is generally the most accurate. They also suggest that advection does not significantly affect the accuracy. When the Lynch/Werner model was run with no advection and $C_f = 44.286$, (E_t, E_c) became (.039,.540) and (.086,.388) for K_1 and M_2 respectively. With $C_f = 19.81$, the same model gave (E_t, E_c) as (.086,.671) and (.112,.517) for K_1 and M_2 respectively. Thus doubling the bottom friction actually improved the model results slightly, but a factor of ten increase made them substantially worse.

Figure 2 shows the K_1 co-amplitude (values around 40 cm.) and co-phase lines obtained with Walter's[4] model and no advection. (The other two models yielded similar contours.) The coastal amplitude minimum corresponds to the trough of a continental shelf wave. (The other major K_1 component is a Kelvin wave.) Using the coastal amplitude variation as a crude indicator, all three models yield a larger amplitude and shorter wavelength than the 5 cm. and 200-450 km. values observed and predicted theoretically by Crawford and Thomson[9]. Though the amplitude estimates are somewhat better with the Walters and Lynch/Werner models, they might be improved further if the finite element grid were extended northward.

DISCUSSION AND FUTURE WORK

There are many sources for the errors shown in Tables 1 and 2. E_c is consistently much larger than E_t because the observed currents were taken at specific depths whereas the model currents were depth averaged values. A fairer comparison can be made with velocities from the three dimensional Lynch and Werner[9] model, and this will be done in the future.

Errors at each of the 24 sites reveal regions where the model performance might be improved. (For example, the M_2 E_t values for both finite element methods were poorest in Barkley Sound (the bay halfway along Vancouver Island coast), whereas with the Flather model they were poorest in Juan de Fuca Strait.) These site by site errors will be analysed more carefully in the future. There are, however, at least three obvious error sources common to all the models.

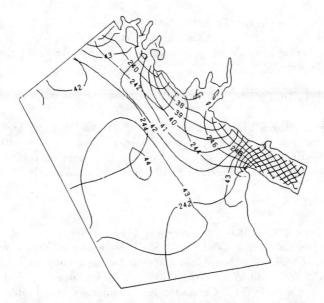

Figure 2: Walter's model K_1 amplitudes(cm) and phases(degrees GMT)

The first is the baroclinic and barotropic nature of tides in the region. As all three models are only barotropic, they cannot be expected to reproduce the observations exactly.

The second common source is model resolution. Both error measures should decrease when the grid or element size is decreased and the coastline and/or bathymetry is better represented in the region of the observation. As the Flather model has a coarser coastline resolution than the finite element models, it is not surprising that its results are less accurate. An improvement can also be expected when observations are compared to interpolated model values rather than to the nearest grid point or node.

The third common error source is the presence of other tidal constituents. M_2 and K_1 only account for about half the total diurnal and semi-diurnal tidal range along the Vancouver Island coast. Through nonlinear interactions, other, presently neglected, constituents can be expected to modify the M_2 and K_1 results from what they were found to be here. (Though the foregoing tests suggest that this modification may not be significant.)

Future work will include more rigorous comparisons of model results with observations, attempts to improve model accuracy in problem regions, the implementation of the three-dimensional Lynch and Werner[9] model, and the inclusion of six more tidal constituents (Q_1, O_1, P_1, N_2, S_2, and K_2) which, together with M_2 and K_1, account for about 90%

of the tidal range along the coast. Some attention will also be given to improving the triangular grid through a northward extension, a modification in Barkley Sound, and a smoothing of some of the rapid changes in triangle size.

ACKNOWLEDGEMENTS

I thank Roger Flather, Dan Lynch, Francisco Werner, and Roy Walters for the use of their models; Falconer Henry for help with his grid generation package; Pat Crean for the harmonic constants from his model; and Howard Freeland for digitized bathymetry.

REFERENCES

1. Flather, R.A. (1988), A Numerical Model Investigation of Tides and Diurnal Period Continental Shelf Waves Along Vancouver Island, submitted to Journal of Physical Oceanography.
2. Lynch, D.R. and W.G. Gray (1979), A Wave Equation Model for Finite Element Tidal Computations, Computers and Fluids, Vol. 7, pp. 207-228.
3. Werner, F.E. and D.R. Lynch (1987), Field verification of Wave Equation tidal dynamics in the English Channel and southern North Sea, Adv. Water Resources, Vol. 10, pp. 115-130.
4. Walters, R.A. (1987), A model for tides and currents in the English Channel and southern North Sea, Adv. Water Resources, Vol. 10, pp. 138-148.
5. Henry, R.F. (1988), Interactive Design of Irregular Triangular Grids, in Proceedings of the 7th Int. Conf. on Computational Methods in Water Resources, Cambridge MA, 1988.
6. Flather, R.A. (1987), A Tidal Model of the Northeast Pacific, Atmosphere-Ocean, Vol. 25(1), pp. 22-45.
7. Crean, P.B. (1983), The development of rotating, non-linear numerical models (GF2,GF3) simulating barotropic mixed tides in a complex system located between Vancouver Island and the mainland, Can. Tech. Rep. Hydrogr. Ocean Sci., No. 31, 65pp.
8. Crawford, W.R. and R.E. Thomson (1984), Diurnal-period continental shelf waves along Vancouver Island: a comparison of observations with theoretical models, Journal of Physical Oceanography, Vol. 14(10), pp. 1629-1646.
9. Lynch, D.R. and F.E. Werner (1987), Three-dimensional hydrodynamics on finite elements: part I: linearized harmonic model, International Journal for Numerical Methods in Fluids, Vol. 7, pp. 871-909.

Computation of Currents due to Wind and Tide in a Lagoon with Depth-Averaged Navier-Stokes Equations (Ulysse Code)
J.M. Hervouet
Electricité de France, Direction des Etudes et Recherches, Laboratoire National d'Hydraulique, 6 Quai Watier, 78400 Chatou, France

INTRODUCTION:

This study was realised on behalf of the S.E.M.E.D. public works company which is intending to build touristic equipments (marina,wharfs,hotels) along the GHAR-EL-MELH lagoon in Tunisia. This 2 km long lagoon is situated in the North of Tunisia, near Farina cape. A harbour and a channel connect the lagoon to the sea. The knowledge of the water exchanges with the open sea is of extreme importance to evaluate the impact on environment and the salubrity of the project. The aim of the study was to determine the currents in the lagoon and the mixing of waters due to tide and wind. The computation was realised on a cray XMP with the curvilinear finite differences code ULYSSE. Tide, wind,turbulence and bottom friction were taken into account for the computation of currents. The non-orthogonal mesh,with approximately 6000 nodes (see picture 1), was designed using elliptic equations, with the "PENELOPE" code.

To follow the sea water in the lake, we used a tracer with initial value 100 in the sea and 0 in the lagoon; this tracer is solution of an advection-diffusion equation.

THE EQUATIONS: The depth-averaged Navier-Stokes equations we used read as follows

Notations:
   u and v : components of the velocity.
      h : water depth
      p : pressure
      k : turbulent energy from the depth-averaged k-ε model (equations not written here for conciseness). k, together with ε, is used to compute the turbulent viscosity v_t.

Continuity equation:

$$\frac{\delta(uh)}{\delta x} + \frac{\delta(vh)}{\delta y} = -\frac{\delta h}{\delta t}$$

Momentum equations:

$$\frac{\delta u}{\delta t} + u\frac{\delta u}{\delta x} + v\frac{\delta u}{\delta y} = -\frac{1}{\rho}\frac{\delta p}{\delta x} + \frac{1}{h}\frac{\delta}{\delta x}(v_t h \frac{\delta u}{\delta x}) + \frac{1}{h}\frac{\delta}{\delta y}(v_t h \frac{\delta u}{\delta y}) - \frac{2}{3}\frac{\delta(hk)}{\delta x} + F_x \text{ wind} + F_x \text{ bot}$$

$$\frac{\delta v}{\delta t} + u\frac{\delta v}{\delta x} + v\frac{\delta v}{\delta y} = -\frac{1}{\rho}\frac{\delta p}{\delta y} + \frac{1}{h}\frac{\delta}{\delta x}(v_t h \frac{\delta v}{\delta x}) + \frac{1}{h}\frac{\delta}{\delta y}(v_t h \frac{\delta v}{\delta y}) - \frac{2}{3}\frac{\delta(hk)}{\delta y} + F_y \text{ wind} + F_y \text{ bot}$$

Tracer:

$$\frac{\delta T}{\delta t} + u\frac{\delta T}{\delta x} + v\frac{\delta T}{\delta y} = \frac{1}{h}\frac{\delta}{\delta x}(v_\tau h \frac{\delta T}{\delta x}) + \frac{1}{h}\frac{\delta}{\delta y}(v_\tau h \frac{\delta T}{\delta y})$$

Wind: The acceleration due to wind was modelled as follows

$$\vec{F}_{wind} = a \frac{\vec{U}_{air}}{h} |\vec{U}_{air}|$$ $\vec{U}_{air}$ is the wind velocity and "a" a coefficient depending of $\vec{U}_{air}$

Bottom friction: We used the Manning-Strickler formula

$$\vec{F}_{bot} = - \frac{g \vec{U} |\vec{U}|}{h^{4/3} K^2}$$ K is the Strickler coefficient ($K = \frac{1}{M}$, M being the Manning coefficient)

The original feature of the computation lies in the unknowns of the momentum equations: usually in the shallow water equations the unknowns are the velocity and the water depth (or the free-surface elevation). Here the water depth is assumed to be known: h follows the local open sea level subject to tide, and varies with respect to time, but is taken uniform in the lagoon, as propagation time is negligible on such a short distance. Therefore the equations are the same as the usual depth-averaged pressure flow equations but two differences:

-A source term $-\delta h/\delta t$ in the continuity equation.

-Variable but known coefficient h in the momentum equations.

Numerical methods used:

The ULYSSE code is based upon a fractional step method and the equations are split into 3 main steps:

- Advection
- Diffusion + source terms
- Continuity

Schemes adapted to every kind of problem may thus be used: method of characteristics for the advection terms, standard finite differences discretization for the elliptic diffusion operator (solved with a S.O.R. method), Runge-Kutta method for the source terms (wind and bottom friction), and finite volumes discretization for the continuity equation (solved with a conjugate residual algorithm) providing a rigorous "numerical" mass conservation.

Results:

2 types of computation were performed:

- With tide only.
- With tide and a 5 m/s West wind.

They showed the importance of wind (see picture 3 and compare with picture 2) which creates long recirculations, and the necessity to bring some modifications to the initial project to suppress stagnation areas.
Picture 4 shows our tracer (sea water) entering the lagoon.

In conclusion, it was found that nearly 5% of the water in the lagoon was renewed during 1 tide with wind, so that the retention time of water was only 10 days. It would have been 13 days without wind.

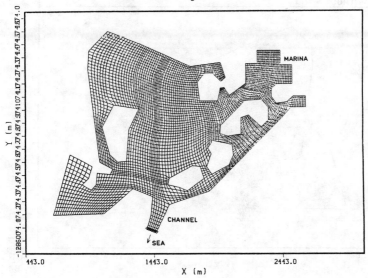

PICTURE 1

Lagune de ghar-el-melh

non orthogonal mesh

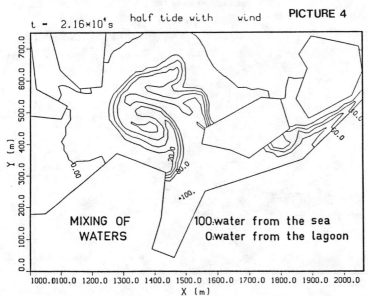

PICTURE 4

Lagune de ghar-el-melh

$t = 2.16 \times 10^4$ s   half tide with wind

MIXING OF WATERS

100: water from the sea
0: water from the lagoon

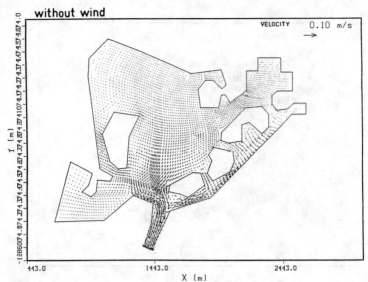

Lagune de ghar-el-melh
3 HOURS AFTER HIGH TIDE
without wind

PICTURE 2

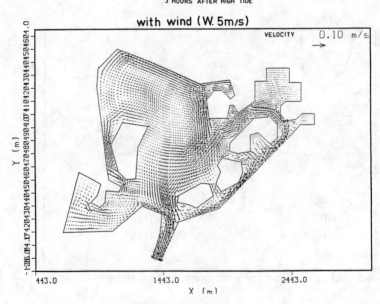

Lagune de ghar-el-melh
3 HOURS AFTER HIGH TIDE
with wind (W. 5m/s)

PICTURE 3

The Shallow Water Wave Equations on a Vector Processor
I.P.E. Kinnmark and W.G. Gray
Department of Civil Engineering, University of Notre Dame, Notre Dame, IN 46556, USA

INTRODUCTION

The shallow water equations are used to determine circulation patterns and maximum and minimum tides at the interior of a fluid body subject to a tidal forcing at the open boundaries of the region. These circulation patterns are of primary interest in a model to determine transport of different species for environmental applications. Another important application is tsunamis. These are waves generated from earthquakes under the floor of the oceans. When these waves propagate into shallow regions their celerities decrease and amplitudes increase, and can cause severe flooding. An interesting possibility for use of shallow water equation models is evaluation of schemes for conversion of tidal energy into commercially available energy. A good description of various proposed tidal energy projects may be found in Charlier[1] (see also Count[2] and Gray and Gashaus[3]). Another possible application is generation of wave heights to be used as input data for determination of loads on offshore structures. Shallow water equation models have also been developed for the purpose of investigating the effect of waves generated from nuclear explosions near the surface of the ocean or under water (see Leendertse[4]).

The advent of computers exhibiting fast scalar execution as well as vector processing capability provides a powerful tool for application of simulation models to large field data sets as well as multiple simulations for parameter identification purposes. The current study will compare overall execution time, scalar vs. vector performance and reprogramming efforts when a computationally intensive finite element wave equation model is utilized on a computer with vector processing capability.

GOVERNING EQUATIONS

The conservation of mass and momentum in a body of fluid, assuming negligible viscosity, long wavelength waves, constant fluid density and good vertical mixing is described by the shallow water equations. The forcing is of tidal and atmospheric type.

Continuity Equation

$$L(\zeta,u,v) \equiv \frac{\partial \zeta}{\partial t} + \frac{\partial (Hu)}{\partial x} + \frac{\partial (Hv)}{\partial y} = 0 \qquad (1)$$

Momentum Equation in Eastward Direction

$$M_x(\zeta,u,v) \equiv \frac{\partial u}{\partial t} + u\frac{\partial u}{\partial x} + v\frac{\partial u}{\partial y} + \tau u - fv + g\frac{\partial \zeta}{\partial x} - A_x/H = 0 \qquad (2)$$

Momentum Equation in Northward Direction

$$M_y(\zeta,u,v) \equiv \frac{\partial v}{\partial t} + u\frac{\partial v}{\partial x} + v\frac{\partial v}{\partial y} + \tau v + fu + g\frac{\partial \zeta}{\partial y} - A_y/H = 0 \qquad (3)$$

where

ζ is elevation above a datum (L)

h   is bathymetry (L)

$H = h+\zeta$ is total fluid depth (L)

(u,v)   is vertically averaged fluid velocity in eastward direction (x) and northward direction (y) (L/T)

τ is the non–linear friction coefficient (1/T)

f   is the Coriolis parameter (1/T)

g   is acceleration due to gravity (L/T^2)

(A_x, A_y) is atmospheric (wind) forcing in eastward direction (x) and northward direction (y) (L^2/T^2)

x   is positive eastward (L)

y   is positive northward (L)

t   is time (T)

This, so called primitive form, of the shallow water equations introduces spurious oscillation, when solved numerically. Therefore Lynch and Gray[5] introduced the Wave Equation formulation of the shallow water equations, which suppresses spurious spatial oscillations without compromising the predictive capability of the model.

Wave Continuity Equation

$$W(\zeta,u,v) \equiv \frac{\partial^2 \zeta}{\partial t^2} + \tau \frac{\partial \zeta}{\partial t}$$

$$-\frac{\partial}{\partial x}[\frac{\partial(Huu)}{\partial x} + \frac{\partial(Huv)}{\partial y} - fHv + gH\frac{\partial \zeta}{\partial x} - A_x]$$

$$-\frac{\partial}{\partial y}[\frac{\partial(Huv)}{\partial x} + \frac{\partial(Hvv)}{\partial y} + fHu + gH\frac{\partial \zeta}{\partial y} - A_y]$$

$$-Hu\frac{\partial \tau}{\partial x} - Hv\frac{\partial \tau}{\partial y} = 0 \quad (4)$$

In the wave equation approach Equation (4) is replacing the primitive continuity equation (1) and solved in conjunction with the momentum equations (2) and (3). The governing equations are discretized spatially using finite elements and nodal integration. The temporal discretization is symmetrical, minimum level type (see Kinnmark[6]), except for the nonlinear convective terms.

APPLICATION

The application is a one dimensional open channel flow problem. The horizontal channel is closed at one end and subject to a sinusoidal forcing of the elevation at the opposite end (see Figure 1). The problem under consideration is therefore, in the wave equation form

$$\frac{\partial^2 \zeta}{\partial t^2} + \tau \frac{\partial \zeta}{\partial t} - gh\frac{\partial^2 \zeta}{\partial x^2} \quad (5)$$

$$\zeta(x_1,t) = \zeta_0 \cos\omega t \quad (6)$$

$$\frac{\partial \zeta(x_2,t)}{\partial x} = 0 \quad (7)$$

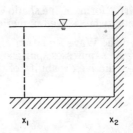

Figure 1. One Dimensional Open Channel Flow
Problem Geometry.

The exact, steady state solution to this problem is presented in Lynch and Gray[7] and given in equation (8)

$$\zeta = \text{Re}\left\{\zeta_0 e^{i\omega t} \frac{\cos[\beta(x - x_2)]}{\cos[\beta(x_1 - x_2)]}\right\};$$

$$\beta^2 = (\omega^2 - i\omega\tau)/gH_o \; ; \; i = \sqrt{-1} \qquad (8)$$

IMPLEMENTATION ON A VECTOR PROCESSOR

The model was run on a CRAY X–MP/48 at the San Diego Supercomputer Center.

The original surface flow code was designed and used extensively in scalar mode at various machines: IBM 3033, IBM 3081 and for smaller problems on an IBM PC AT. The code was originally not designed with a vectorizing capability in mind. It was therefore not entirely surprising that turning the vectorizing option on or off in the compiler showed virtually no difference (less than 0.01%) in execution time for the original code. It is a well known fact that implicit finite element codes consume most of the computational effort in only two parts of the program: 1. the integration (setup of the matrix and the right hand side); 2. the solution of the resulting system of linear equations. For the medium size prototype problem investigated (283 equtions) the two timing methods (the SECOND routine and the FLOWTRACE compiler option) gave the same result to within 2 significant digits: 58% of total execution time when running 300 time steps was spent in the subroutine setting up the matrix.

Reprogramming effort was therefore directed towards the matrix setup routine. The main impediment to vectorization was twofold: scalar intermediate quantities and data dependency. The data dependency was broken through splitting the main do–loop into a set of do–loops, thus completing the computation of one set of vectors prior to utilizing them in the following calculation. The intermediate scalar quantities were eliminated at the cost of increased storage: a substantial number of new arrays were introduced. All in all this "vectorization" effort required partial or complete change of nearly every line in a 400 line subroutine − a rather massive effort. The resulting speedup of this subroutine was a factor 3.7 due to reprogramming and vectorization alone. Further improvement of maybe another factor 2 should be possible through reorganization of the array storage procedure. Due to the difference between node versus element numbering of a finite element grid the current algorithm gathers array elements at every computation and scatters them after each individual computation. An initial gathering of all vectors at the beginning of the subroutine, completion of all calculations without gather/scatter and a final scattering procedure at the end of the routine should lead to improved computational speed. Initial gathering is accomplished through converting the conventional node number ordering of the arrays into element ordering.

Secondly speedup of the matrix solution procedure was considered. The original flow code was utilizing the YALE Sparse Matrix package available at the SDSC CRAY X−MP/48 under the name SMPAK. The Summer Institute notes[8] (Section Math Libraries: table on page 34) indicates a vectorization speedup of at least a factor 10 for the vector size in this prototype example. Contrary to this the actual runs indicated a speedup of approximately 10%. Most of all this probably indicates the need to benchmark the SMPAK on sparse matrices with different structures to try to get a handle on which kind of matrix equations exhibit a significant vector speedup using the SMPAK.

Overall the total original flow code experienced a speedup of approximately a factor 2 using the vector processor capability of the CRAY X−MP/48. In addition another factor 6 speedup was obtained in scalar mode on the CRAY X−MP/48 compared to the IBM 3033 on which the code was originally run. Thus altogether a speedup of a factor 12 was obtained which will allow the code to run on much larger field data sets and even allow multiple runs for parameter identification purposes.

CONCLUSIONS

Implementation of a finite element shallow water wave equation solver on a CRAY X−MP/48 gave speedups of over an order of magnitude compared to an IBM 3033. To achieve significantly higher speedups the basic numerical algorithms must be designed specifically with a vector processor in mind, thus necessitating the need to create a new code largely from scratch.

ACKNOWLEDGEMENT

The first author wishes to acknowledge the San Diego Supercomputer Center for use of their CRAY X−MP/48 in performing the calculations described in this paper.

REFERENCES

1. Charlier, R.H. (1982). Tidal Energy, Van Nostrand Reinhold Company, New York.

2. Count, B. (Editor) (1980). Power From Sea Waters, Academic Press, London.

3. Gray, T.J. and Gashaus, O.K. (Editors) (1972). Tidal Power, Plenum Press, New York.

4. Leendertse, J.J. (1967). Aspects of a Computational Model for Long Period Water–Wave Propagation, Rand Memorandum RM–5294–PR, Santa Monica.

5. Lynch, D.R. and Gray, W.G. (1979). A Wave Equation Model for Finite Element Tidal Computations, Computers and Fluids, Vol. 7, pp. 207–228.

6. Kinnmark, I.P.E. (1986). The Shallow Water Wave Equations: Formulation, Analysis and Application, Lecture Notes in Engineering 15, Springer–Verlag, Berlin.

7. Lynch, D.R. and Gray, W.G. (1978). Analytic Solutions for Computer Flow Model Testing, Journal of the Hydraulics Division, Vol. HY10, pp. 1409.

8. Pfeiffer, W. (1987). San Diego Supercomputer Center Summer Institute 1987–Lecture Notes, SDSC, San Diego.

Testing of Finite Element Schemes for Linear Shallow Water Equations

S.P. Kjaran and S.L. Hólm
Vatnaskil Consulting Engineers, Reykjavik, Iceland
S. Sigurdsson
Science Institute, University of Iceland, Reykjavik, Iceland

INTRODUCTION

We consider the problem of testing finite element schemes for shallow water free surface flow, including the effects of friction, windstress, and Coriolis force.

By neglecting non-linear terms in the shallow water equations, they can be reduced to a Helmholtz equation by considering only periodic or time independent solutions. In the case when only effects of friction and windstress are included there exist analytical solutions for simple geometries, as derived by eg. Lynch and Gray (1978), that lend themselves very well for comparisons with solutions obtained by finite element schemes. When the effect of Coriolis force is also present, analytical solutions, however, become much more complicated even in simple geometries due to awkward boundary conditions. In this case we therefore propose the use of a simple finite difference scheme for comparisons with the finite element solutions. Since all the difference approximations can be made to be of second order for simple geometries, a posteriori error estimates for the finite difference solutions are available using extrapolation.

We demonstrate how this approach has been used to test a new staggered finite element scheme presented in a separate paper by Sigurdsson et.al. (1988).

BASIC EQUATIONS

The shallow water equations in wave equation formulation may be expressed as follows:

$$\frac{\delta^2 \eta}{\delta t^2} + \beta \frac{\delta \eta}{\delta t} + \nabla \cdot (\frac{\delta \eta}{\delta t}\overline{V}) = g\nabla \cdot (H\nabla\eta) - f(\nabla \times H\overline{V})_z + H\nabla\beta \cdot \overline{V} - \nabla \cdot \overline{W} + \nabla \cdot (H(\overline{V} \cdot \nabla)\overline{V}) \quad (1)$$

and

$$\frac{\delta \overline{V}}{\delta t} + (\overline{V} \cdot \nabla)\overline{V} = -g\nabla\eta + \frac{1}{H}\overline{W} - A\overline{V} \tag{2}$$

$$A = \begin{Bmatrix} \beta & -f \\ f & \beta \end{Bmatrix} \tag{3}$$

where $(\nabla \times (a_x, a_y))_z$ denotes $\delta a_y/\delta x - \delta a_x/\delta y$ and η is the free surface elevation above mean sea level, D is the depth corresponding to mean sea level, H is the total depth, $\overline{V}$ is the vertically averaged fluid velocity, $\overline{W}$ is the windstress, β is friction coefficient, g is the acceleration of gravity, f is the Coriolis parameter.

By neglecting the convective terms, linearizing the equations and assuming that the friction coefficient and the windstress are constant eq. 1 and 2 are simplified as follows:

$$\frac{\delta^2 \eta}{\delta t^2} + \beta \frac{\delta \eta}{\delta t} = g\nabla \cdot (D\nabla\eta) - f(\nabla \times D\overline{V})_z \tag{4}$$

$$\frac{\delta \overline{V}}{\delta t} = -g\nabla\eta + \frac{1}{D}\overline{W} - A\overline{V} \tag{5}$$

and the boundary conditions are given by head boundary condition and the following no flow boundary condition:

$$g \frac{\delta \eta}{\delta n} = \frac{1}{D} \overline{W} \cdot \overline{n} + f \overline{V} \cdot \overline{t} \tag{6}$$

where $\overline{n}$ is normal to the boundary and $\overline{t}$ is the tangential vector.

DERIVATION OF TEST SOLUTIONS

Assuming periodic and steady state solutions we can write

$$\eta(x,y,t) = \eta_o(x,y) + \eta_1(x,y)e^{i\omega t} \tag{7}$$

$$\overline{V}(x,y,t) = \overline{V}_o(x,y) + \overline{V}_1(x,y)e^{i\omega t} \tag{8}$$

We first note that eq. 5 is satisfied provided:

$$\overline{V}_o = \frac{-g}{\beta^2 + f^2} \begin{Bmatrix} \beta & f \\ -f & \beta \end{Bmatrix} (\nabla \eta_o - \frac{1}{gD} \overline{W}) \tag{9}$$

$$\overline{V}_1 = - \frac{g}{\alpha^2+f^2} \begin{Bmatrix} \alpha & f \\ -f & \alpha \end{Bmatrix} \nabla \eta_1 \qquad (10)$$

where $\alpha = \beta + i\omega$

Substituting into eq. 4 we get

$$\nabla \cdot (D\nabla \eta_0) + \frac{f}{\beta} (\nabla D \times \nabla \eta_0)_z = 0 \qquad (11)$$

$$\nabla \cdot (D\nabla \eta_1) + \frac{f}{\alpha} (\nabla D \times \nabla \eta_1)_z - i (1+(\frac{f}{\alpha})^2) \frac{\alpha \omega}{g} \eta_1 = 0 \qquad (12)$$

where $((a_x, a_y) \times (b_x, b_y))_z$ denotes the scalar $a_x b_y - a_y b_x$

Working in (x,y)-coordinates no-flow boundary conditions on a boundary where x is constant result in the boundary conditions.

$$\beta \frac{\delta \eta_0}{\delta x} + f \frac{\delta \eta_0}{\delta y} = \frac{1}{gD} (\beta W_x + f W_y), \quad \alpha \frac{\delta \eta_1}{\delta x} + f \frac{\delta \eta_1}{\delta y} = 0 \qquad (13)$$

for eq. 11 and 12 respectively, whereas no-flow boundary conditions where y is constant result in the following equation:

$$-f \frac{\delta \eta_0}{\delta x} + \beta \frac{\delta \eta_0}{\delta y} = \frac{1}{gD} (-f W_x + \beta W_y), \quad -f \frac{\delta \eta_1}{\delta x} + \alpha \frac{\delta \eta_1}{\delta y} = 0 \qquad (14)$$

for eq. 11 and 12 respectively.

Working in polar coordinates, r and θ, analogous boundary conditions apply on no-flow boundaries where r is constant or θ is constant, with $\frac{\delta \eta}{\delta x}$ and $\frac{\delta \eta}{\delta y}$ being replaced by $\frac{\delta \eta}{\delta r}$ and $\frac{1}{r} \frac{\delta \eta}{\delta \theta}$.

In the absence of the Coriolis force, i.e. setting f = 0 eq. 11 and 12 reduce to Laplace and Helmholtz equations and eq. 13 and 14 to simple Neumann conditions. In this case analytical solutions have been derived using separation of variables by eg. Lynch and Gray (1978) for a variety of simple geometries in both (x,y)- and polar coordinates and a variety of depth patterns. In these examples we have a periodic Dirichlet boundary condition along one of the boundaries and no-flow boundary conditions along the remaining ones. After analytical expressions have been obtained for η_0 and η_1, $\overline{V}_0$ and $\overline{V}_1$ can be evaluated using eqs. 9 and 10. When the Coriolis force is present the technique of separation of variables breaks down due to the presence of the tangential derivative in the boundary conditions described by eqs. 13 and 14. However for simple geometries in (x,y)- or polar coordinates, simple second order difference approximations are available for eqs. 11 and 12 as well as for the boundary conditions in eqs. 13 and 14 rendering a uniformly second order difference approximation for η_0 and η_1. Thus by obtaining approximations for two different grid spacings, h and h/2 say, and denoting the corresponding approximate values at a common gridpoint P by $\eta_p^{(h)}$ and $\eta_p^{(h/2)}$, the extrapolated value $\eta_p^{(e)} = \eta_p^{(h/2)} + (\eta_p^{(h/2)} - \eta_p^{(h)})/3$.

gives an improved approximation and the correction term serves as an a posteriori error estimate at P.

This is the approach taken in the following tests when $f \neq 0$. We illustrate the construction of the second order difference approximation in the case of eq. 12 in (x,y)- coordinates with boundary conditions given by eqs. 13 and 14. Other cases are treated in an analagous manner.

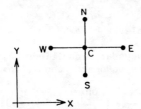

If the gridpoint C in the figure is an interior point of the difference grid it is associated with the following $O(h^2)$ difference approximation to eq. 12.

$$D(\eta_N + \eta_S + \eta_W + \eta_E - 4\eta_C) + \frac{h}{2}\frac{\delta D}{\delta x}[\eta_E - \eta_W + \frac{f}{\alpha}(\eta_N - \eta_S)]$$

$$+ \frac{h}{2}\frac{\delta D}{\delta y}[\eta_N - \eta_S - \frac{f}{\alpha}(\eta_E - \eta_W)] - ih^2(1+(\frac{f}{\alpha})^2)\frac{\alpha w}{g}\eta_C = 0 \qquad (15)$$

h denotes the grid spacing and the subscript the point to which the approximate value applies.

If points S, C ad N lie along a no-flow boundary for which x is constant, and W is the exterior point the approximation associated with C is obtained by first introducing the $O(h^2)$ centered difference approximation for the boundary condition in eq. 13.

$$\alpha(\eta_E - \eta_W) + f(\eta_N - \eta_S) = 0 \qquad (16)$$

and hence substituting $\eta_E + \frac{f}{\alpha}(\eta_N - \eta_S)$ for η_W into eq. 15.

Similarily if points W, C and E lie along a no-flow boundary for which y is constant and N is the exterior point the approximation associated with C is obtained from eq. 14

$$-f(\eta_E - \eta_W) + \alpha(\eta_N - \eta_S) = 0 \qquad (17)$$

i.e. by substituting $\eta_S + \frac{f}{a}(\eta_E - \eta_W)$ for η_N into eq. 15.

Finally if C is a cornerpoint between two such boundaries with W and N being exterior points, in order for both eqs. 16 and 17 to hold, we substitute η_E for η_W and η_S for η_N into eq. 16.

The resulting linear system of equations in the unknown η-approximations at interior points and no-flow boundaries is solved by applying a Gauss elimination routine for systems with complex coefficients. Finally approximations to the velocities are obtained from eqs. 9 and 10, using a second order centered difference approximation for $\nabla\eta$.

TESTS APPLIED TO STAGGERED FINITE ELEMENT SCHEME

We demonstrate in this final section how test solutions of the above type have been used in order to test a new finite element scheme for eq. 4, 5 and 6. The scheme itself is described in a separate paper by Sigurdsson et. al. (1988) and is called a staggered finite element scheme. The geometrical configuration and boundary conditions for the tests are shown in fig. 1. The rectangular mesh has 289 nodes and the element size is 2500 m. In the polar case there are 90 nodes and the elements are divided radially at 5^0 intervals.

For pure tidal forcing with tidal amplitude 1 m without Coriolis effect and wind stress results for the rectangular case are shown in fig. 2 for waterlevel and velocity vs. time for a location in the middle of the area. The friction coefficient is $6.7 \cdot 10^{-5} s^{-1}$, and the depth variation is parabolic ranging from 20 m at the open end to 10 m at the closed end. Table 1 summarizes the results for the polar case giving the radial varation in waterlevel and velocity. The depth varies between 25 m and 4 meters. The theoretical results for the examples above are taken from Lynch and Gray (1978).

TABLE 1 Radial variation of waterlevel and velocity

Distance from mouth m	Waterlevel		Velocity	
	theo-retical cm	calcul-ated	theo-retical cm/s	calcul-ated
3750	78.1	78.3	8.45	8.04
11250	76.7	76.8	9.18	8.66
18750	75.1	75.2	9.35	9.03
26250	73.6	73.6	7.64	7.25

Results for pure wind setup are shown in figs. 3 and 4. The wind direction is shown in the figures. The wind velocity is 10 m/s and the windstress is $2.9 \cdot 10^{-4} m^2/s^2$. Fig. 3 compares calculated results with theoretical results taken from Lynch and Gray (1978). The depth varies radially as $r^{1/2}$ between 14 and 22 meters. Fig. 4 compares calculated results from the finite element method and the finite difference solution to the Helmholz equation, eq. 11, with constant depth equal to 10 m.

Finally figs. 5 and 6 show results when the Coriolis force has been taken into account with Coriolis parameter $1.3 \times 10^{-4} s^{-1}$. The calculations are compared with finite difference solution to Helmholtz equation (eq.12). Fig. 5 shows time variation of waterlevel and velocity for a location in the middle of the rectangular area and fig. 6 shows area variation of the waterlevel.

ACKNOWLEDGEMENTS

This work was supported by the National Science Foundation in Iceland, Science Institute of the University of Iceland and Vatnaskil Consulting Engineers Ltd.

REFERENCES

Lynch, D.R., Gray, W.G., (1978): Analytical Solutions for Computer Flow Model Testing, J.of the Hydr. Div., 1409-1428.

Sigurdsson, S., Kjaran, S.P., Tómasson, G.G., (1988): A Simple Staggered Finite Element Scheme for Simulation of Shallow Water Free Surface Flows. VII International Conference on Computational Methods in Water Resources, Massachusetts Institute of Technology.

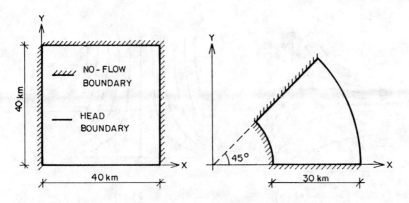

FIG. 1 Geometrical configuration and boundary conditions

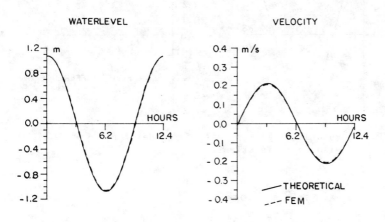

FIG. 2 Waterlevel and velocity for tidal forcing

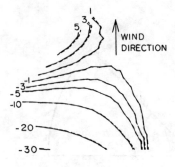

FIG. 3 Wind setup, waterlevel in mm

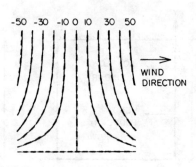

FIG. 4 Wind setup, waterlevel in mm

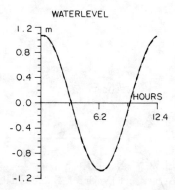

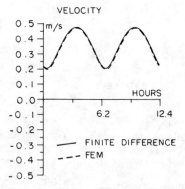

FIG. 5 Waterlevel and velocity for tidal forcing with Coriolis effect

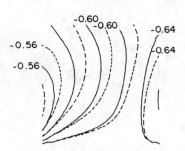

FIG. 6 Coriolis effect, waterlevel in m

INVITED PAPER
Long Term Simulation and Harmonic Analysis of North
Sea/English Channel Tides
D.R. Lynch and F.E. Werner
*Thayer School of Engineering, Dartmouth College, Hanover, N.H.
03755, USA*

INTRODUCTION

The results contained herein represent an extension of the calculations described by Werner and Lynch (1987). In that paper, a finite element model of the North Sea/English Channel system was forced for a period of 3 days (corresponding to March 15-17, 1976) by eleven tidal constituents (O_1, K_1, M_2, N_2, S_2, K_2, M_4, MS_4, MN_4, M_6 and $2MS_6$) and the results compared to time-series at 11 coastal stations and 8 current meter stations collected during JONSDAP '76 (Ramster, 1977). Similarly, the model results were Fourier-decomposed (using a standard FFT subroutine) and the results of the *lumped* diurnal, semi-diurnal, quarter-diurnal and sexto-diurnal constituents were obtained. The above calculations represented the proposed method of model-comparison that was discussed at the Tidal Forum at the VIth Conference on Finite Elements in Water Resources held in Lisbon (in June 1986) and which is being continued at the present Conference. These computations have provided an excellent benchmark for comparing models with each other, including to date three finite element models (Gray et.al. 1987; Walters 1987; Werner and Lynch 1987) and a finite difference model (Ozer and Jamart, 1987). However, the comparison with field observations has been less clear and has generated considerable international discussion. The present paper is a contribution to this latter, critical discussion, in which we utilize a longer-term simulation to distinguish among the various harmonic constituents which were lumped together in the earlier report.

PROCEDURE

In order to retrieve the 11 constituents used in forcing the model, a long-term simulation needs to be implemented. The length of the simulation is dictated, in this case, by the need to separate the K_2 from the S_2 constituent. Specifically, if $\sigma_{S_2} = 2.0000\,cycles/day$ and $\sigma_{K_2} = 2.0055\,cycles/day$ are the frequencies of the S_2 and K_2 constituents respectively, the length of the record ($2N$) needed to separate the two frequencies is $2N = 1/(\sigma_{K_2} - \sigma_{S_2})$ (Godin, 1972; p. 204), or approximately 182 days. We therefore ran our model for the time period which started on 15 March 1976 (0000 hours) and ended on 20 September 1976 (2300 hours). The first 5 days were discarded (for spin-up reasons), i.e., through 20 March (0000 hours), leaving a record length of 184 days and 23 hours. The run-time in 32-bit precision on a VaxStation II (LINPACK rating of 0.17 MFLOPS) is 62.5 hours of CPU for the 190 days of real-time simulation; required memory is 0.85 Mbytes. The mesh appears in Fig. 1.

The model formulation used is described in Werner and Lynch (1987):

$$\frac{\partial^2 H}{\partial t^2} + \tau_0 \frac{\partial H}{\partial t} - \nabla \cdot [\nabla \cdot (H\mathbf{vv}) + gH\nabla\zeta + \mathbf{f} \times H\mathbf{v} + (\tau - \tau_0)H\mathbf{v} - H\Psi] = 0 \quad (1)$$

$$\frac{\partial \mathbf{v}}{\partial t} + \mathbf{v} \cdot \nabla \mathbf{v} + g\nabla\zeta + \mathbf{f} \times \mathbf{v} + \tau\mathbf{v} = \Psi \quad (2)$$

where the notation is standard. Equation (1) is the "generalized" Shallow Water Wave Equation (Lynch and Gray, 1979; Kinnmark and Gray, 1985) and equation (2) is the vertically integrated momentum equation in nonconservative form. As in Werner and Lynch (1987), the atmospheric forcing Ψ was zero, the Chezy bottom friction parameter was kept constant at 65 $m^{1/2}s^{-1}$ throughout the model, the time-step was 5 minutes, the minimum bathymetric depth was 15 meters, and $\tau_0 = 2 \times 10^{-4} s^{-1}$. No adjustment of free-parameters took place; note also there are no horizontal viscous stresses in this model.

Once the simulation time series were generated, we decomposed them using the tidal analysis software developed by Foreman (1977). With the 184-day record we were able to fit with confidence to 51 tidal constituents as shown in figure 2. This provided an extra opportunity to test the hypothesis that the field data was indeed properly captured with only the original 11 components provided, with interesting results.

RESULTS

Table 1 contains the observed amplitude (in meters) and phase (in degrees; relative to Greenwich), as well as the numerical results, for the 11 tidal components and field stations provided by Drs. G.Verboom and C. Le Provost for the first Tidal Forum in Lisbon (1986). We regard these as very good. Other sources of data include the co-tidal and co-range charts of Chabert D'Hieres and Le Provost (1979), Fornerino (1982), Le Provost and Fornerino (1985), and Prandle (1980). We have constructed similar charts and find very satisfying agreement. Table 1 also permits a direct comparison with the harmonic model results of Walters (1987).

Of the 51 constituents obtained in the analysis, most of the energy is indeed recovered in the 11 constituents used in the forcing; however the non-linear interactions in the model have put some energy into other harmonics. Examples we have found include the $2MS_2$ and $2MN_2$ (which the tidal analysis package does not distinguish from the μ_2 and L_2). To illustrate, we include the full spectrum of the model response at Dover in figure 2. Fig. 2a contains *all* 51 component amplitudes; fig. 2b focuses on the 40 components *not* used in the forcing, i.e. those generated entirely by nonlinear internal interactions (note the change in scale of the y-axis). The $2MS_2$ has an amplitude at this station of about 13 centimeters, and the $2MN_2$ is 6.5 centimeters.

With regard to the tidal-elevation time-series at the 11 coastal stations, our original (1987) comparisons were not as satisfying as the harmonic comparisons reported here. The raw simulation results contain the full spectrum of frequencies – those forced on the boundary and those generated internally through non-linear interactions. Thus, some of the original "discrepancies" between the time-stepping solution and the data might be due to missing energy in constituents *not* provided to reconstruct the field observations. In other words, *the data ought to be considered as filtered*, and the most appropriate comparison is with simulation results passed through the identical filter.

We have therefore revisited this comparison armed with the full harmonic decomposition of the simulation results. In figures 3-13 we compare three signals for each of the 11 tidal stations:
  a) the field data, reconstructed from the 11 components provided ("Data"; solid line);
  b) the simulation results, reconstructed from the same 11 components only ("Filtered"; long dash line);
  c) the raw, unfiltered time series from the simulation ("Unfiltered"; short dash line).

Items (a) and (c) are identical to those reported previously. A simple visual comparison reveals that the filtered simulation results generally agree better with the data than do the unfiltered results. In particular, notice the considerable improvement at Dieppe, Calais, Boulogne, Zeebrugge and Dover. A slight amplitude deterioration (overestimate) may be found in St. Malo and Cherbourg, and a phase shift in Lowestoft.

A second source of discrepancy in the time-series comparisons arises from the limited spectrum used to force the model at the open boundaries. Essentially all but 11 components were forced to vanish on these boundaries. Since the system exhibits resonances at some of the common frequencies, relatively small amplitude forcing at the boundaries could be amplified at some of the tidal stations. As a result we presently cannot compare the spectra as in figure 2b with field data – the data would combine the effects of internal generation and boundary excitation. We view this as the next logical step in this exercise.

ACKNOWLEDGEMENTS

This work was supported by the National Science Foundation, Grant No. CEE-835-2226; and by the Digital Equipment Corporation. We thank M.G.G. Foreman, R.A. Walters, B.M. Jamart, C. Le Provost, and G.K. Verboom for a stimulating discussion of these results; J.T.C. Ip for running the long-term simulations; and J. Drolet for constructing the finite element mesh.

REFERENCES

Chabert D'Hieres, M.M.G. and C. Le Provost (1979). Atlas des composantes harmoniques de la maree dans La Manche. *Annals Hydrographiques*, 6, 5-36.

Foreman, M.G.G. (1977). *Manual for Tidal Heights Analysis and Prediction.* Pacific Marine Science Report 77-10, Institute of Ocean Sciences, Patricia Bay, Sidney, B.C. 97pp., Unpublished Manuscript.

Fornerino, M. (1982). Modelisation des courants de maree dans La Manche. These de Docteur-Ingenieur, L'Institut National Polytech. de Grenoble, 267pp.

Gray, W.G., J. Drolet, I.P.E. Kinnmark (1987). A simulation of tidal flow in the southern part of the North Sea and the English Channel. *Advances in Water Resources*, **10**, 131-137.

Godin, G. (1972). *The Analysis of Tides*. Liverpool Univ. Press, 264pp.

Kinnmark, I.P.E. and W.G. Gray (1985). A generalized wave equation formulation of tidal circulation. *Proc. 4th Int. Conf. on Num. Meths. in Laminar and Turbulent Flows*, Taylor, et al., eds., Pineridge Press, Swansea, U.K., 1312-1324.

Le Provost, C. and M. Fornerino (1985). Tidal spectroscopy of the English Channel with a numerical model. *J. Phys. Oceanogr.*, **15**, 1009-1031.

Lynch, D.R. and W.G. Gray (1979). A wave equation model for finite element computations. *Computers and Fluids*, **7**, 207-228.

Ozer, J. and B.M. Jamart (1987). Computation of tidal motion in the English Channel and southern North Sea: comparison of various model results. *MUMM's contribution to BSEX*, **Tech. Rep. TR07**, first draft, July 1987.

Prandle, D. (1980). Co-tidal charts for the Southern North Sea. *Deutsch. Hydrogr. Zeitschr.*, **33**, 68-81.

Ramster, J.W. (1977). Development of cooperative research in the North Sea. The origins, planning and philosophy of JONDSAP '76. *Marine Policy*, October 1977, 318-325.

Walters, R.A. (1987). A model for tides and currents in the English Channel and southern North Sea. *Advances in Water Resources*, **10**, 138-148.

Werner, F.E. and D.R. Lynch (1987). Field verification of wave equation tidal dynamics in the English Channel and southern North Sea. *Advances in Water Resources*, **10**, 115-130.

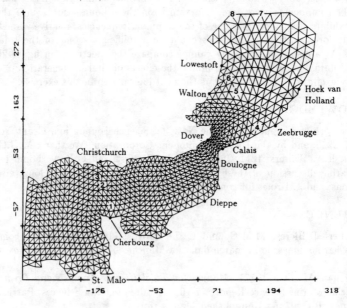

Figure 1. Computational mesh with 990 nodes and 1762 linear elements.

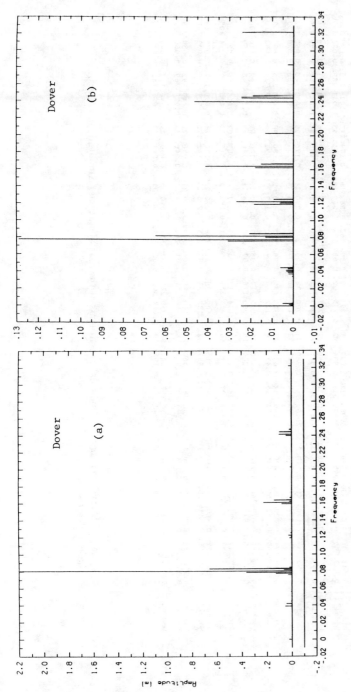

Figure 2. Spectrum of computed response at Dover: (a) all 51 constituents in the following order: Z_0 (the residual), SSA, MM, MSF, MF, α_1, $2Q_1$, Q_1, O_1, τ_1, β_1, NO_1, P_1, K_1, ϕ_1, J_1, SO_1, OO_1, v_1, ϵ_2, μ_2, N_2, M_2, MKS_2, L_2, S_2, K_2, MSN_2, η_2, MO_3, M_3, SO_3, MK_3, SK_3, MN_4, M_4, SN_4, MS_4, MK_4, S_4, SK_4, $2MK_5$, $2SK_5$, $2MN_6$, M_6, $2MS_6$, $2MK_6$, MSK_6, $3MK_7$, M_8; and (b) same but with the 11 forcing components removed.

Station	O1 A	O1 ph	K1 A	K1 ph	N2 A	N2 ph	M2 A	M2 ph	S2 A	S2 ph	K2 A	K2 ph	MN4 A	MN4 ph	M4 A	M4 ph	MS4 A	MS4 ph	M6 A	M6 ph	2MS6 A	2MS6 ph
St. Malo	0.079	344.	0.093	96.	0.740	162.	3.740	178.	1.470	229.	0.416	226.	0.111	267.	0.276	279.	0.207	335.	0.025	352.	0.029	39.
Model	0.093	347.	0.110	96.	0.716	166.	3.877	177.	1.421	230.	0.388	227.	0.137	285.	0.270	296.	0.252	351.	0.008	292.	0.009	0.
Cherbrg.	0.064	354.	0.091	106.	0.365	210.	1.870	230.	0.690	273.	0.197	269.	0.045	334.	0.155	353.	0.078	49.	0.025	101.	0.027	135.
Model	0.070	4.	0.104	110.	0.368	209.	1.974	226.	0.692	272.	0.184	271.	0.042	347.	0.128	5.	0.078	57.	0.028	95.	0.028	134.
Dieppe	0.041	36.	0.075	123.	0.570	291.	3.080	311.	1.020	1.	0.297	359.	0.093	162.	0.262	185.	0.174	242.	0.029	299.	0.022	351.
Model	0.049	45.	0.095	131.	0.510	293.	3.020	311.	0.934	1.	0.255	2.	0.087	172.	0.269	188.	0.171	245.	0.034	289.	0.038	336.
Boulogne	0.039	77.	0.046	131.	0.526	310.	2.930	331.	0.960	21.	0.273	21.	0.109	199.	0.325	223.	0.218	276.	0.064	93.	0.068	136.
Model	0.045	86.	0.064	139.	0.481	311.	2.899	329.	0.885	21.	0.243	22.	0.110	210.	0.324	226.	0.206	280.	0.099	83.	0.096	134.
Calais	0.050	147.	0.030	68.	0.440	325.	2.460	345.	0.780	38.	0.200	38.	0.090	226.	0.250	249.	0.170	306.	0.050	146.	0.060	193.
Model	0.056	140.	0.022	100.	0.393	317.	2.425	338.	0.721	29.	0.198	31.	0.079	228.	0.224	249.	0.131	297.	0.094	106.	0.090	155.
Zeebrug.	0.110	174.	0.020	339.	0.370	355.	1.590	14.	0.430	72.	0.120	72.	0.060	2.	0.090	26.	0.080	98.	0.080	325.	0.090	13.
Model	0.111	169.	0.057	350.	0.236	349.	1.622	15.	0.435	66.	0.122	70.	0.046	20.	0.159	36.	0.093	96.	0.067	330.	0.082	30.
Hoek v H	0.100	180.	0.070	346.	0.110	29.	0.770	57.	0.190	116.	0.050	117.	0.060	79.	0.160	104.	0.100	161.	0.040	146.	0.040	100.
Model	0.119	175.	0.070	347.	0.117	29.	0.972	55.	0.224	110.	0.065	116.	0.077	90.	0.226	112.	0.116	166.	0.066	46.	0.069	103.
Christc.	0.040	347.	0.094	123.	0.062	256.	0.420	285.	0.125	312.	0.034	312.	0.033	19.	0.107	41.	0.083	102.	0.016	66.	0.023	122.
Model	0.045	359.	0.093	118.	0.108	217.	0.387	227.	0.213	266.	0.055	259.	0.060	11.	0.184	26.	0.115	83.	0.095	59.	0.094	105.
Dover	0.060	172.	0.060	46.	0.410	309.	2.230	332.	0.710	23.	0.180	25.	0.130	199.	0.270	220.	0.083	273.	0.070	104.	0.070	149.
Model	0.054	169.	0.044	58.	0.365	310.	2.197	331.	0.661	22.	0.182	22.	0.082	215.	0.226	233.	0.142	281.	0.101	96.	0.099	146.
Walton	0.120	183.	0.070	348.	0.290	311.	1.350	331.	0.360	25.	0.100	25.	0.040	299.	0.080	316.	0.050	24.	0.040	292.	0.050	339.
Model	0.130	171.	0.101	351.	0.184	292.	1.178	321.	0.324	6.	0.088	11.	0.019	284.	0.084	312.	0.046	39.	0.058	269.	0.060	321.
Lowest.	0.140	166.	0.080	321.	0.150	234.	0.680	257.	0.200	296.	0.050	296.	0.020	311.	0.040	334.	0.040	23.	0.040	116.	0.040	160.
Model	0.135	160.	0.114	334.	0.116	232.	0.603	266.	0.196	300.	0.049	303.	0.015	324.	0.038	319.	0.044	20.	0.025	181.	0.051	247.

Table 1. Observed versus computed tidal constituents at the coastal stations. Amplitude in meters; phase in degrees.

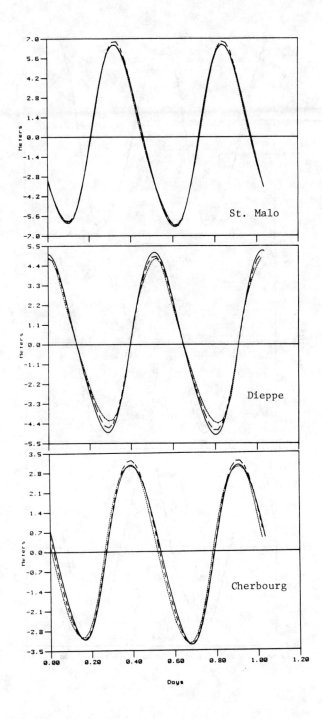

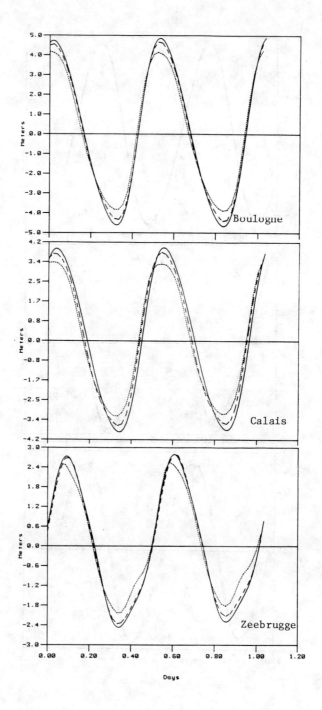

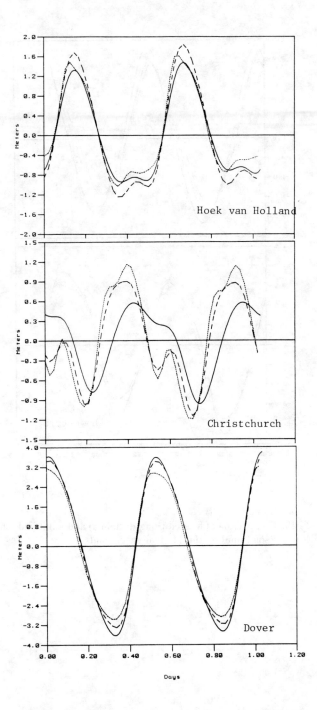

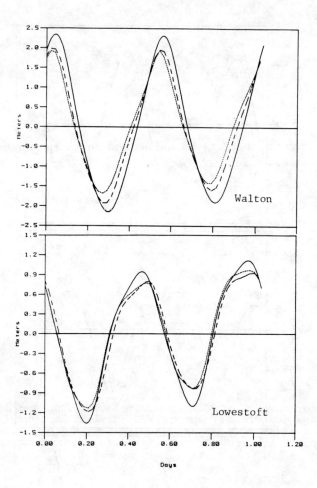

Figures 3 - 13. Comparison of filtered (long-dash line) and unfiltered (short-dash line) with data (solid line) at the 11 tidal checkpoints.

Tidal Motion in the English Channel and Southern North Sea: Comparison of Various Observational and Model Results

J. Ozer and B.M. Jamart

MUMM (Management Unit of the Mathematical Models of the North Sea and the Scheldt Estuary), Institut de Mathématique, Avenue des Tilleuls 15, B-4000 Liège, Belgium

Abstract

A conventional finite-difference, time-stepping (semi-implicit) model is entered in the model intercomparison exercise known as the "Tidal Forum". The results of that model are compared to those obtained by Walters[7] and by Werner and Lynch[8].

A distinction is proposed between two kinds of simulation: i)"time domain simulation", *i.e.*, a time-stepping computation starting with the water at rest and covering a period of only a few days; ii)"spectral approach", in which the harmonic constituents are calculated, either directly or indirectly. The first kind of simulation is represented by the first entry of Werner and Lynch and by our own; the spectral approach is followed by Walters and by Werner and Lynch in their second computation.

The reference or "sea truth" data proposed in the Tidal Forum exercise consist of a harmonic reconstitution of "observed" tidal heights and time series of currents at several locations. When evaluated against *that* benchmark, the results of the spectral approach are more satisfactory than those of the time domain simulations.

The comparison of the Tidal Forum reference data set with observations available from other sources suggests that some effort to tighten the definition of the sea truth would be worthwhile.

1. Introduction

The present study is a contribution to the exercise of model assessment and comparison known as the "Tidal Forum". The aim of the exercise is to perform similar calculations with various numerical models and to compare the model results between themselves and with "reality". The models under consideration are those solving the classical vertically-integrated, "2-D" equations of shallow water wave motion.

The test problem of the Tidal Forum has been set up under the leadership of Drs. W.G. Gray and G.K. Verboom. The problem consists of computing the motion due to tides in the English Channel and in the southern part of the North Sea. A gridded bathymetric data set and the harmonic components of several tidal constituents along the two open boundaries constitute the database provided to the participants. A detailed description of the reference computation and of the area under study, as well as a number of results, are given by Werner and Lynch[8], Gray et al.[3], and Walters[7].

We have performed the reference calculation with a finite-difference model and we have compared the results with those obtained, using finite element codes, by Werner and Lynch[8] and by Walters[7]. The purpose of this paper is to summarize our findings to date.

A brief overview of the three models, henceforth referred to as Werner's model, Walters' model and MU-model, respectively, is given in section 2. The various runs performed with the models are also described in this section. We shall distinguish between two kinds of simulations : "time domain simulation" and "spectral approach", respectively.

In a time domain simulation, the model, starting with the sea at rest, is run for a relatively short period of time. After a warm-up of a few days, the time-dependent model results are compared to a signal which will be referred to as "the observation". In the Tidal Forum experiment, the observations are obtained by a harmonic recomposition including the same constituents as those introduced along the open boundaries.

In the spectral approach, either the governing equations are solved in the frequency domain (e.g., Pearson and Winter[6], Jamart and Winter[4,5], Walters[7]), or the model is run for a sufficiently long period of time that the various constituents of interest can be identified using a classical harmonic analysis technique. In that case, the time evolution, for a particular day, will be computed using the same harmonic constituents as those introduced in the observations.

Section 3 is devoted to the analysis of the model results. Using as a benchmark the time series calculated from harmonic constituents, it appears that the results of the time domain simulations are less satisfactory than those of the spectral simulations. This point is discussed at some length.

The reference data is analyzed in section 4. In order to check the data set used in the Tidal Forum to validate the models, we have compared the amplitudes and phases of the constituents at various coastal stations as they can be found in several data banks. The discrepancies between these data sets, at some stations, can be of the same order of magnitude as the differences between model results and "observations".

Preliminary conclusions and suggestions are discussed in the last section.

2. Overview of the models and of the simulations

For the three models, the governing equations are the vertically-integrated non-linear shallow water waves equations. The bottom friction is parameterized by a conventional quadratic law. The horizontal stress is neglected. Along the open boundaries, the time evolution of the sea surface elevation is prescribed using a harmonic reconstitution of 11 constituents (O1, K1, M2, S2, N2, K2, M4, MS4, MN4, M6 and 2MS6). In the experiment, it is proposed to run the models for a period of 72 hours starting with the sea at rest on the 15th of March 1976, at 0:00 GMT. The comparisons are to be made for the 17th of March.

The first model included in the comparison (Werner and Lynch[8]) uses a finite element method to solve the equations in a Cartesian coordinate system. The resolution is performed using a time-stepping procedure, with $\Delta t = 5$ min. The numerical grid has 990 nodes. The minimum depth is set equal to 15 m. The bottom friction coefficient is taken as $2.32 \ 10^{-3}$. In a first run (Werner (r1)), the recommendations mentioned above have been followed exactly. A second, much longer (182 days) run (Werner (r2)) has also been performed, from which the amplitudes and phases of the 11 constituents introduced at the boundaries have been computed everywhere in the domain. For our comparison purposes, the time series corresponding to the latter simulation are calculated from the computed harmonic constants.

The second model (Walters[7]) also uses the finite element method to solve the governing equations in a Cartesian coordinate system. In Walter's approach, the equations are solved in the frequency domain by an iterative procedure. The numerical grid and the bathymetry are the same as in Werner's model except that the minimum depth is set equal to 5 m. In two runs performed with the model, the bottom friction coefficient is either equal to $2.32 \ 10^{-3}$ (Walters (CA)) or to $2.04 \ 10^{-3}$ (Walters (CB)).

The third model, the MU-model, uses finite difference analogs on a uniform staggered grid (Arakawa C-grid) to solve the equations in a spherical coordinate system. The time-stepping scheme is a semi-implicit, alternate direction method patterned after that proposed by Beckers and Neves[1]. The timestep is 20 min. The grid spacing is equal to 1/12 degree latitude and to 1/8 degree longitude. The number of "wet points" is equal to 1386. A time domain simulation was performed with the model using a bottom friction coefficient equal to $2.32 \ 10^{-3}$.

3. Analysis of the results

Only tidal elevations are discussed hereafter. The model responses are compared to the "observations" for the 17th of March 1976. In an attempt to present a general overview of the abilities of the models, two time-independent estimators are introduced. The first is the maximum absolute error, *i.e.*, the maximum difference between observed and computed elevations. The second estimator is the root mean square error (RMS) over the 24 hour period. The RMS errors corresponding to the five simulations under consideration are listed in Table 1.

The results of these calculations, and the visual inspection of time series at the reference coastal stations, indicate that the largest discrepancies between models and observations occur at station "Christchurch" (see map in, *e.g.*, Werner and Lynch[8]). The responses of all the models appear very bad, though consistent, at this location. This may be due to the local influence of the Isle of Wight which is not taken into account in the finite element models and very schematized in the finite difference model.

The maximum absolute errors show that the discrepancies between observations and model results can be rather large during the course of an integration (up to 1 m). This particular measure of the quality of fit of the models may well not be the one of most concern in some applications. Indeed, depending on the purposes of the model, the time, during the tidal cycle, at which the largest differences occur can be more important. This point is not discussed here. We subjectively consider such errors as "large", especially taking into account the relatively small size of the area of interest.

STATION	MU-MODEL	WERNER (r1)	WERNER (r2)	WALTERS (CA)	WALTERS (CB)
SAINT MALO	20.4	47.1	15.0	19.4	16.0
CHERBOURG	13.8	12.7	13.3	13.7	10.8
DIEPPE	44.2	54.1	17.3	25.1	11.8
BOULOGNE	61.4	41.6	15.6	32.5	28.9
CALAIS	61.8	46.1	37.7	34.7	24.7
ZEEBRUGGE	36.6	30.3	9.4	21.3	20.5
HOEK VAN HOLLAND	21.2	14.8	21.0	12.8	26.1
LOWESTOFT	12.1	12.8	15.1	13.8	14.0
WALTON	24.4	47.0	36.4	50.6	42.3
DOVER	31.4	33.6	15.6	22.5	17.1
CHRISTCHURCH	52.2	56.5	48.2	56.6	51.6
Mean RMS	34.5	36.0	22.2	27.5	23.9
Standard deviation	18.1	16.4	12.5	14.7	12.9

Table 1 : Comparison between model results and the Tidal Forum database: RMS errors, in cm, for March 17, 1976.

The more general measure of fit, the root mean square error, indicates that a distinction can indeed be made between two approaches, as discussed in section 1. On the average, we find that the RMS values are nearly 1.5 times larger in the "time domain simulations" ($\sim$ 0.35 cm) than in the simulations based on the spectral approach ($\sim$ 0.25 cm).

Concerning the time domain simulations, the results of the two models (Werner (r1), and MU) are very similar. The largest values of the RMS error are not observed at the same locations but they are of the same order of magnitude. The mean values of the RMS are equal. For Werner's model, the errors at the stations Saint Malo and Walton are perhaps surprising because these stations are not far from the open boundaries. At the Saint Malo station, the error seems to be due to a small phase shift in the model response. In the MU-model results, the RMS error seems to be more "organized" in the sense that it increases with the distance from the open boundaries. The largest values are found at Boulogne and Calais. In this region, the numerical grid of the MU-model is relatively coarse as compared to the grid of the finite element model. The values of the RMS might indicate that the numerical grid of the MU-model needs to be refined or that the configuration of the solid boundary must be modified in this region.

At this time, we can only speculate on the reasons why the "time domain simulations" do not compare as well as the "spectral simulations" to the "observations". Spectrally speaking, the model response in a time domain simulation may comprise more modes than the harmonic reconstitution used as the reference. The nonlinear terms in the model equations generate, in the interior of the domain, several harmonics of the various constituents introduced along the open boundaries. These harmonics are probably not negligible everywhere in the interior of the domain. As long as the observations at the test stations are defined using the same harmonic reconstitution as along the open boundaries, the RMS error will probably be always larger in the "time domain simulations" than in the spectral computations. On the other hand, since the neglected harmonics are clamped along the open boundaries (no radiation condition has been specified), they cannot be correctly estimated in the interior, unless their amplitude is indeed very small at the open boundaries.

The values of the RMS in the spectral simulations are quite similar. In Walters' simulations, the best results are obtained with the smallest bottom friction coefficient. However, these results are very close to those obtained by Werner with a larger bottom friction coefficient. This aspect needs to be investigated further. As a starting point, and in line with the above observations, we remark that Walters' model has fewer constituents than can be extracted from Werner's long simulation results.

4. Analysis of the data

The differences between model results and observations noted in the preceding section are, from our point of view, relatively important. It could thus be concluded that a lot of work ("tuning") remains to be done in order to improve the model responses. However, such an effort would only be worthwhile if the data set used to compute the reference is sufficiently reliable.

To address this question, we have used the same statistical approach as in the preceding section to compare time series based on the combination of 11 tidal constituents obtained from three sources: the Tidal Forum database; data obtained from the British Marine Information and Advisory Service (MIAS); data recently used at the Delft Hydraulics Laboratorium (DHL) to calibrate the Dutch tidal model (G.K. Verboom, personal communication).

The DHL data set is, in general, very close to that of the Tidal Forum. However, at Cherbourg, Dieppe and Boulogne, the RMS errors (6.8, 5.4, and 10.9 cm, respectively) are not small compared to those calculated for Werner's model (r2) and Walters' model (CB). As for the MIAS data set, the results at Boulogne (RMS = 169.5 cm) and Christchurch (RMS = 123.7 cm) are disconcerting. This may be due to a coding error. For some other stations, the RMS error is again of the same order of magnitude as that found for the best model simulations (*e.g.*, Calais: 18.4 cm, Zeebrugge: 8.0 cm).

On the basis of these comparisons, it seems necessary to verify the reference data set before ascribing the discrepancies to problems inherent to the models. It may also be useful to investigate the reliability of the boundary conditions.

A second question which needs to be addressed is the number of constituents that are necessary and sufficient for a "good" simulation of the tidal motion in the area. We have mentioned the difficulties encountered in comparing the results of a time domain simulation to a harmonic reconstitution including only the same constituents as the boundary forcing. One way to circumvent such difficulties is to include in the calculation of the reference a larger number of constituents. At this writing, we can only do this for Zeebrugge, a Belgian coastal station for which we have a one year long series of observations. Using Foreman[2]'s package, 114 constituents have been extracted from that time series. If the 11 constituents considered so far are compared to the Tidal Forum database as we did for the MIAS and DHL data, we find a maximum difference of 10.7 cm and a RMS of 6.5 cm. If we now use all 114 constituents to recompute the time evolution of the elevation on the 17th of March, 1976, we find that the tidal range is not significantly different from that obtained from the Tidal Forum database. However, it is puzzling to note that the peculiar behaviour of the tidal curve shortly after low tide observed in the results of the "time domain simulations" (see Fig. 9 in Werner and Lynch[8]) is also obvious when 114 constituents are used whereas this feature is completely obliterated when only 11 harmonics are included. This indicates the influence, at this particular time and place, of constituents excluded from the intercomparison exercise.

The influence of the number of constituents included in a harmonic reconstitution can be also evaluated by computing day by day, on a long series, the values of the global estimators for i) a reconstitution with 11 constituents; and, ii) a series calculated with 114 constituents. We have performed such a computation, using the actual observations as the reference. The influence of phenomena not taken into account in the harmonic analysis (atmospheric forcing, ...) appears clearly in these results. More importantly for the present discussion, we note that both errors (maximum error and RMS error) can be up to 30% smaller when all 114 constituents are used in the harmonic reconstitution. Considering that the area under study is relatively small and shallow, our provisional conclusion is that the number of modes included in the computation (and, hence, specified at the boundaries) might have to be increased to improve the responses of the models. If these constituents cannot be analyzed from the relatively short time series available at the offshore stations, they can perhaps be provided by numerical models covering a larger area.

5. Concluding remarks

The results of three models, a finite difference model and two finite element models, included in the Tidal Forum experiment have been compared.

At this stage, it seems that the way selected to perform a particular calculation is more important than the numerical approach chosen to solve the governing equations. Under the same conditions, both numerical techniques (finite differences and finite elements) seem to provide results which are mostly undistinguishable. On the other hand, the results obtained following a spectral approach are better than those given by a time domain simulation. The fact that the reference is defined only by a harmonic reconstitution of the constituents used to drive the system explains partially the differences between the two kinds of calculations.

Another potential problem for the success of the intercomparison experiment is that the information used to compute "the observation" (amplitude and phase of the constituents) differs from one data bank to another. Differences between data sets can be comparable to the discrepancies observed for some model responses. From this point of view, an intercomparison experiment between various data analysis techniques should be instructive.

Finally, we wish to assert our conviction that the comparative assessment of numerical models, in the spirit of the Tidal Forum experiment, is a worthwhile exercise. We submit that a similar experiment, performed with three-dimensional codes rather than with vertically-integrated models, would also be of interest to the modelling community.

6. Acknowledgments

This work was partially supported by Det norske Veritas within the frame of an industry research and development project called the Bottom Stress Experiment funded by Fina Exploration Norway.

We thank G.K. Verboom, R.A. Walters, and F.E. Werner for fruitful discussions and early communication of unpublished results. We also thank C. Coolen and A.F. Lucicki for assistance with the text processing.

7. References

1) Beckers P.M. and Neves R.J. (1985), A semi-implicit tidal model of the North European Continental Shelf, Applied Mathematical Modelling, Vol.9, No.6, pp. 395-402.
2) Foreman, M.G.G. (1977), Manual for Tidal Heights Analysis and Prediction. Pacific Marine Science Report 77-10, Institute of Ocean Sciences, Patricia Bay, Sydney, B.C., 97 pp.
3) Gray W.G., Drolet J. and Kinnmark I.P.E. (1987), A simulation of tidal flow in the south part of the North Sea and The English Channel, Advances in Water Resources, Vol.10, pp. 131-137.
4) Jamart B.M. and Winter D.F. (1978), A new approach to the computation of tidal motions in estuaries. In Hydrodynamics of Estuaries and Fjords, (Ed. Nihoul J.C.J.), pp.261-281, Elsevier, Amsterdam.
5) Jamart B.M. and Winter D.F. (1982), Finite element solution of the shallow-water wave equations in Fourier space, with application to Knight Inlet, British Columbia. Chapter 8, Finite Elements in Fluids, (Ed. Gallagher R.H., Norrie D.H., Oden J.T. and Zienkiewicz O.C.), Vol.4, pp.157-177, Wiley-Interscience, London.
6) Pearson C.E. and Winter D.F. (1977), On the calculation of tidal currents in homogeneous estuaries, Journal of Physical Oceanography, Vol.7, No.4, pp. 520-531.
7) Walters, R.A. (1987), A model for tides and currents in the English Channel and southern North Sea, Advances in Water Resources, Vol.10, pp. 138-148.
8) Werner F.E. and Lynch D.R. (1987), Field verification of wave equation tidal dynamics in the English Channel and southern North Sea, Advances in Water Resources, Vol.10, pp. 115-130.

Experiments on the Generation of Tidal Harmonics

R.A. Walters
U.S. Geological Survey, 1201 Pacific Avenue, Suite 450, Tacoma, WA 98402, USA

F.E. Werner
Thayer School of Engineering, Dartmouth College, Hanover, NH 03755, USA

ABSTRACT

The dependent variables in the shallow water equations, sea level and velocity, are expanded in a time-average plus the sum of periodic components whose frequencies are known. Following harmonic decomposition, the equations are solved for sea level resulting in an elliptic equation for sea level and the two components of the momentum equation. Time-stepping methods are not used; rather, the dependent variables are the complex amplitudes. In this study, we examine the effects of various approximations to the nonlinear terms, namely wave transport, advection, and bottom friction.

INTRODUCTION

A frequency-domain, finite element model for the shallow water equations was developed in order to study circulation and solute transport in estuaries over long time periods. The spatial part of the governing equations is approximated using finite element methods with triangular elements and linear basis functions. The temporal part is approximated using harmonic decomposition so that the dependent variables are characterized by an amplitude and a phase for each frequency component rather than by a time series. A time series representation can be obtained through synthesis of the components.

With harmonic decomposition, the linear terms in

the governing equations only have an effect on the principal frequency constituent of the particular equation under consideration. However, the nonlinear terms generate contributions at sums, differences, and multiples of the principal constituents and hence act as complicated source terms. For the nonlinear advection and wave transport, the frequency difference terms provide the source terms for the fortnightly and longer period motions, whereas the frequency sums provide the source terms for the overtides and compound tides. The treatment of the quadratic bottom friction terms is difficult because of their nonlinearity.

Our purpose in the numerical experiments described herein, is to assess the ability of the frequency-domain model to reproduce the nonlinear source generation terms. The method used here entails making calculations for a simplified set of constituents in an irregular network, and examining the effects of the various source terms. The results are compared to those from a time-domain model.

MODELS

The frequency-domain model used here is described in detail in references 1 and 2. Only a brief summary is presented for completeness. The governing equations are the shallow water equations that are derived from the depth-integration of the Navier-Stokes equations with a free surface. Horizontal stresses are not included. The dependent variables, sea level η and horizontal velocity $\mathbf{u}$, are expanded as a time-averaged component (subscript 0) plus a sum of periodic components (subscript n) whose frequencies are known. Note that boldface type will denote vector quantities. The various components are separated by harmonic decomposition[3,2].

After this decomposition, the continuity equation and the two momentum equations are solved for η_n, the complex sea level amplitude for component n, to arrive at

$$-i\omega_n \eta_n + \nabla \cdot \left\{ \left[\frac{gH}{q^2 + f^2} \right] [q(\nabla \eta_n - \mathbf{T}_n) - f \times (\nabla \eta_n - \mathbf{T}_n)] \right\} - \nabla \cdot \mathbf{W}_n = 0 \quad (1)$$

where $q = -i\omega_n + \tau_n$, τ_n is the linear part of the bottom friction, and W_n and gT_n are the nonlinear terms from the continuity equation and the momentum equations, respectively. Once sea level is known, the velocity components are back-calculated using the momentum equations. The system of equations exhibits an oscillatory convergence that is damped with an underrelaxation factor applied to velocity.

Because the nonlinear terms are the focus of this study, they need to be described in more detail. The nonlinear wave transport term from the continuity equation is

$$W_n = -\frac{1}{2}\sum_{i,j}(\eta_i u_j) \qquad (2)$$

and the nonlinear term from the momentum equations is

$$T_n = -\frac{1}{2g}\sum_{i,j} u_i \cdot \nabla u_j - \frac{\tau'_b}{g} \qquad (3)$$

subject to the constraint that $\omega_i + \omega_j = \omega_n$, or the sum or difference in frequency must be the same as the frequency with index n. Both τ_n and τ'_b depend on the details of the approximation for bottom friction in the following way. Here bottom friction is given by

$$\tau_b = \frac{k|u|u}{H + \eta} \qquad (4)$$

where k is taken as a constant with a value of 0.00232 in this study (corresponding to a Chezy coefficient of 65 $m^{1/2}s^{-1}$). Following Snyder et al[3], $|u|$ is separated into a time-independent and a time-dependent part, and expanded in a Taylor series about the latter. The first term in the series is

$$\tau_n = \frac{k}{H}\left(\frac{1}{2}\right)\left[\sum_j u_j \cdot u_{-j} + 4u_0^2\right]^{\frac{1}{2}} \qquad (5)$$

the steady part of τ_b. The next two terms in the series are

$$\tau'_b = \frac{k}{8H\lambda}\sum_{\substack{ijm \\ i\neq -j}}(u_i \cdot u_j)u_m - \frac{k}{128H\lambda^3}\sum_{\substack{ijpqm \\ i\neq -j \\ p\neq -q}}(u_i \cdot u_j)(u_p \cdot u_q)u_m \qquad (6)$$

where $\lambda = (H/k)\tau_n$, the rms component of velocity,

$\omega_i + \omega_j + \omega_m = \omega_n$ in the first term on the right, and $\omega_i + \omega_j + \omega_p + \omega_q + \omega_m = \omega_n$ in the second term. Different combinations of the constituents in equations (2), (3), and (6) combine to form the source terms for the overtides as discussed below.

The time-domain model is the explicit wave equation model as described in Werner and Lynch[4]. This model treats the nonlinear terms directly without approximation and is therefore used as a basis for comparison with the frequency-domain model.

The network used in this study encompasses the English Channel and southern North Sea and includes 990 nodes and 1762 elements. The network and an observational data set was made available through the Tidal Flow Forum associated with this conference series. The data set includes harmonic constants for 11 constituents with values along the open boundaries and at 11 sea level and 8 current meter stations. Results using the full 11 constituent data set are reported by Walters[2] and by Werner and Lynch[4] for the frequency-domain model and the time-domain model, respectively. In general, the results agree reasonably well with the data. The time-domain model[5] appeared to have slightly better results than the frequency-domain model for the highest frequency constituents (sexto-diurnal tides) when compared to the results cited by Le Provost and Fornerino[6], although there is considerable uncertainty in the spatial distribution of these short wavelength components.

In order to simplify this study, yet retain the essence of the nonlinear couplings, we reduced the constituents to M_2, M_4, and M_6. In this manner, there is one constituent in each frequency band such that they are integer multiples which eliminates the fine structure in the spectrum and hence additional complications with closely-spaced constituents.

RESULTS

In this region, the tides are dominated by the M_2 constituent which has a maximum amplitude of about 4 m. The M_4 tidal amplitude is about 8% and the M_6 is about 2% that of the M_2 when averaged over a number of observation stations (Le Provost and Fornerino[6]). Thus the problem can be characterized as a dominant wave that supplies the nonlinear forcing for the higher harmonics. In the following discussion, the indices 1, 2, and 3 will refer to the M_2, M_4, and M_6 constituents, respectively.

The resultant amplitudes for the M_2 constituent where almost identical between the two models in this study, but larger than those for the full constituent set. This happens because of the absence of damping from the excluded constituents as discussed in reference 6. The only significant nonlinear term that appears is $k(\mathbf{u}_1 \cdot \mathbf{u}_1)\mathbf{u}_{-1}/(8H\lambda)$ from eq. (6). This term scales to be 50% of the friction term $\tau_1 u_1$ when the tidal ellipse is a straight line and zero when it is a circle. In effect, it provides a correction to increase the bottom friction.

The resultant amplitudes for the M_4 were also very close between the two models. This constituent is forced primarily by the wave transport and advection terms, $-(1/2)(\eta_1 \mathbf{u}_1)$ and $-(\mathbf{u}_1 \cdot \nabla \mathbf{u}_1)/(2g)$, contained in equations (2) and (3). Although the triangular elements with linear bases provide a low order approximation to these terms, the favorable comparison with the field data suggests that the approximation is adequate here. One additional term of importance is $k[(\mathbf{u}_1 \cdot \mathbf{u}_2)\mathbf{u}_{-1} + (\mathbf{u}_{-1} \cdot \mathbf{u}_2)\mathbf{u}_1]/(4H\lambda)$ which has properties similar to the corresponding correction term for the M_2 as discussed above.

The source terms for the M_6 constituent are relatively complicated because of the presence of both two-frequency and three-frequency interactions. Note that for bottom friction, the dominant term in the expansion of eq. (4) is $\tau_3 u_3$ and the terms in eq. (6) provide a correction. However, for the generation terms the first term is $k(\mathbf{u}_1 \cdot \mathbf{u}_1)\mathbf{u}_1/(8H\lambda)$ from eq. (6) which is the second term in the expansion of eq. (4). Thus, the next term in the series must usually be included as it provides about a 12% correction to the term above. In effect, more terms in the series expansion need to be retained in the generation term for equivalent accuracy than with the bottom friction term.

In addition to the source terms discussed above, both advection and wave transport are important processes. The corresponding terms involve coupling between the M_2 and M_4 constituents and are $-(\mathbf{u}_1 \cdot \nabla \mathbf{u}_2 + \mathbf{u}_2 \cdot \nabla \mathbf{u}_1)/(2g)$ from the advection terms, and $-(\eta_1 \mathbf{u}_2 + \eta_2 \mathbf{u}_1)/2$ from the wave transport terms. Without these terms, the sea level amplitude was too small in the North Sea and too large in the English Channel by about 50%, and the spatial distribution was poor in the far western end of the network.

Including all the terms in the expansion as noted above, the results for the two models were in good agreement. The greatest differences occurs in the western part of the English Channel where there is an amphidrome.

CONCLUSIONS

Based upon the observations and results from another model, the frequency-domain model is capable of reproducing the correct source terms for the overtides. However, a practical limit in frequency when using this approach seems to be the sexto-diurnal constituents because of the increasing complexity of the source terms with frequency. This limit is tempered by the fact that there is little energy in the higher frequency constituents.

REFERENCES

1. Walters, R.A. (1986) A finite element model for tidal and residual circulation, Communications in applied numerical methods, Vol.2, pp. 393-398.

2. Walters, R.A. (1987) A model for tides and currents in the English Channel and southern North Sea, Advances in Water Resources, Vol.10, pp. 138-148.

3. Snyder, R.L., Sidjabat, M., and Filloux, J.H. (1979) A study of tides, set-up, and bottom friction in a shallow, semi-enclosed basin. Part II: Tidal model and comparison with data, Journal of Physical Oceanography, Vol.9, pp. 170-188.

4. Werner, F.E., and Lynch D.R. (1987) Field verification of wave equation tidal dynamics in the English Channel and southern North Sea, Advances in Water Resources, Vol.10, pp. 115-130.

5. Lynch, D.R., and Werner, F.E. (1988) Long-term simulation and harmonic decomposition of English Channel tides, Unpublished data.

6. Le Provost, C., and Fornerino, M. (1985) Tidal spectroscopy of the English Channel. with a numerical model, Journal of Physical Oceanography, Vol.15, pp. 1009-1031.

A 2D Model for Tidal Flow Computations

C.S. Yu, M. Fettweis and J. Berlamont
Katholieke Universiteit te Leuven, Civil Engineering Department,
Laboratory of Hydraulics, de Croylaan 2, Bus 4, 3030 Heverlee, Belgium

ABSTRACT

A two-dimensional depth-averaged numerical model for tidal flow computations is presented. The model applies the finite difference method together with a falsified alternating direction implicit scheme to solve the shallow water equations. The possibility of using the technique of grid refinement has been included in the model. A North-West European Continental Shelf model is presented.

INTRODUCTION

The finite difference method (FDM) is commonly used to solve shallow water equations for two-dimensional (2D) flow problems in estuarine and coastal waters. Implicit finite difference schemes have been widely used after *Leendertse*[1] first proposed, because the implicit scheme is unconditionally stable which is so attractive from an economic point of view. A disadvantage of FDM is the difficulty of describing an area of the computational field in detail without increasing the total number of computational grid points.

A 2D depth-averaged numerical simulation model for tidal flow computations has been developed. The model is based on a finite difference solution for the shallow water equations by using a falsified alternating direction implicit(FADI) scheme. The model not only saves some computer time, but also reduces the possible non-linear instability as the derivatives are centered.

Another great advantage is that the model includes the possibility of applying grid refinement to any desired region in the model area without increasing the total number of the computational grid points. The variables at the fine grid points are calculated simultaneously as the variables at the coarse grid points. This model is suitable especially for the shelf sea dynamics. Shallow areas or areas with strong varying bathymetry can be described in detail with a good computing efficiency; this cannot be achieved by schemes which are restricted by the CFL criterion.

A brief out-line of the model is presented. The treatment of the grid refinement technique is described. Tidal flow computations of the North-West European Continental Shelf are discussed.

GOVERNING EQUATIONS

The 2D depth-averaged shallow water equations are in the following form :
Continuity equation

$$\frac{\partial z}{\partial t} + \frac{\partial (Hu)}{\partial x} + \frac{\partial (Hv)}{\partial y} = 0 \qquad (1)$$

Equations of horizontal motion in x direction

$$\frac{\partial u}{\partial t} + u\frac{\partial u}{\partial x} + v\frac{\partial u}{\partial y} + g\frac{\partial z}{\partial x} - fv + \tau_{bx} - \tau_{sx} = 0 \qquad (2)$$

and, in y direction

$$\frac{\partial v}{\partial t} + u\frac{\partial v}{\partial x} + v\frac{\partial v}{\partial y} + g\frac{\partial z}{\partial y} + fu + \tau_{by} - \tau_{sy} = 0 \qquad (3)$$

where the notation is
  t : time (seconds)
  x : horizontal coordinate, positive eastward (m)
  y : horizontal coordinate, positive northward (m)
  z : water surface elevation above a datum (m)
  h : bathymetry, measured from mean-sea-level (m)
  H : (= h + z), total depth of water (m)
  u : depth mean velocity in x direction (m/s)
  v : depth mean velocity in y direction (m/s)
  g : the acceleration due to gravity
  f : the Coriolis parameter ($= 2\Omega \sin \phi$, where Ω is the angular speed
      of the earth's rotation)
  τ_{bx}, τ_{by} : bottom friction components in x and y direction respectively
  τ_{sx}, τ_{sy} : wind stress components on the water surface in x and y
      direction respectively.

The bottom friction terms were assumed in the following quadratic law:

$$\tau_{bx} = g\frac{u\sqrt{u^2+v^2}}{C^2 H}, \tau_{by} = g\frac{v\sqrt{u^2+v^2}}{C^2 H} \qquad (4)$$

where C is the Chézy coefficient, and the wind stress is assumed in the form

$$\tau_{sx} = \frac{C_D \rho_a W_u \sqrt{W_u^2 + W_v^2}}{\rho_w H}, \tau_{sy} = \frac{C_D \rho_a W_v \sqrt{W_u^2 + W_v^2}}{\rho_w H} \qquad (5)$$

where C_D is the drag coefficient of the air-water interface, ρ_a is the air density, ρ_w is the water density and W_u, W_v are the local wind velocities.

NUMERICAL MODEL

FDM has been applied to solve the differential equations (Eqs. 1-3) numerically. The variables are described on a spatially staggered grid as shown in Fig.1. In order to gain some computer time and to save some memory spaces, the ADI scheme of the *Leendertse* type has been modified. A method of the splitting-up operator approach was applied to split the difference form of the continuity equation into two falsified forms for the two alternating directions. For more details see Abbott et al.[2] and Yu et al.[3]. The consistency and stability of this method were examined by *Yanenko*[4]. In each direction the FADI scheme contains only one momentum equation and a falsified continuity equation. These equations have the following difference form:

At odd time-steps (x direction)

$$\frac{z_{i,j}^{n+1} - z_{i,j}^{n-1}}{2\Delta T} + \frac{Hu_{i+1/2,j}^{n+1} - Hu_{i-1/2,j}^{n+1}}{\Delta X} = 0 \quad (6)$$

$$\frac{u_{i+1/2,j}^{n+1} - u_{i+1/2,j}^{n-1}}{2\Delta T} + u_{i+1/2,j}^{n+1} A_x + \bar{v} A_y + g\frac{z_{i+1,j}^{n+1} - z_{i,j}^{n+1}}{\Delta X} - f\bar{v} + \tau_{bx} - \tau_{sx} = 0 \quad (7)$$

where

$$\bar{v} = \frac{1}{4}(v_{i,j-1/2}^n + v_{i,j+1/2}^n + v_{i+1,j-1/2}^n + v_{i+1,j+1/2}^n) \quad (8)$$

$$\tau_{bx} = g\frac{u_{i+1/2,j}^{n+1}\sqrt{(u_{i+1/2,j}^{n-1})^2 + \bar{v}^2}}{C^2 H_{i+1/2,j}^n} \quad (9)$$

where i, j are the finite coordinates in x and y direction, respectively; ΔT is the time-step size (sec); ΔX is the spatial grid size (m); n is the time level ($n-1, n+1,...$ for odd time-steps, $n, n+2,...$ for even time-steps) and A_x, A_y are the second order upstream difference forms of the advective terms. From Eqs.6-9, z_{odd} and u are solved implicitly. The second stage has similar difference equations for solving z_{even} and v in y direction at even time-steps.

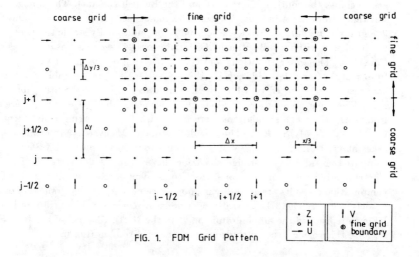

FIG. 1. FDM Grid Pattern

GRID REFINEMENT

The possibility of using grid refinement is included in the model. The fine grid pattern is shown in Fig. 1. The computational order of the equations at both odd and even time-steps was pre-managed into a coefficient matrix. In order to make a continuous implicit computation of each line, both grid systems were linked together by the following principles:

- The calculation starts from the left boundary, the variables in the coarse grid are calculated until the left boundary points of the fine grid are reached.
- The boundary points of the fine grid are situated on water level points of the coarse grid, as shown in Fig.1.
- The values of the coarse grid points which overlap the boundary points of the fine grid are used as the boundary values for the fine grid.
- The values of the fine grid boundary points which are situated between two coarse grid points are interpolated from these two values.
- The calculation of the fine grid variables proceeds in the same way as that of the coarse grid until a closed boundary or a coarse grid boundary is reached.
- The computation continues with the following coarse grid points.

In such a way the variables situated on both coarse and fine grid are solved simultaneously. Detailed coding were described by Moelans et al.[5].

MODEL APPLICATION

The verification of the model was discussed in Ref.3 where the model results were compared with the analytical solutions of an idealized example and applied to several practical cases. The model has been applied to calculate the tidal flows of the North-West European Continental Shelf. The shelf model represent an area from $12°W$ longitude to the Baltic Sea, and from $48°N$ to $61°N$ latitude. The fine grid includes the southern North Sea and the English Channel. The topography was taken from the Admiralty Charts. The coarse grid size is $24*24km^2$ and the fine grid size is $8*8km^2$. Since the scheme is unconditionally stable the time-step can be chosen without restriction by CFL criterion. 447 seconds was chosen as the time-step. The water elevations and the momentum of the model were started from rest. After a few tidal cycles the sea surface reaches a stable condition. The results were stored in a file which is used as the initial condition for later runs. The open boundary data along the continental shelf edge were generated from $IAPSO^6$. The simulations were done with 6 tidal forcing constituents (i.e. $O_1, K_1, N_2, M_2, S_2, K_2$). The calculation of one M_2 cycle (= 100 time-steps) takes about 16 min on a HP-VECTRA and about 30 sec on a VAX-8600. The computed co-tidal and co-range chart for the M_2 tide is illustrated in Fig. 2. The computed tidal currents for 6 tidal components are shown in Fig. 3. The computational results are generally in good agreement compared to those of other models, e.g. see the 3D spectral model done by *Davies*[7]. The grid size is too large to reproduce the amphidromic point in the North Channel of the Irish Sea because this region is not included in the fine grid computations.

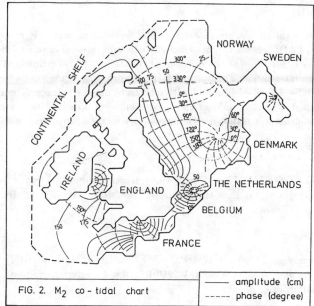

FIG. 2. M$_2$ co-tidal chart

— amplitude (cm)
----- phase (degree)

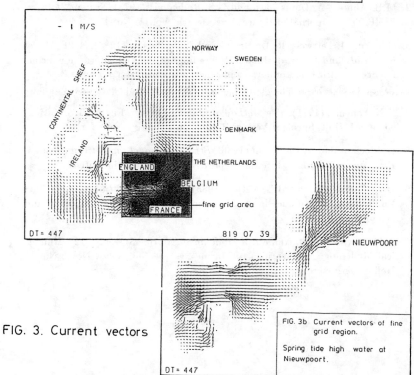

FIG. 3. Current vectors

FIG. 3b Current vectors of fine grid region.

Spring tide high water at Nieuwpoort.

CONCLUSION

A 2D numerical simulation model for tidal flow computations is presented. The proposed FADI scheme is more efficient than conventional implicit schemes, because of the falsification of the differential equations. A grid refinement technique has been introduced. This increased the applicability of the model to areas which have to be described in detail without loss of much efficiency. The model has been applied to the North-West European Continental shelf. The advantages of the grid refinement are shown. The results are in good agreement with published data.

ACKNOWLEDGEMENT– Part of this research is supported by the Belgian program "Scientific research on the Antarctic" (Services of the Prime Minister - Science Policy Office).

REFERENCES

1. J.J. Leendertse (1970), *A Water Quality Simulation Model for Well-mixed Estuaries and Coastal Seas*, RM-6210-RC, Rand Corporation, Santa Monica, California.

2. M.B. Abbott, A. Damsgaard and G.S. Rodenhuis (1973), *System 21, "JUPITER"*, Journal of Hydraulic Research, Vol.11, No. 1.

3. C.S. Yu, M. Fettweis, R. De Bruyn and J. Berlamont (1988), *A 2D Model for Steady and Unsteady Flows*, Proceedings of the "1st international conference in Africa on computational methods and water resources" Vol.2 Comp. Mech. Publ., Boston and Springer-Verlag, Berlin.

4. N.N. Yanenko (1971), *The Method of Fractional Steps*, English Translation, ed. by M. Holt, Springer-Verlag, Berlin - Heidelberg - New-York.

5. D. Moelans and R. De Bruyn (1986), *Stromingsmodel van het Continentaal Plat van de Noordzee*, Internal Report No. 28-HY-10. (in Dutch)

6. D.E. Cartwright and B.D. Zetler (1985), *Pelagic Tidal Constant*, IAPSO Publication Scientifique No. 33, The International Union of Geodesy and Geophysics, Paris.

7. A.M. Davies (1987), *Numerical Modelling of Marine Systems*, 1-24, Numerical Modelling: Applications to Marine Systems, ed. by J. Noye, North-Holland, Amsterdam.

SECTION 3B - LAKE AND ESTUARY MODELS

A Coupled Finite Difference - Fluid Element Tracking Method for Modelling Horizontal Mass Transport in Shallow Lakes

P. Bakonyi and J. Józsa

Research Centre for Water Resources Development, VITUKI, Budapest, Hungary

ABSTRACT

Wind-induced horizontal transport phenomena have been analysed in shallow lakes with a coupled 2-D circulation-fluid element tracking model. Advective currents are calculated by a depth-integrated ADI-type finite difference model. In subdomains the influence of small scale bottom topography on the velocities is taken into account by local grid refinement. To avoid numerical diffusion and to obtain qualitative results fast and easily, transport simulation is based on the Lagrangian Fluid Element Tracking method, modelling the paths of marked fluid elements in the advective-diffusive velocity field. Diffusive velocity fluctuations are generated by Monte Carlo technique. As possible applications the simulation of storm-induced displacement of large pollution clouds, water exchange of bays and the mixing of inflow water are given in Lake Balaton.

INTRODUCTION

One of the most important factors in the global water quality of lakes is the mass exchange between areas having different water quality. In shallow lakes with insignificant throughflow this process is determined mainly by the wind-induced horizontal currents. In some cases the shallowness allows the the 2-D depth-integrated calculation of the unsteady advective velocity field, which is used then to simulate advection-diffusion processes. To treat the latters grid methods such as finite differences or elements are traditional interpreted in Eulerian frame. They present, however, numerical difficulties (numerical diffusion and/or oscillation) when being applied for advection dominated problems or to handle point sources, steep concentration and velocity gradients, furthermore the information about the origin and the path of the transported masses is totally lost. These difficulties do not arise in Lagrangian system in which the water is interpreted as a finite set of marked fluid elements owning some fixed physical properties. The numerical description of the transport process then consists of tracking the independent fluid elements by a suitable numerical time integration of the unsteady velocity field (Maier-Reimer and Sündermann[1]). At any time the element positions will give a global picture about the process. The advective part of the velocity comes from the circulation model, turbulent diffusion is taken into account by generating random velocity fluctuations and superimposing them onto the advective velocity field. In the present paper the Fluid Element Tracking (FET) method is used as an unconditionally stable and efficient tool for the qualitative description of different transport phenomena including the transport of dissolved or suspended passive material moving with the water.

CIRCULATION MODEL

Advective currents are calculated by a 2-D depth-integrated flow model based on the complete shallow water equations. The solution is achieved with an ADI-type finite difference method on orthogonal staggered grid. At areas of special interest locally refined grid is applied to obtain finer scale advective velocity resolution reflecting the influence of small scale bottom and shore line topography. The connection of the different grids is assured through discharge-type boundaries presenting no significant numerical problems.

FLUID ELEMENT TRACKING

Considering the velocity as the sum of an advective $\underline{u}_a$ and a diffusive $\underline{u}'$ part the position of a fluid element at time $t+\Delta t$ from that at t can be given by:

$$\underline{x}(t+\Delta t) = \underline{x}(t) + \underline{u}(\underline{x}(t),t)\Delta t \qquad (1)$$

where $\underline{x}(t+\Delta t)$ and $\underline{x}(t)$ are the 2-D position vectors and:

$$\underline{u} = \underline{u}_a + \underline{u}' \qquad (2)$$

Advective velocities are determined from the flow model using e. g. bilinear interpolation function.

Velocity fluctuations can be modelled by a Monte Carlo technique. Taking the random numbers from tophat distribution Maier-Reimer and Sündermann[1] give the relation between the band width U of the fluctuation, the diffusion coefficient D and the timestep Δt:

$$U = (6D/\Delta t)^{1/2} \qquad (3)$$

At closed boundaries perfect reflection is prescribed, at outflow boundaries elements leaving the domain are omitted for further computation. The timestep is limited only by accuracy requirement.

APPLICATION

The applicability of the method is demonstrated in Lake Balaton, the largest shallow lake in Central-Europe (Fig.1.). The western end of the lake has been investigated in detail presenting particularly severe water quality problems caused mainly by the highly polluted inflow and the insignificant water exchange. N-NW storms dominate the region inducing seich motion and strong circulatory currents as well. A typical velocity field corresponding to the relatively steady stage of a two days N-NW wind is shown in Fig.2. The flow calculations have been performed on a coarse grid for the whole lake and on a three times finer one for the part inside the frame in Fig.1. To estimate the real water exchange between the two western bays an appropriate control cross-section could not be chosen. That is why fluid parcels were marked at grid nodes at the beginning of the storm and tracked during the two days period. In Fig.3. the dashed line shows the separation line, on the right hand side of which initially fluid elements were marked stronger. It can be seen that despite the rather high velocities the storm has hardly induced water exchange. Using a north-south separation not presented here the final situation shows an intensive exchange in the cross direction. Another aspect is to mark fluid elements along the contour of large pollution clouds. Then supposing a strong mixing caused by waves in the vertical during the early stage of the storm, the FET method is applicable to determine their displacement by pure advection and their final position at the end of the storm. Fig.4. shows the new position of three initially circular clouds after one day in the above mentioned storm. The last example demonstrates the tracking of the inflow river water in the storm. This problem requires to mark continuously the water masses entering the bay. Releasing one element in each time step the final positions can be seen in Fig.5.

reflecting the influence of the unsteady conditions and diffusion on the mixing.

SUMMARY

A simple and efficient way has been given for the qualitative simulation of various transport processes in shallow lakes based on the tracking of marked fluid elements and dissolved or suspended passive material parcels in an unsteady 2-D advecticve-diffusive velocity field. The FET has been applied to Lake Balaton in several aspects. Working in Lagrangian system one can track arbitrary sets of elements to illuminate the transport phenomena from various points of view. The accuracy can be improved using higher order time integration and velocity interpolation. Quantitative results can be obtained by increasing the number of elements resulting in a linear increase in CPU time. Further improvements should include the 3-D extension and the direct use of velocity fluctuation data (Zannetti[2]).

1. Maier-Reimer E. and Sündermann J. (1982), On Tracer Methods in Computational Hydrodynamics. In Engineering Application of Computational Hydraulics, (Ed. Abbott M.B. and Cunge J.A.), Vol.1, pp.198-217, Pitman.

2. Zannetti P. (1984), New Monte Carlo Scheme for Simulating Lagrangian Particle Diffusion with Shear Effect, Applied Mathematical Modelling, Vol.8, pp.188-192.

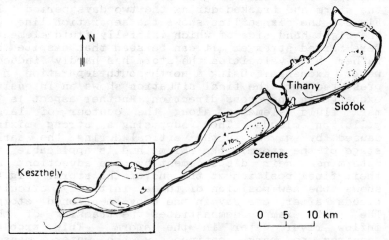

Figure 1. Lake Balaton. Depth contours in meters. Refined grid area falls inside the frame.

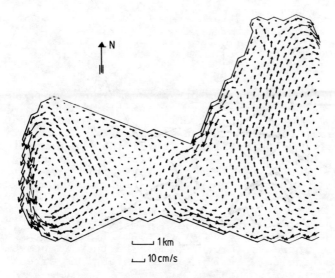

Figure 2. Typical flow pattern induced by a N-NW wind in the western end of the lake.

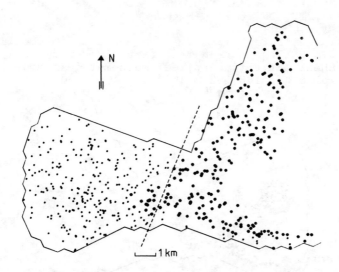

Figure 3. Marked fluid element positions after a two days N-NW storm. Initial separation according to the dashed line, initial positions at the nodes with stronger mark on the right hand side. $D = 0.2$ m²/s.

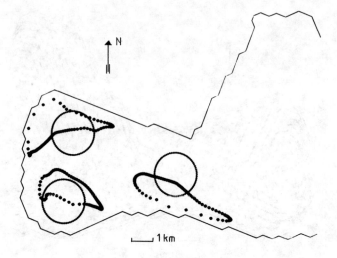

Figure 4. Advection of circular pollution clouds. Positions after one day in the N-NW storm. Contours are described with 80 elements.

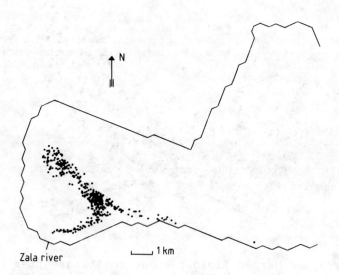

Figure 5. Mixing of the inflow water due to the two days N-NW storm. One element is released in each time step. $D = 0.2$ m^2/s.

Hydrodynamics and Water Quality Modeling of a Wet Detention Pond

D.E. Benelmouffok and S.L. Yu
Department of Civil Engineering, University of Virginia, Charlottesville, VA 22901, USA

ABSTRACT

A vertically averaged two-dimensional hydrodynamic model was developed to simulate the flow and pollutant transport in small scale basins and to assess the pollutant removal performance of an urban wet detention pond located in Charlottesville. The differential equations were numerically solved on a space staggered grid using a two-level time ADI integration scheme.

INTRODUCTION

Most of the models developed for estimating the performance of a detention basin are based either on a plug flow or completely mixed systems which seldom occurs in reality. The shallow water equations and the two-dimensional advective-dispersive equations have been widely used for the simulation of flow in large scale systems where the driving forces are either the wind or the tide, but no attempt has been made to use this approach in smaller scale systems where the main driving forces are the flows in and out of the basin.

GOVERNING EQUATIONS

For two-dimensional flow, vertical integration of the momentum and continuity equations yields the following basic equations:

$$\frac{\partial U}{\partial t} + U\frac{\partial U}{\partial x} + V\frac{\partial U}{\partial y} + g\frac{\partial \varsigma}{\partial x} + \frac{\tau_x^b}{\rho H} = 0 \quad (1)$$

$$\frac{\partial V}{\partial t} + U\frac{\partial V}{\partial x} + V\frac{\partial V}{\partial y} + g\frac{\partial \varsigma}{\partial y} + \frac{\tau_y^b}{\rho H} = 0 \quad (2)$$

$$\frac{\partial \varsigma}{\partial t} + \frac{\partial (HU)}{\partial x} + \frac{\partial (HV)}{\partial y} = 0 \quad (3)$$

The variables U and V are the vertically averaged fluid velocity components, and x, y are the cartesian coordinates in the horizontal plane, g in the acceleration of gravity, τ_x^b, τ_y^b are the components of bottom stress coefficients in the x, y directions, ζ is the free surface displacement, $H = h + \zeta$ is the total depth in the basin, ρ is the water density and t is the time.

The variations of pollutant concentrations with space and time are simulated with the vertically averaged, two-dimensional mass balance equation.

$$\frac{\partial(HC)}{\partial t} + \frac{\partial(HUC)}{\partial x} + \frac{\partial(HVC)}{\partial y} - \frac{\partial(HD_x \frac{\partial C}{\partial x})}{\partial x} - \frac{\partial(HD_y \frac{\partial C}{\partial y})}{\partial y} - HS_A = 0 \quad (4)$$

In equation (4), the first term represents the rate of change of the pollutant A, the second and third terms represent the advective fluxes and the fourth and fifth terms represent the dispersive fluxes. S_A represents the sink term associated with pollutant A, D_x and D_y are the dispersion coefficients.

SOLUTION OF THE GOVERNING EQUATIONS

The technique used to solve these equations is the well-known Alternating Direction Implicit finite-difference method (ADI) on a space staggered grid pioneered by Leendertse[1]. The ADI finite difference method used in this work uses fractional (half) and whole time steps for time differencing the momentum, continuity and pollutant transport equations. The basis of the ADI scheme is the same as the classical ADI since the marching in time involves two half time steps, the only difference is the differencing of the momentum and continuity equations. The solution of the shallow water equations coupled with the pollutant transport equation, as described by Cheng and Casulli[2] (1982) is illustrated in Fig. 1.

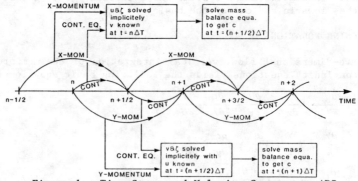

Figure 1. Time Staggered Velocity Components ADI

In the first operation at time level n (going from t to t + 1/2Δt, the x-momentum equation which is time-differenced between t = (n + 1/2)Δt and (n - 1/2)Δt, and the continuity equation which is time-differenced between t = nΔt and (n + 1/2) Δt, are solved first for the water levels ζ and velocities U, at time level n + 1/2. Thus U and ζ are solved implicitly because the matrix equations for U-ζ take a tridiagonal form. These results are then used in the pollutant transport equation to obtain the constituent concentration, $C_i^{n+1/2}$ at time level n + 1/2. The results of this first operation are then used at time level n + 1/2 (going from t + 1/2 Δt to t + Δt) to determine the unknowns in the second half time step. This time the y-momentum equation, which is time differenced between t = (n + 1/2)Δt and (n + 1)Δt, and the continuity equation which is time-differenced between t = (n + 1/2)Δt and (n + 1)Δt, are solved for the water levels ζ and the V velocities, at time level n + 1. This new information is then used in the pollutant transport equation to obtain the constituent concentration, C_i^{n+1}, at time level n+1. This procedure is then repeated for each succeeding full time step.

DIFFERENCING OF THE NON-LINEAR TERMS

When using a central-difference of the non-linear advective term a static instability occurred. A static instability means that the error grows monotonically. On the other hand a dynamic instability creates oscillatory fields and can be removed by decreasing the time step Δt. A static instability cannot be removed by a decrease in Δt, but by using some other finite difference scheme (Roache[3]).

An upwind scheme presented by Smith and Cheng[4] (1987) was substituted for the central difference formulation and is shown below for the non-linear term of x-momentum equation.

$$U\frac{\partial U}{\partial X} = \begin{cases} U_{i+1,j}^{n+1/2} \times \frac{1}{\Delta X}(U_{i+1,j}^{n-1/2} - U_{i-1,j}^{n-1/2}) & U_{i+1,j}^{n-1/2} > 0 \\ \\ U_{i+1,j}^{n+1/2} \times \frac{1}{\Delta X}(U_{i+3,j}^{n-1/2} - U_{i+1,j}^{n-1/2}) & U_{i+1,j}^{n-1/2} < 0 \end{cases} \quad (5)$$

In differencing the mass transport equation, a central difference scheme was first used for the non linear terms. The use of central differences leads to negative concentration and consequently to the nonconservation of pollutant mass. An upwind scheme was adapted to this inflow driven situation and is presented in Equation 6 for the term $\partial(HUC)/\partial X$.

In a space staggered grid, we have on each side of a mesh two velocities. The sign of these velocities determines, by upwind differencing, which cell values of C and U to be used. This scheme is called the second upwind differencing or donor cell

$$\frac{\partial(HUC)}{\partial X} = \begin{cases} \frac{HU_{i-1,j}^{n+1/2}}{\Delta X}\left[C_{i,j}^{n+1/2} - C_{i-2,j}^{n+1/2} \right] & U_{left} > 0 \\ \\ \frac{HU_{i+1,j}^{n+1/2}}{\Delta X}\left[C_{i+2,j}^{n+1/2} - C_{i,j}^{n+1/2} \right] & U_{right} < 0 \end{cases} \quad (6)$$

method. These two conditions for the upwind differencing, of the non-linear terms in the transport equation, are easily incorporated in the algorithm by using the equation below, shown for the x-component.

$$U\frac{\partial C}{\partial X} = \frac{1}{2\Delta X}\left[U_{i+1,j}^{n+1/2}C_{i+2}^{n+1/2} - U_{i+1,j}^{n+1/2}C_i^{n+1/2} + U_{i-1,j}^{n+1/2}C_i^{n+1/2} - U_{i-1,j}^{n+1/2}C_{i-2}^{n+1/2} \right] -$$

$$\frac{1}{2\Delta X}\left[|U|_{i+1}^{n+1/2}C_{i+2}^{n+1/2} - |U|_{i+1}^{n+1/2}C_i^{n+1/2} - |U|_{i-1}^{n+1/2}C_i^{n+1/2} + |U|_{i-1}^{n+1/2}C_{i-2}^{n+1/2} \right] \quad (7)$$

One of the main disadvantages of upwind differencing is the introduction of numerical viscosity as described by Roache[3] (1972). Numerical viscosity cannot be reduced by decreasing the time step but is on the other hand strongly dependent on the mesh size ΔX. Thus it can be reduced by using a finer mesh. In the modeling of small scale basin numerical viscosity was not observed due to the required fine mesh used. In order to give the best approximation of the real situation, the mesh size should be equal or close to the width of the channel inflow, because the driving forces which dicates the hydrodynamic behavior of such systems are the inflows into the basin. On the other hand the effects of these inflow velocities are negligible in large scale basin.

The small mesh size ($\Delta X = \Delta Y = 4m$ for the detention pond modeled) will increase the computer time but will avoid the introduction of numerical viscosity. For one hour of simulation the central processor time is around 140 cps for a Δt of 10 seconds which is eight times the time step required for explicit schemes (working at CFL conditions).

The use of upwind differences drastically improved the performance of the model, and the model performed adequately without numerical instability and the mass of water and pollutant were conserved throughout the simulation.

STUDY AREA AND SAMPLING PROGRAM

In February 1985, the University of Virginia entered into an agreement with the Virginia Department of Conservation and Historic Resources, Division of Soil and Water Conservation, for conducting a demonstration project on selected Best Management Practices (BMP) for nonpoint source pollution control in the Albemarle/Charlottesville area. One of the BMP's tested is an existing wet detention pond. The pond, known as the Four Seasons Detention Pond, has a surface area of 1.67 acres and an average depth of 7.6 feet. The detention pond has two inflow locations because the whole catchment is hydrologically divided into two subcatchments.

Three complete storm events have been sampled at the three locations of the pond to monitor inflow and outflow fluxes of pollutant. Eight pollutants were analyzed, NO_3, ortho-phosphate, ammonia, total kjeldahl nitrogen (TKN), total phosphorus, lead, zinc and total suspended solids. Samples were taken at inflow 1 (FS1), at inflow 2 (FS2) and at the outflow (FS3) by means of automatic samplers.

HYDRODYNAMIC AND POLLUTANT TRANSPORT SIMULATIONS

Using the quantity and quality field data at FS1 and FS2 as input for the developed model, outflow hydrographs and pollutographs at the FS3 inflow are generated. The computed surface elevation at the riser is inputed in a previously calibrated depth-discharge curve to compute the outflow hydrographs.

The quality data consist pollutographs at the inflows (FS1 and FS2) and at the outflow FS3. The samples were not filtered hence the total concentration measured at different times includes the dissolved and particulate part. Nevertheless some of the pollutants are known to be mainly in dissolved form like NO_3, NH_3 and orthophosphate. In order to test the quality part of the model, simulations were made assuming the equalization process. In this work we define the equalization process as the transport and dispersion of pollutant without any sink term. We assume that all the pollutants have the same transport patterns as the water itself.

Using the three storms selected previously, simulations were made for all the pollutants analyzed. The first observation made from the equalization process is that there is a much better agreement between the computed and observed values for NO_3, NH_3 and orthophosphate than for the other constituents. These constituents are know to be mainly in a dissolved form. This is depicted in Figure 2 for NH_3 for the May 23, 1985 event and also in Table 1 which shows the computed and trapping efficiencies for these three constituents.

On the other hand simulation results show larger deviations from the observed values for TSS, TP, and TKN respectively. The larger deviations were as expected for total suspended solids.

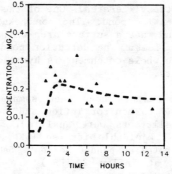

Figure 2. NH_3 - May 23, 1985

TABLE 1 Observed and Computed Removal Efficiencies

	May 23 Obs.	May 23 Comp.	June 10 Obs.	June 10 Comp.	Oct. 20 Obs.	Oct. 20 Comp.
NO3	58.1	59.5	51.2	62.8	52.1	57.3
H3	49.6	32.2	53.1	51.2	51.8	34.3
OP	48.9	53.5	50.2	55.5	--	53.8

CONCLUSIONS

A model based on the shallow water equations and a two-dimensional advection dispersion equation can be applied to analyze the flow and pollutant transport in small scale basin such as detention pond.

REFERENCES

1. Leendertse, J.J. and E.C. Gritton (1971). A Water Quality Simulation Model for Well Mixed Estuaries and Coastal Seas. Vol. II Computation Procedures. The Rand Corporation. Report R-708-NYC.

2. Cheng, R.T. and V. Casulli (1982), On Lagrangian Residual Currents with Applications in South San Francisco Bay, California, Water Resources Research, 18(6), 1652-1662.

3. Roache, P.J. Computational Fluid Dynamics. Hermosa Publishers. Albuquerque, NM.

4. Smith, L.H. and R.T. Cheng (1987), Tidal and Tidally Averaged Circulation Characteristics of Suisun Bay, California, Water Resources Research, 23(1), 143-155.

Solving the Transport Equation using Taylor Series Expansion and Finite Element Method

C.L. Chen
Systech Engineering, Inc., 3744 Mt. Diablo Blvd., Suite 101, Lafayette, CA 94549, USA

ABSTRACT

A numerical scheme is presented to obtain the solution of the depth-averaged convection-diffusion (transport) equation together with the shallow water equations. Galerkin finite element method is applied to the spatial approximation of the transport equation. The direct use of Taylor series expansion to the second-order derivative in the time domain results in a simple forward time-stepping scheme. This scheme introduces balancing tensor diffusivities into the transport equation to compensate for the truncation error due to forward differencing in time, and it precludes spurious oscillations. Computational efficiency can be obtained by lumping the mass matrix and using one-point quadrature. The scheme is conditionally stable and second-order accurate.

INTRODUCTION

The simulation of transport problems in a shallow water basin with a free surface requires the solution of the depth-averaged transport equation. The two-dimensional form of this equation can be written as

$$\partial Q/\partial t + \underline{V} \cdot \nabla Q + \nabla \cdot (HD \cdot \nabla Q)/H - S = 0 \qquad (1)$$

where Q is the depth-averaged transport variable; t is the time; $\underline{V}$ is the velocity vector; ∇ represents the spatial gradient; H is the water depth; D is the eddy diffusivity; S represents the source/sink of the transport variable.

If the velocity and water depth are also to be modeled, the depth-averaged momentum equations and continuity equation, called the shallow water equations, will be solved together.

The shallow water equations are

$$\partial H/\partial t + \nabla \cdot (H\underline{V}) = 0$$

$$\partial \underline{V}/\partial t + \underline{V} \cdot \nabla \underline{V} + g\nabla z + f \times \underline{V} + \tau \underline{V} - \underline{W}/H = 0 \quad (3)$$

Where z is the elevation of the free surface relative to a reference datum; g is the gravity; f is the Coriolis parameter; τ is the bottom stress parameter; W represents atmospheric forcing function.

In finite difference methods, the upwind differencing on the convection term has long been the mainstream algorithm of solving the transport equation. In finite element methods, the weighted upwind approach is the mainstream algorithm. Among the weighted upwind approaches, Brooks and Hughes'[1] streamline upwind/Petrov-Galerkin approach precludes the excessive crosswind diffusion and eliminates artificial diffusion.

Recently, Gresho, et al.[6] introduced a modified Galerkin finite element method (MDGFEM) for solving the Navier-Stokes equations (NS) and advection-diffusion equation (CD). Their scheme includes an interesting feature: the time stepping is simple forward, but a balancing tensor diffusivity (BTD) is introduced into the transport equation to compensate for the error of temporal approximation. The scheme is conditionally stable, second-order accurate, and cost-effective. It does not have the crosswind diffusion, although it does send out wiggle signals for problems too difficult for the selected mesh.

Chen[2] developed a numerical scheme similar to MDGFEM for solving the depth-averaged transport equation. In that the Taylor series expansion is applied to temporal approximation. He also developed a modified leapfrog scheme for solving the primitive shallow water equations. In both schemes the Galerkin finite element method is applied to the spatial approximation. Mass matrices are all lumped for computational efficiency. This paper will discuss the behavior of both numerical schemes.

FINITE ELEMENT SOLUTIONS OF THE TRANSPORT EQUATION

Consider the one-dimensional linearized transport equation given by

$$\partial Q/\partial t + U\partial U/\partial x - D\partial^2 Q/\partial x^2 = 0 \quad (4)$$

where U is the depth-averaged velocity in the x direction; D is the diffusion coefficient. U and D are assumed constants. Applying the Galerkin finite element method to equation (4) and lumping the mass matrix result in the following numerical representation at any interior node i:

$$Q_{i,n+1} = Q_{i,n} - \Delta t U(Q_{i+1,n} - Q_{i-1,n})/2\Delta x + D(Q_{i+1,n} - 2Q_{i,n} + Q_{i-1,n})/\Delta x^2 \quad (5)$$

One Fourier component of the general solution of equation (5) can be represented as

$$\hat{Q} = Q^* \exp(ibt + idx) \qquad (6)$$

The substitution of (6) into (5) yields

$$e = 1 - 2aR - icB \qquad (7)$$

where $e=\exp(ib\Delta t)$, is the complex amplification factor; $c=U\Delta t/\Delta x$, is the Courant number, $a=D\Delta t/\Delta x$, is the diffusion parameter; $B=\sin(d\Delta x)$, and $R=1-\cos(d\Delta x)$.

The stable condition requires that $|e| \leq 1$ for all phase angles. Leonard[7] gave the necessary and sufficient condition for stability in the sense of von Neumann as

$$c^2 \leq 2a \leq 1 \qquad (8)$$

The Taylor series expansion in time for Q yields

$$Q_{n+1} = Q_n - \Delta t U \partial Q/\partial x + (D\Delta t + \frac{1}{2}\Delta t^2 U^2) \partial^2 Q/\partial x^2 \qquad (9)$$

In an analogue of above Fourier analysis, the stability constraint of the Taylor series expansion scheme is

$$c^2 \leq 2a + c^2 \leq 1 \qquad (10)$$

the left inequality gives $a \geq 0$, and the right one gives

$$c \leq (\sqrt{1+p^2} - 1)/p \qquad (11)$$

where $p=c/a$, is the grid (cell) Peclet number. Following Pinder and Gray[10], the number of time steps required for the analytic wave to propagate a wavelength, N, can be defined as

$$N = L/(\Delta x c) \qquad (12)$$

where L is one wavelength of the analytic wave. The ratio of the computed to analytic amplitude is given by

$$r = [|e|/\exp(-4\pi^2 a(\Delta x/L)^2)] \qquad (13)$$

The phase lag, A, after one complete wavelength is

$$A = -kN - 2\pi \qquad (14)$$

where k is the phase angle of the numerical wave.

Pinder and Gray[10] examined several numerical schemes for the solution of the transport equation. Their work is very instructive. However, in their stability analysis (pp158-159), the convection and diffusion parameters do not meet the stability criteria of explicit schemes. That leads them to a statement: the explicit scheme is found to be unstable for the given value of c and a and any reasonable spatial mesh. In contrast, the remaining two (implicit) schemes are stable for arbitrary mesh spacing.

In this paper, I will examine the performance of two explicit schemes: FTCS (forward time, centered space) and FTCSBTD (FTCS + BTD), and one implicit scheme: CN (Crank-Nicholson scheme, without lumping). The eigenvalues, amplitude ratios, and phase lags of these three schemes are plotted in Fig. 1. Note that the transport parameters, c and a all meet the stability criteria (8) and (10). It is seen that the two explicit schemes are conditionally stable, and have better performance when the transport is convection dominant.

If the velocity U is not a constant, the finite Taylor series expansion for Q is

$$Q_{n+1} = Q_n + \frac{\partial Q}{\partial x}(-U\Delta t - \frac{\Delta t^2}{2}\frac{\partial U}{\partial t} + \frac{\Delta t^2}{2}U\frac{\partial U}{\partial x} - D\frac{\Delta t^2}{2}\frac{\partial U}{\partial x^2}) + \frac{\partial^2 Q}{\partial x^2}(D\Delta t + \frac{\Delta t^2}{2}U^2) \quad (15)$$

the second term in the right hand side of equation (15) contributes to the phase speed of numerical waves. The effect of each of that term on phase speed has not been quantified.

According to Gresho et al.[6] the local error of the FTCSBTD scheme is $O(\Delta t(\Delta x^2 + D t + \Delta t^2)]$, If $D=0$, global one is $O(\Delta x^2 + \Delta t)$. If $D \neq 0$ and fixed, the global error is also $O(\Delta x^2 + \Delta t^2)$ when the scheme is stable. So this scheme is considered as second-order accurate in space and time. If the velocity derivatives in equation (15) are retained, the velocity terms should be at least second-order accurate such that they do not deteriorate the accuracy of the scheme. This requirement can be met by selecting an appropriate algorithm for solving the shallow water equations.

FINITE ELEMENT SOLUTION OF THE SHALLOW WATER EQUATIONS

Gray[3] reviewed various numerical schemes for the finite element solutions of the shallow water equations. They also presented a wave equation model. In the wave equation model the continuity equation is transformed into wave form such that the shortest numerical wave ($2\Delta x$ in 1D problem) can be suppressed with central differencing scheme. Chen[2] presented a modified leapfrog scheme for solving the primitive shallow water equations. The scheme is less effective to suppress short waves than the wave equation model. However, it is about 30 percent more efficient than the wave equation model in terms of CPU time (based on the computer code presented by Lynch and Gray[6]). Both the primitive model and the wave equation model are conditionally stable and second-order accurate. They are suitable for incorporation with Taylor series expansion approach in solving the transport equation.

One important issue in solving the shallow water equations is the treatment of boundary conditions. Lynch[9] pointed out that the elimination of the continuity or wave equation along boundaries where elevation is specified produces a mass imbalance which is first order in the mesh spacing; while retention of these equations enforces the global balance. He presented a simple way to correct the mass imbalance in which the difference wave equations or continuity equations are all retained. Once the water depths have all been computed, the equations on boundaries yield the fluxes, which in turn serve as boundary conditions for the velocity calculation. Similar ideas and procedures have been introduced by Gresho et al.[5] in solving NS equations.

The other issue is regarding the determination of the normal direction along the boundary of finite element grid. Various algorithms have been presented in literatures. The algorithm presented by Gray[4] is employed in Chen's model.

REFERENCES:

1. Brook A.N. and Hughes T.J.R. (1982), Streamline Upwind /Petrov-Galerkin Formulations for Convection Dominated Flows with Particular Emphasis on the Incompressible NS Equations, Computer Methods in Applied Mechanics and Eng. Vol.32, pp.199-259.
2. Chen C.L. (1985), Simulation of Hydrodynamics and Water Quality in a Well-Mixed Estuary by Using Finite Elemenet Methods Dr.Eng. Thesis, University of Wisconsin-Milwaukee.
3. Gray W.G. (1982), Some Inadequacies of Finite-Element Models as Simulators of Two-Dimensional Circulation, Advance Water Resources, Vol.5, pp.171-177.
4. Gray W.G. (1984), On Normal Flow Boundary Conditions in Finite Element Codes for Two-Dimensional Shallow Water Flow, Int. J. Num. Methods in Fluids, Vol.4, pp.99-104.
5. Gresho P. Lee R. and Sani R. (1980), On the Time-Dependent Solution of the Incompressible NS Equations in Two and Three Dimensions, Recent Advances in Numerical Methods in Fluids, Vol.1, Pineridge Press, Swansea, U.K., pp.1-27.
6. Gresho P. Chan S. Lee R. and Upson C. (1984), A Modified Finite Element Method for Solving the Time-Dependent, Incompressible NS Equations, Part 1: Theory, International J. For Numerical Methods In Fluids, Vol.4, pp.557-598.
7. Leonard B.P. (1980), Note on the von Neumann Stability of the Explicit FTCS Convective Diffusion Equation, Applied Mathematic Modelling, Vol.4, pp.401-402.
8. Lynch D.R. and Gray W.G. (1980), An Explicit Model for Two-Dimensional Tidal Circulation Using Triangular Finite Elements: WAVETL User's Manual, NTIS 5, PB 80-226046.
9. Lynch D.R. (1984), Mass Balance in Shallow Water Simulations, 5th Int. Sym. on FE Problems, University of Texas at Austin.
10. Pinder G.F. and Gray W.G. (1977), Finite Element Simulation in Surface and Subsurface Hydrology, Academic Press, New York.

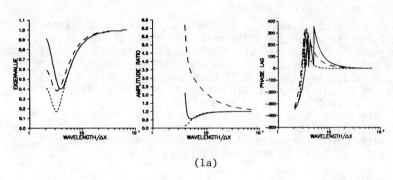

(1a)

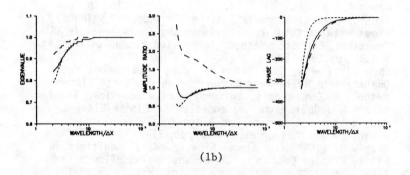

(1b)

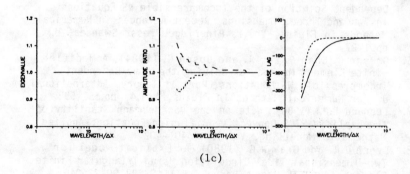

(1c)

Fig. 1. The behavior of three numerical schemes in solving the finite element discretization of the transport equation: FTCSBTD (———), FTCS (———), CN (----); (1a) c=0.4, a=0.4, p=1; (1b) c=0.2, a=0.02, p=10; (1c) c=0.01, a=0.0001, p=100.

Cooling-Induced Natural Convection in a Triangular Enclosure as a Model for Littoral Circulation

G.M. Horsch and H.G. Stefan
St. Anthony Falls Hydraulic Laboratory, CME Department, University of Minnesota, Minneapolis, Minnesota 55414, USA

1. INTRODUCTION

In most studies of natural convection, known temperatures are imposed at fixed boundaries (walls). In many environmental water resources applications, however, the flow is excited by heat transfer through the surface of the water body. An example is convective circulation in sidearms of cooling lakes (e.g. Brocard and Harleman[2]). The incentive for the computations which are presented herein is the need to estimate the flow that develops near shore during cooling of lakes, ponds, bays, and similar water bodies. The convective flow is driven by a horizontal temperature (density) gradient which forms because regions of progressively larger depth are subjected to approximately the same rate of surface cooling. The exchange between littoral and open waters induced by this flow is of importance, e.g. to the understanding of the removal of dissolved substances in the littoral waters by aquatic plant communities. In the convective littoral flow the horizontal temperature gradient develops naturally in response to heat transfer in the vertical. Because of this extra degree of freedom, both the velocity scale and the horizontal temperature gradient are dependent on the imposed parameters of the problem: the surface cooling rate, the geometry of the domain, and the fluid properties of water. The dimensionless independent variables of the problem are the Rayleigh number (or Grashof number), the Prandtl number and the bottom slope. Intermittent convective cells similar to Bénard cells can be present during the development of the convective circulation, whereas flows with imposed temperature gradient, may have no cells at all. (Poulikakos and Bejan[1]).

2. GOVERNING EQUATIONS

The simulation model is formulated in terms of the equations of continuity, momentum and energy, all expressed in polar coordinates, in which the domain can be fitted naturally.

The momentum equations:

$$\frac{\partial u}{\partial t} + v\frac{\partial u}{\partial r} + \frac{u}{r}\frac{\partial u}{\partial \theta} + \frac{uv}{r} = -\frac{1}{r}\frac{\partial P}{\partial \theta} + \nu\left[\frac{1}{r}\frac{\partial}{\partial r}\left(r\frac{\partial u}{\partial r}\right)\right.$$
$$\left. + \frac{1}{r^2}\frac{\partial^2 u}{\partial \theta^2}\right] + \nu\left[-\frac{u}{r^2} + \frac{2}{r^2}\frac{\partial v}{\partial \theta}\right] + g\beta \cos\theta(T_o - T) \quad (1)$$

$$\frac{\partial v}{\partial t} + u\frac{\partial v}{\partial r} + \frac{u}{r}\frac{\partial v}{\partial \theta} - \frac{u^2}{r} = -\frac{\partial P}{\partial r} + \nu\left[\frac{1}{r}\frac{\partial}{\partial r}\left(r\frac{\partial v}{\partial r}\right)\right.$$
$$\left. + \frac{1}{r^2}\frac{\partial^2 v}{\partial \theta^2}\right] - \nu\left[\frac{v}{r^2} + \frac{2}{r^2}\frac{\partial u}{\partial \theta}\right] + g\beta \sin\theta(T_o - T) \quad (2)$$

Continuity equation:

$$\frac{1}{r}\frac{\partial}{\partial r}(rv) + \frac{1}{r}\frac{\partial u}{\partial \theta} = 0 \quad (3)$$

Energy equation:

$$\frac{\partial T}{\partial t} + v\frac{\partial T}{\partial r} + \frac{u}{r}\frac{\partial T}{\partial \theta} = \alpha\left[\frac{1}{r}\frac{\partial}{\partial r}\left(r\frac{\partial T}{\partial r}\right) + \frac{1}{r^2}\frac{\partial^2 T}{\partial \theta^2}\right] \quad (4)$$

where $u=u(r,\theta)$ = tangential velocity, $v=v(r,\theta)$ = radial velocity, T_o = initial temperature, ν = kinematic viscosity, β = thermal expansivity, g = acceleration of gravity, α = thermal diffusivity. In these equations the Boussinesq approximation has been used, and in the body force the density has been expressed as a linear function of the temperature. This formulation of the density has implications for the occurrence of a quasi-steady state as discussed below.

The boundary conditions are shown in Figure 1, where k = conductivity, H = imposed surface heat loss; and R, θ_T specify the size of the domain.

The detailed scaling of natural convection problems by Patterson and Imberger[10] and Bejan[1] can be used to select the appropriate scales in order to make the equations nondimensional. Ostrach[7] in reviewing low aspect ratio rectangular enclosures with differentially heated endwalls, noted a great variety of scales. A source of difficulty in the choice of the approximate scales in natural convection problems is the inability to predict a priori the flow pattern. This difficulty is aggravated when the horizontal temperature gradient which drives the flow is not imposed, but rather develops naturally.

The scales used in the present study are:

Length scale $\quad h_s = R \sin \theta_T \quad$ (5)

Velocity scale $\quad v_s = Gr^{1/2} \nu/h_s = (g\beta H/k)^{1/2} h_s \quad$ (6)

Time scale $\quad t_s = Gr^{-1/2} h_s^2/\nu = (k/g\beta H)^{1/2} \quad$ (7)

Temperature scale $\quad \Delta T_s = H h_s/k \quad$ (8)

The nondimensional independent parameters of the problem are: Grashof number, $Gr = g\beta H h_s^4 \nu^{-2} k^{-1}$; Prandtl number, $Pr = \nu/\alpha$; and slope of the enclosure, $S = \tan \theta_T$. The Raleigh number, $Ra = Gr\, Pr$, can be used instead of Gr. If the dependence on the Prandtl number is included, the proper velocity scale is $v_s = Gr^{1/2} Pr^{-1/2} \nu/h_s$. Using this scale, the results show that the velocity maxima are of order one.

3. QUASI-STEADY STATE PROBLEM

Because the domain is cooled from the surface while the bottom and the side are insulated, the water temperatures will continue to drop and no steady state in terms of temperatures can be achieved. Because the fluid properties (ν, α, k, β) are represented as constants and the density is expressed as a linear function of temperature, the velocities and the temperature gradients (instead of the temperature) become steady. This is referred to as a quasi-steady state at which all temperatures drop at the same rate:

$$\frac{dT}{dt} = - \frac{2H}{\rho c R \theta_T} \quad (9)$$

where c = specific heat. Among the quantities to be extracted from the numerical solutions are the strength of the convective circulation Q and the total radial heat transfer $\overline{H}$. The latter is defined as:

$H(r) = \int_0^{\theta_T} (\rho c T - k \frac{\partial T}{\partial r}) r\, d\theta$ and made dimensionless as $Nu = \overline{H}/Hh$. Nu is a form of Nusselt number; a simple heat balance shows that at steady state: $Nu(r') = r'(1 - r' \sin \theta_T)$, where $r' = r/h_s$ (Fig. 2).

4. THE NUMERICAL METHOD

The conservation equations were discretized in primitive variables and solved numerically using the SIMPLE algorithm (Patankar[9]). The main features of the method include preservation of the conservation properties of the equations, the power-law interpolation for the combined convection-diffusion fluxes, and fully implicit time-marching. This method has been

implemented in the code of Patankar[9]. The code was modified to
allow calculation of unsteady flow. Minor modifications were
implemented to make possible vectorization of the code on the
Cray 2, although the Gauss elimination which cannot vectorize,
was left intact. The resulting code ran twice as fast as the
original one.

The grid for the calculation of natural convection should be
fine enough for all pertinent lengthscales to be discretized,
and also ensure that false diffusion does not become excessive.
Important lengthscales of the problem have been identified
elsewhere (Horsch and Stefan[6]). For discretization of the surface thermal boundary layer, the grid is made progressively
finer near the surface. In the radial direction the grid is
uniform.

Although the power-law scheme is a very close, efficient
approximation to the exact solution of the one-dimensional
convection-diffusion problem, false diffusion can still arise
in a multi-dimensional flow (De Vahl Davies and Mallinson[4];
Pantankar[8]). A remedy is to align the gridlines, whenever
possible, with the flow and make the grid fine enough. To the
extent possible, the polar coordinates achieve the first
requirement (the underflow and the return flow move in the
radial direction). The formation, however, of intermittent
Bénard-like cells (thermals) at high Rayleigh numbers can cause
severe false diffusion at their ephemeral boundaries. False
diffusion was estimated using the formula of De Vahl Davis and
Mallinson[4]. For the $Ra = 10^4$ solution false diffusion was less
than 4 percent of the actual viscosity. For the 10^6 solution
false diffusion was up to 40 percent of the physical viscosity
but only at the turning points where the flow intersects gridlines at an angle, and much smaller elsewhere. This run took 3
hours on the Cray 2 (5×10^{-4} CPU secs/iter/timestep/grid point).

5. RESULTS

Numerical solutions at $Ra = 10^4$, 10^6, 10^8 and $Pr = 7$ (corresponding to water at 20°C), with $S = 0.2$ were extracted. The
initial condition is an isothermal, quiescent (zero velocity)
body of water. The isotherms must curve near the pointed end
of the domain to meet the adiabatic boundary perpendicularly.
The local temperature gradient thus created induces a cell at
the corner (Figure 3a). In response, a weaker, counterrotating
cell slowly transports cool surface water towards the deep end,
where the thermal boundary layer thickness, and eventually a
cell of finite strength forms (Fig. 3a). This may include an
aspect ratio ($1/\sin\theta_T$) influence in the evolution of the flow
(Busse[3]). The flow initially develops similarly at the two
ends for all three Ra numbers, but the subsequent stages (Figs.
3b and 3c) are heavily dependent upon Ra. (The streamfunction
ψ is made dimensionless using $\psi_s = v_s h_s$.) All numerical values
shown are dimensionless.

Only three instances from the evolution of the Ra = 10^6 simulation are shown in Fig. 3. It can be seen that many cells develop in the transition period. The steady state consists of one main cell with a smaller cell near the surface of the deep-end side. The evolution of velocities in time and at several locations is shown in Fig. 4. At dimensionless times on the order of 400 to 500, temperatures have reached a steady state for practical purposes while velocities are still in an oscillatory mode which is later damped out. In applications where the littoral zone is vegetated an apparent viscosity which accounts for the added resistance offered to the flow by the plants should be used. Then Ra values of the order 10^6 become relevant to the analysis of natural systems.

The evolution of the Ra = 10^8 run is considerably more complicated. It is possible that at Ra = 10^8 a steady state is reached only in a time-averaged sense.

6. ACKNOWLEDGEMENTS

This work was supported by the Legislative Commission on Minnesota Resources, St. Paul, MN, and by the Supercomputer Institute of the University of Minnesota, Minneapolis, MN, and by an Alexander S. Onassis Foundation Scholarship.

REFERENCES

1. Bejan A. (1984) Convection Heat Transfer, John Wiley.
2. Brocard, D. N. and Harleman, D.R.F. (1980) Two-Layer Model for Shallow Horizontal Convective Circulation, J. Fluid Mechanics, Vol. 100:129.
3. Busse F. H. (1981) Transition to Turbulence in Rayleigh-Benard Convection, in Hydrodynamic Instabilties and the Transition to Turbulence, Springer-Verlag.
4. De Vahl Davies G. and Mallinson, G. D. (1976) An Evaluation of Upwind and Central Difference Approximations by a Study of Recirculating Flow, in Computers and Fluids, Vol. 4:29-43.
6. Horsch, G. M. and Stefan, H. G. (1986) Convective Currents on Sloping Boundaries, Proceedings, Int'l. Symposium on Buoyant Flows, Frame Publishing Company.
7. Ostrach. S. (1982) Natural Heat Transfer in Cavities and Cells, in Heat Transfer, Proceedings, Seventh Int'l. Heat Transfer Conference, Hemisphere Publ.
8. Patankar, S. V.(1980) Numerical Heat Transfer, Hemisphere.
9. Patankar S. V. (1982) A General Purpose Computer Program for Two-Dimensional Elliptic Situations, Mechanical Engineering Dept., Univ. of Minnesota, Minneapolis.
10. Patterson, J. and Imberger, J. (1980) Unsteady Natural Convection in a Rectangular Cavity, J. Fluid Mechanics, Vol. 100:65-86.
11. Poulikakos, D. and Bejan, A. (1983) The Fluid Mechanics of an Attic Space, J. Fluid Mechanics, Vol. 131:251-269.

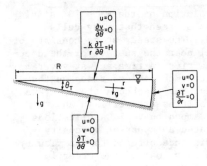

Fig. 1. Computation domain and boundary conditions.

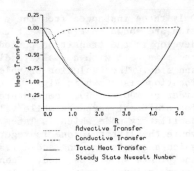

Fig. 2. Dimensionless horizontal heat transfer Nu at $Ra = 10^6$, $S = 0.2$, and $Pr = 7$.

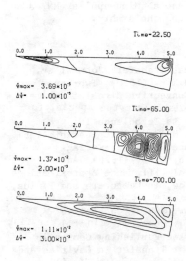

Fig. 3. Streamlines at three times for $Ra = 10^6$, $S = 0.2$, and $Pr = 7$.

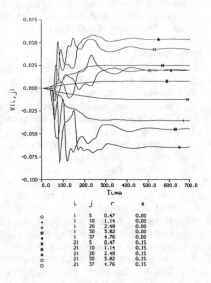

Fig. 4. Evolution of radial velocities at several locations in the domain for $Ra = 10^6$, $S = 0.2$, and $Pr = 7$.

System Identification and Simulation of Chesapeake Bay and Delaware Bay Canal Hydraulic Behavior
B.B. Hsieh
Maryland Department of Natural Resources, Annapolis, MD 21401, USA

ABSTRACT

The Kalman filter combined with one-dimensional shallow water wave equation is qpplied to describe the tidal hydraulic behavior of the Chesapeake Bay and Delaware Bay canal (C&D canal). This model joins two separate estuarine systems and one branched in the Elk River area. The field measurements from surface water elevations and tidal currents are used to calibrate this stochastic-deterministic model. The agreement between field observations and filtered estimate is better than field and the purely numerical solution. The model performance is evaluated by the statistical methods. The considerations for extending the modeling process to the system noise and the measurement noise provide much better predictions and can be used to simulate the special forcing events.

INTRODUCTION

The increase of shipping activities through the waterway and the concern of baywide resources, revealed the important issue of the dynamic of the C&D canal. Particularly, the correct estimation of tidal flux from this intersystem could address much more transport behavior of the distribution of a contaminant or a fish egg and larvae. Hydrodynamic equations have been successfully used by many researchers(Gardner and Pritchard (1974); Hunter(1975); Rives and Pritchard (1978), among others) for constructing a deterministic numerical solution of the C&D canal. However, a substantial error could be produced by the purely stochastic meteorological forcing. This contrast to the purely stochastic model is incapable of providing predictions over the spatially-varying distribution of long wave propagation. Budgell(1981) developed a stochastic-deterministic model for applying to the Great Bay estuary. This type of approach considers the system noise, such as the specified cross-sectional area, flow field, boundary conditions, and the measurement noise is mainly caused by instrument sensitivity. Budgell(1982) indicated without corrections, this modeling error could be propagated through time and space by the deterministic model. A means of computing the optimal correction to be applied to the computed values at each time step is through the use of the Kalman Filter.

In this paper, the stochastic filter derived from the Kalman filter is constructed around the one-dimensional equations governing motion in an open channel. The boundary conditions are using the surface water elevation at Reedy Point at its eastern end and at Old Town Point as its western end. One tidal measurement station at Chesapeake City and the current meters at Summit Bridge are used for model calibration and filtered estimate calculation. The application of this model can be extended to design future monitoring programs and to simulate the storm event which may cause abnormal tidal fluctuations.

MODEL DESCRIPTION

The one-dimensional hydrodynamic equations(the continuity equation and momentum equation) combined with a set of equations defining flow conditions at junctions and boundaries constitute the deterministic component of the model. The matrix form may be expressed as

$$A(n,n+1) X(n+1) = B(n,n+1)X(n) + G(n,n+1) U(n+1) \quad (1)$$

where $A(n,n+1)$ and $B(n,n+1)$ are 2Nx2N coefficients matrices which depend on the schemes and N grid points.

$X^T(n+1) = (u_1, \eta_1, \ldots, u_n, \eta_n)$ 2Nx1 velocity and surface elevation vectors at time level (n+1).

$X(n)$ = 2Nx1 velocity and surface elevation vector at time level n.

$U(n+1)$ = vector containing boundary conditions at time level (n+1).

$G(n,n+1)$ = Nxr matrix which specifies which equations are r boundary conditions.

Equation (1) can be expressed in an alternative form using the inverse of $A(n,n+1)$.

$$X(n+1) = \Phi(n,n+1) X(n) + G^*(n,n+1) U(n+1) \quad (2)$$

where $\Phi(n,n+1) = A^{-1}(n,n+1) B(n,n+1)$
$G^*(n,n+1) = A^{-1}(n,n+1) G(n,n+1)$

Jawinski(1970) developed the algorithm of the discrete system for computing an optimal filter based on observations. The prediction step consists of

$$X(n+1/n) = \Phi(n,n+1) X(n/n) + G^*(n,n+1) U(n+1) \quad (3)$$

$$P(n+1/n) = \Phi(n,n+1) P(n/n) \Phi^T(n,n+1) + \tau Q(n+1) \tau^{*T} \quad (4)$$

The measurement update step consist of

$$X(n+1/n+1) = X(n+1/n) + K(n+1)(Z(n+1) - H X(n+1/n)) \quad (5)$$

$$P(n+1/n+1) = P(n+1/n) - K(n+1) H P(n+1/n) \qquad (6)$$

where K(n+1)   = the Kalman Gain
       Q(n+1)   = system noise covariance
       τ = transition matrix of the system noise
       R(n+1)   = measurement noise covariance
       Z(n+1)   = observations
       H   = measurement transition matrix
       X(n+1/n) = one-step ahead prediction vector
       P(n+1/n) = the error covariance of X(n+1/n)
       X(n+1/n+1) = the filtered estimate of X(n+1)
       P(n+1/n+1) = the error covariance of X(n+1/n+1)

Equation (3) to (6) constitute the discrete form of the Kalman filter for specified system noise covariance and measurement noise covariance. This part is linked with equation (1) and (2) for solving the entire canal system by each time step.(Figure 1)

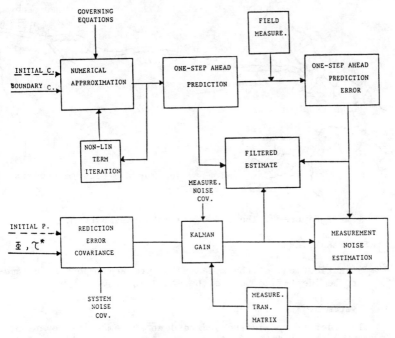

Figure 1. Block diagram for a stochastic-deterministic model.

The C&D CANAL STUDY

In order to calculate the tidal elevation and current velocity from the hydrodynamic equations, this system is divided into a number of river reach over a branched estuarine waterway.(Figure 2) The upper portion of the Elk River is considered as the extending tidal behavior area. After the cubic spline interpolation process and the estimation of current velocity from 6 mooring

meters at three levels, a data set with 3 tidal heights, one
freshwater inflow and cross-sectional mean velocity is created.
This data file covers the period of June 14, 1984 to July 5, 1984
with half-hourly basis.

Four points implicit finite-difference scheme with weighting
factor equal to 0.55 is used to solve the matrix system. Twenty
pairs of continuity equations and momentum equations plus three
boundary conditions and three junction equations form a 46x46
matrix system. The instability for inverse this matrix is usually
obtained due to band width of matrix is produced. The technique
for decomposing a matrix into upper triangle and lower triangle
can avoid direct inverse computation. The non-linearity of the
momentum equation is solved by the Newton-Raphson method. (Mickle
and Sze, 1972)

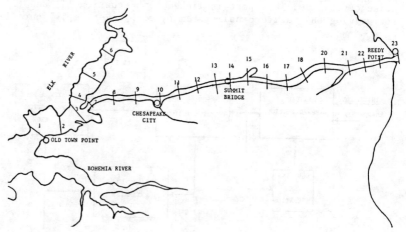

Figure 2. The C&D Canal Estuarine System.

SYSTEM PARAMETERS

The specifications of the model structure are important parameters to be identified before the model is run.

(1) system noise covariance

In order to obtain the independent system noise covariance
the data set is divided into two data files. First set(10
tidal cycles) is used to generate the system noise covariance while the second set is used to calibrate the model.
The ARIMA process is used to generate the white noise from
the field data for both variables. It is assumed that the
system noise covariance for current velocity remains the same
in all grid systems The system noise covariance for tidal
heights will remain the constant for some certain grid points
only.
$$Q_{1-6} = 0.0011 \text{ m}^2 \qquad Q_{7-14} = 0.0038 \text{ m}^2$$

$Q_{15-23} = 0.0036 \text{ m}^2$ $Q_u = 0.0028 \text{ m}^2/\text{sec}^2$

(2) measurement noise covariance
$R = 0.000225 \text{ m}^2$ $R_u = 0.0001 \text{ m}^2/\text{sec}^2$

(3) friction factor (initial) = 0.020

(4) initial error covariance $P(n/n)_\eta = 0.09$ m $P(n/n)_u = 0.09$ m

(5) initial conditions $\eta(x,0) = 0.0$ m $U(x,0) = 0.0$ m/sec

MODEL CALIBRATION AND EVALUATION

The stochastic component of the model cannot be introduced until the deterministic component along with friction factor are determined. It is more practical to use several tidal cycles (3-5) data for calibrating the friction factor instead of using the entire file to do this adjustment. The best solution of friction factor is 0.021. The criterion of this calibration is to select the minimum sum of square error between the observations and model computed values.

The filtered estimate of surface elevations and current velocities are calculated by the filtering process. The comparison from hour 200 to 245 between the observed values and filtered prediction are shown by Figure 3(a) and 3(b). The improvements are made by this calculation. The more significant advantages of the filtered estimate for current velocities is attributed to the noisy system of the velocity variables. An even better performance could be made if the data set is selected by the winter month.

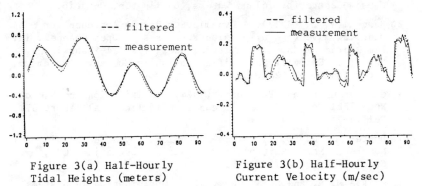

Figure 3(a) Half-Hourly Tidal Heights (meters)

Figure 3(b) Half-Hourly Current Velocity (m/sec)

The performance of the model is evaluated by the statistical methods. The harmonic analysis is used to examine the consistency of tidal constitutes between field measurements and the numerical solution. The amplitude and phase angle for major components of diurnal and semi-diurnal frequency bands are calculated. It is found that the dominant components, such as M2, S2, and N2, do not vary too much (5-10 percent). The F-test is used to search how the stochastic-deterministic approach is superior to purely numerical solution. Both tidal elevations (at Chesapeake City) and tidal currents (at Summit Bridge) show their significance at 0.01 level. (Table 1)

STATION	DEGREE OF FREEDOM	STO-DETER MODEL	NUMERICAL MODEL	F-VALUES
CHESAPEAKE CITY	558	0.01589	0.06938	4.36 **
SUMMIT BRIDGE	558	0.00419	0.000938	2.24 **

Table 1. Estimated model prediction error covariance.

CONCLUSIONS

The one-dimensional hydrodynamic equations combined with the Kalman filter is used to modeling the tidal hydraulic behavior of the C&D canal. The filtered estimate from the computation algorithms for the solution is better than the numerical approximation only. The extension work involved more sampling stations for both surface water elevations and current velocities and can identify a better picture of the C&D canal waterway.

REFERENCES

1. Budgell W.P. (1981), A Stochastic-Deterministic Model for Estimating Tides in Branched Estuaries, Manuscript Report Series No. 10. Ocean Science and Surveys Central Region, Canada Centre for Inland Waters, Burlington, Ontario.

2. Budgell W. P. (1982), A Dynamic-Stochastic Approach for Modeling Advection-Dispersion Processes in Open Channel. In: Time Series Methods in Hydrosciences, (Ed. El-Shaarwi and Esberty), Elsevier Scientific Publishing Company, Amsterdam, Netherlands, pp. 244-263.

3. Gardner G. B. and Pritchard (1974), Verification and Use of a Numerical Model of the C&D canal, CBI, Technical Report 87., Ref 74-7.

4. Hunter, J. R. (1975), A One-Dimensional Dynamic and Kinematic Numerical Model Suitable for Canals and Estuaries, CBI, Special Report 47, Ref. 75-10.

5. Mickle M. H. and Sze T. W. (1972), Optimization in Systems Engineering, Intext Educational Publishers, Scranton, Pennsylvania.

6. Rives, S. R. and Pritchard D. W. (1978), Adaptation of J. R. Hunters' One-Dimensional Model to the C&D Canal System, CBI, Special Report 66, Ref 78-6.

A Layered Wave Equation Model for Thermally Stratified Flow
J.P. Laible
Department of Civil and Mechanical Engineering, The University of Vermont, Burlington, Vermont 05405, USA

1. Introduction. The dynamic behavior of stratified bodies of water is dramatically different from that of homogeneous bodies. It is well known that large amplitude internal waves exist in lakes in temperate climates when thermal stratification develops in the summer months. This motion can induce substantial flows under the right wind conditions and hence can play a significant role in the transport process in the main body of a lake. It can also induce exchange processes with the adjacent coves and bays.

Internal waves have periods that are often orders of magnitude larger than the gravity waves based on the celerity of the same body under homogeneous conditions. The dynamic response of a thermally stratified body of water consists of the superposition of the homogeneous type waves and the waves arising from stratification. As will be demonstrated, both waves are important to the solution and hence the solution techniques must not mask or inadvertently filter out what might otherwise appear to be troublesome higher frequency components of the response. Because of this situation it is imperative that the solution technique be as noise free as possible and not require excessive damping to control stability.

The work of *Lynch* and *Gray*[1] ,*Kinnmark* and *Gray*[2] , *Werner* and *Lynch*[3,4] , *Laible*[5,6,7] and *Gray* and *Kinnmark*[8] , thoroughly document the virtues of several variations of the wave equation form of the differential equations for homogeneous bodies of water. The ability of these models to control potentially spurious $2\Delta X$ waves and allow for a free choice of friction parameters, trumpets the prospect of success at simulating the dynamics of stratified flow using a wave equation approach. On that note the extension of the wave model to a two layered system was initiated. In what follows, the form of the general equations will be described, and a special (X-Z problem) case will be detailed. The resulting model was applied to Lake Champlain, Vermont in order to predict the motion of the thermocline. The solution of the resulting matrix equations was obtained by both direct numerical integration and *modal* analysis. A successful simulation of actual field data encourages continued development.

2. The Governing Equations. The fundamental equations of conservation of mass and momentum for horizontal flow are symbolized as:

$$L = \frac{\partial H}{\partial t} + \vec{\nabla} \cdot (H\vec{V}) = 0 \qquad (2.1)$$

$$\vec{M}_c = \frac{\partial (H\vec{V})}{\partial t} + \vec{\nabla}(H\vec{V}\vec{V}) + \vec{f} \times H\vec{V} + \frac{1}{\rho} H\vec{\nabla} P + \vec{\tau}_b - \vec{\tau}_t = 0 \qquad (2.2)$$

$$\vec{m} = \frac{\partial \vec{v}}{\partial t} + \vec{\nabla}\vec{v} \cdot \vec{v} + \vec{f} \times \vec{v} + \frac{1}{\rho}\vec{\nabla} P - \frac{\partial}{\partial z}\left(\epsilon \frac{\partial \vec{v}}{\partial z}\right) = 0 \qquad (2.3)$$

Equation 2.1 is the vertically integrated continuity equation where : H is the total depth (thickness of a layer in the stratified model); $\vec{V}$ is the two dimensional horizontally averaged velocity vector $\vec{V} = U(x,y,t)\vec{i} + V(x,y,t)\vec{j}$; $\vec{\nabla} = \frac{\partial}{\partial x}\vec{i} + \frac{\partial}{\partial y}\vec{j}$; $\vec{i}$ and $\vec{j}$ are unit vectors in the east and north direction; $\vec{f}$ is the Coriolis vector and t=time. *Equation 2.2* is the vertically integrated conservative momentum equation where : ρ = mass density ; P is pressure and τ_b , τ_t are the stresses at the bottom and top respectively. *Equation 2.3* is also a two dimensional momentum equation but retains the z dependence of the flow where : $\vec{v} = u(x,y,z,t)\vec{i} + v(x,y,z,t)\vec{j}$; ϵ is the vertical eddy viscosity and z is the depth coordinate (positive *downward*) from the mean water level at the surface.

The *generalized wave equation* may be formed from the vertically integrated equations 2.1 and 2.2 as:

$$W = \frac{\partial L}{\partial t} - \vec{\nabla} \cdot \vec{M}_c + GL \qquad (2.4)$$

where G is a stationary constant. Substitution of equations 2.1 and 2.2 yields the wave equation as given by *Gray* and *Kinnmark*[8]. Here we follow the same lines of reasoning for each of two layers with density assumed constant in each layer. In order to resolve the vertical structure of the flow in each layer however, equation 2.3 is used in lieu of equation 2.2 for computing the velocities. The equations for solution are therefore:

$$\frac{\partial^2 H_n}{\partial t^2} + G\frac{\partial H_n}{\partial t} - \vec{\nabla} \cdot (\vec{M}_n + \vec{\nabla} \cdot (H_n \vec{V}_n)\vec{V}_n) + G\vec{\nabla} \cdot (H_n \vec{V}_n) = 0 \qquad (2.5)$$

$$\frac{\partial \vec{v}_n}{\partial t} + \vec{\nabla}\vec{v}_n \cdot \vec{v}_n + \vec{f} \times \vec{v}_n + \frac{1}{\rho}\vec{\nabla} P_n - \frac{\partial}{\partial z}\left(\epsilon_n \frac{\partial \vec{v}_n}{\partial z}\right) = 0 \qquad (2.6)$$

where $n = 1, 2$ for the upper and lower layers respectively. The upper layer of a stratified lake is taken to extend through the epilimnion to the thermocline, located in the metalimnion. The lower layer extends to the bottom from the thermocline. The total depths may be expressed in terms of the initial thicknesses h_1, h_2 and the change in the surface elevations ς_1 and ς_2, each of which are measured as positive when downward from the initial mean water level and the initial datum at depth h_1:

$$H_1 = h_1 + \varsigma_2 - \varsigma_1 \qquad H_2 = h_2 - \varsigma_2 \qquad (2.7)$$

The pressure in each layer reflects the difference in the layer densities:

$$P_1 = \rho_1 g(z - \varsigma_1) \qquad P_2 = \rho_1 g(h_1 - \varsigma_1 + \varsigma_2) + \rho_2 g(z - h_1 - \varsigma_2) \qquad (2.8)$$

where g is gravity. The pressure gradients $\vec{\nabla} P_n$ therefore become:

$$\frac{1}{\rho_1}\vec{\nabla} P_1 = -g\vec{\nabla}\varsigma_1 \qquad \frac{1}{\rho_2}\vec{\nabla} P_2 = -g\vec{\nabla}\varsigma_\rho \qquad (2.9)$$

where:

$$\varsigma_\rho = \frac{\Delta\rho}{\rho_2}\varsigma_2 + \frac{\rho_1}{\rho_2}\varsigma_1 \qquad \Delta\rho = \rho_2 - \rho_1 \qquad (2.10)$$

The terms $\vec{M}_n$ in the wave equation 2.5 are given by:

$$\vec{M}_n = -H_n\frac{\partial \vec{V}_n}{\partial t} = H_n(\vec{\nabla}\vec{V}_n \cdot \vec{V}_n) + \vec{f} \times H_n\vec{V}_n + \frac{1}{\rho_n}H_n\vec{\nabla} P_n + \vec{\tau}_{bn} - \vec{\tau}_{tn} \qquad (2.11)$$

3. Solution Procedure and Finite Element Approximations.

The procedure for solving equations 2.5 and 2.6 by the finite element method requires a rather lengthy description that will be illustrated by a somewhat simpler X-Z problem in section 5. For the general form, a brief description of the process is as follows. Equations 2.7 - 2.10 are substituted into 2.5 to obtain equations in terms of ς_1 and ς_2. The resulting equations are discretized by the Galerkin finite element method (*GFEM*) in the horizontal domain. Nodal integration is used to diagonalize the *mass* and *damping* matrices. The form of the resulting element and global matrix equation is:

$$\begin{bmatrix} M & 0 \\ 0 & \frac{\Delta\rho}{\rho_1}M \end{bmatrix} \begin{Bmatrix} \frac{\partial^2 \varsigma_1}{\partial t^2} \\ \frac{\partial^2 \varsigma_2}{\partial t^2} \end{Bmatrix} + \begin{bmatrix} G & 0 \\ 0 & \frac{\Delta\rho}{\rho_1}G \end{bmatrix} \begin{Bmatrix} \frac{\partial \varsigma_1}{\partial t} \\ \frac{\partial \varsigma_2}{\partial t} \end{Bmatrix}$$

$$+ \begin{bmatrix} K_1 + \frac{\rho_1}{\rho_2}K_2 & \frac{\Delta\rho}{\rho_2}K_2 \\ \frac{\Delta\rho}{\rho_2}K_2 & \frac{(\Delta\rho)^2}{\rho_1\rho_2}K_2 \end{bmatrix} \begin{Bmatrix} \varsigma_1 \\ \varsigma_2 \end{Bmatrix} = \begin{Bmatrix} R_1 \\ R_2 \end{Bmatrix} \qquad (3.1)$$

The momentum equation 2.6 is treated by *GFEM* first in the vertical direction. This is done for each of the two equations. Applying *GFEM* and nodal integration in the horizontal plane decouples the equations and produces a matrix equation for each node. The order of the matrix equation is only equal to the number of degrees of freedom used to define the vertical structure of the flow in the two layers. The scalar equations in the x and y directions for each layer n are of the form:

$$H_n[\alpha]\{\tfrac{\partial X_n}{\partial t}\} + [K_n]\{X_n\} - fH_n[\alpha]\{Y_n\} = \{rx_n\} \qquad (3.2)$$

$$H_n[\alpha]\{\tfrac{\partial Y_n}{\partial t}\} + fH_n[\alpha]\{X_n\} + [K_n]\{Y_n\} = \{ry_n\} \qquad (3.3)$$

The terms X and Y are the nodal parameters on a vertical line that define the structure of the flow. Details of these equations, using Hermitian polynomial basis functions, are given in *Laible*[5,6].

The wave equation is solved first, producing new values of ς_1 and ς_2. These in turn are used in the momentum equations to obtain the next set of X and Y. These terms define the flow at the interface and at the bottom and hence can be used to determine the shear stresses. These stresses are required for equation 2.11 for use in the wave equation. The vertically averaged flows are obtained by integration over the depth and are likewise used in equation 2.11.

4. Time Integration and Modal Analysis. Implicit integration schemes (*Gray* and *Kinnmark*[8]) , were used to directly integrate 3.1 - 3.3 over the time domain. This feature allows for large time steps.

Another possible way to solve the wave equation is to use *modal analysis*. The eigenvalue problem is first solved using the homogeneous form of 3.1 omitting the damping terms. This produces the periods of motion and the mode shapes of the body of water. The actual response due to an arbitrary loading is obtained as a superposition of the independent mode shapes Φ_m (eigenvectors) factored by the response of each q_m. Each q_m is obtained as the response of a one degree of freedom oscillator, with damping G and frequency ω_m (eigenvalues):

$$\ddot{q}_m + G\dot{q}_m + \omega_m^2 q_m = r_m \qquad (4.1)$$

The physical response is now :

$$\{\varsigma\} = \sum_{m=1}^{\varsigma ndof} \{\Phi_m\} q_m \qquad (4.2)$$

The advantage of this approach is that the fundamental mode shapes and frequencies provides some insight into the way the body of water will respond. The largest periods of motion are measurable in the field and hence may be used to verify that the discrete model accurately represents the actual body of water. Although it is often possible in structural dynamics to obtain the physical response by only considering the response of a few of the modes , this has not been found to be the case for these equations. Some of the high frequency modes are essential to obtaining the true physical response as discussed in section 6.

5. An X-Z Model Program.

Here we consider the case of a long narrow body of water such as Lake Champlain, Vermont. In this problem the vertically integrated equations will be used. The Coriolis term and the convective terms are neglected. The x axis runs along the *Talweg* (valley way) of the lake, (figure 1) and z is as before, positive *downward*.

LAKE CHAMPLAIN & Finite Element Nodes
Grand Isle = 17   Burlington Bay = 11   Kingsland Bay = 5

Talweg Length = 73.6 km

Figure 1. *The Talweg and Node Locations.*

The counterparts of equations 2.1 (continuity) and equation 2.2 (momentum) for each layer are:

$$b_1 \frac{\partial \varsigma_1}{\partial t} - b_2 \frac{\partial \varsigma_2}{\partial t} = \frac{\partial (A_1 U_1)}{\partial x} \qquad b_2 \frac{\partial \varsigma_2}{\partial t} = \frac{\partial (A_2 U_2)}{\partial x} \qquad (5.1)$$

$$\frac{\partial (U_1 A_1)}{\partial t} = g \frac{\partial \varsigma_1}{\partial x} A_1 + b_1 \tau_w W^2 - b_2 \tau_1 (U_1 H_1 - U_2 H_2) - \tau_{s1} U_1 A_1$$

$$\frac{\partial (U_2 A_2)}{\partial t} = g \frac{\rho_1}{\rho_2} \frac{\partial \varsigma_1}{\partial x} A_2 + g \frac{\Delta \rho}{\rho_2} \frac{\partial \varsigma_2}{\partial x} A_2$$

$$- b_b \tau_2 H_2 U_2 + b_2 \tau_1 (U_1 H_1 - U_2 H_2) - \tau_{s2} U_2 A_2 \qquad (5.2)$$

where: τ_w, τ_1, τ_2 are wind, interface and bottom friction parameters respectively; τ_{s1} and τ_{s2} are friction parameters for sloping sides; b_1, b_2, b_b are the lake widths at the surface, the level of the thermocline and the bottom respectively; A_n and H_n are the cross-sectional areas and mean depths above and below the thermocline. H_n are as given by equation 2.7. A_1 and A_2 also vary with time i.e., $A_1 = A_{1i} + \varsigma_2 b_2 - \varsigma_1 b_1$ and $A_2 = A_{2i} - \varsigma_2 b_2$. A_{1i} and A_{2i} are the initial cross-sectional areas. All of the friction stresses are assumed to be proportional to the relative average flow.

The *generalized wave equations* for each layer are obtained by applying equation 2.4 , resulting in :

$$b_1 \frac{\partial^2 \varsigma_1}{\partial t^2} + Gb_1 \frac{\partial \varsigma_1}{\partial t} - g \frac{\partial}{\partial x} \left[(A_{1i} + \frac{\rho_1}{\rho_2} A_{2i}) \frac{\partial \varsigma_1}{\partial x} + \frac{\Delta \rho}{\rho_2} A_{2i} \frac{\partial \varsigma_2}{\partial x} \right] =$$

$$g \frac{\partial}{\partial x} \left[(\varsigma_2 b_2 - \varsigma_1 b_1 - \frac{\rho_1}{\rho_2} \varsigma_2 b_2) \frac{\partial \varsigma_1}{\partial x} - (\frac{\Delta \rho}{\rho_2} \varsigma_2 b_2) \frac{\partial \varsigma_2}{\partial x} \right]$$

$$+ \frac{\partial}{\partial x} \left[b_1 \tau_w W^2 - \tau_{s1} U_1 A_1 - \tau_{s2} U_2 A_2 - \tau_{2b_b} U_2 H_2 + GA_1 U_1 + GA_2 U_2 \right]$$
(5.3)

$$b_2 \frac{\Delta \rho}{\rho_1} \frac{\partial^2 \varsigma_2}{\partial t^2} + Gb_2 \frac{\Delta \rho}{\rho_1} \frac{\partial \varsigma_2}{\partial t} - g \frac{\partial}{\partial x} \left[\frac{\Delta \rho}{\rho_2} A_{2i} \frac{\partial \varsigma_1}{\partial x} + \frac{(\Delta \rho)^2}{\rho_1 \rho_2} A_{2i} \frac{\partial \varsigma_2}{\partial x} \right] =$$

$$g \frac{\partial}{\partial x} \left[-(\frac{\Delta \rho}{\rho_2} \varsigma_2 b_2) \frac{\partial \varsigma_1}{\partial x} - (\frac{(\Delta \rho)^2}{\rho_1 \rho_2} \varsigma_2 b_2) \frac{\partial \varsigma_2}{\partial x} \right]$$

$$+ \frac{\Delta \rho}{\rho_1} \frac{\partial}{\partial x} \left[b_2 \tau_1 (U_1 H_1 - U_2 H_2) - \tau_{s2} U_2 A_2 - \tau_{2b_b} U_2 H_2 + GA_2 U_2 \right] \quad (5.4)$$

Linear finite elements and nodal integration were used to obtain equations of the form of 3.1 and 3.2.

6. Application to Lake Champlain, Vt.. The main body of Lake Champlain was modeled as shown in figure 1. The equilibrium depth to the thermocline during late summer was observed to be about 20 m. At each node the cross-sectional areas above and below the thermocline and the other geometric parameters were computed from lake charts. Wind data was collected at the three stations shown in figure 1. An analysis of this data revealed that the wind stress on the lake increases from south to north. The north-south wind stress that develops at a nearby airport (where continuous records are kept) differed from the lake sites by factors of .3 , .6, and 1.3 at nodes 5 , 11 and 18 respectively. The wind stress values ranged from 1×10^{-5} - $2 \times 10^{-4} m^2/s^2$. Field measurements of the motion of the thermocline were taken during August 8 - August 27 , 1987. The density difference was $1 \times 10^{-3} kg/m^3$.

With the lake initially at rest, wind stress values were read into the program at 1 hour increments , starting from August 1 , 1 am. A comparison of the simulated motion of the thermocline ς_2 and the observed motion is shown in figure 2.

The observation point (st.51) is between nodes 17 and 18. The figure illustrates that the modeled and measured response are in good agreement. The modeled amplitudes were less than the observed. The friction parameters used were $G = 2 \times 10^{-7} sec^{-1}$, $\tau_1 = 1 \times 10^{-5} sec^{-1}$,

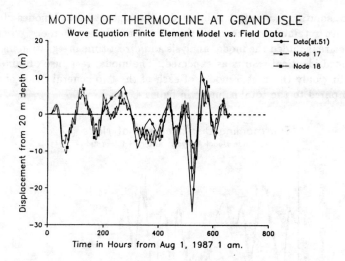

Figure 2. *Thermocline Motion Lake Champlain Vt.*

$\tau_2 = 4 \times 10^{-6} sec^{-1}$, $\tau_{s1} = 2 \times 10^{-6} sec^{-1}$, $\tau_{s2} = 2 \times 10^{-6} sec^{-1}$. Adjustment of these parameters could presumably provide a better fit. Note that the amplitude of the motion is quite large, reaching 20-25 meters. This large motion followed a strong wind event and resulted in tremendous horizontal exchanges of fluid.

The modal analysis revealed that the fundamental period of motion is 3.51 days. Frequency analysis of the thermocline motion data reveals a period of 3.6 days. The mode shapes of two layered bodies are generally of two different types as illustrated in figure 3.

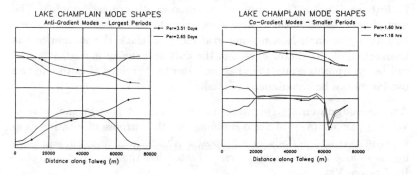

Figure 3. *Mode Shapes Anti-gradient and Co-gradient Shapes.*

The modes of highest period have surface and interface shapes of opposite slope. All of these modes for a discretized model are of higher period than the second class of modes which have parallel slopes. The correct solution can only be obtained if the highest period modes of each class are retained by the numerical time marching scheme. When all modes are included the modal analysis and direct time integration method produced identical results as expected. The modal response obtained by including only the first 5 modes of each of the two general mode shapes, is compared to the total solution in figure 4.

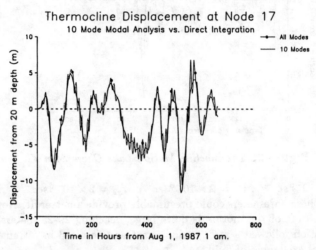

Figure 4. *Partial Mode Analysis vs. Direct Integration Response.*

When only the first 5 large period anti-gradient modes are used, the results are unreasonable and eventually become unstable.

7. Future Work. The results of this pilot model are encouraging. The next phase of development will now focus on the three dimensional model. The ability to include a moving grid mesh to track the motion of the intersection of the thermocline and the bottom is being investigated and will be incorporated into the final model so that regions of both one and two layers can be simultaneously studied.

Acknowledgements. This work was supported in part by the State of Vermont, Department of State Buildings for the purpose of locating deep water intakes and outfalls for a proposed fish hatchery. Field data involving motion of the thermocline was collected by Aquatec. Inc. of South Burlington, Vt.

References

[1] Lynch , D.R. and Gray , W.G. , (1979) *A Wave Equation Model for Finite Element Tidal Computations*. Computers in Fluids, Vol. 7, 207-228 .

[2] Kinnmark, I.P.E. and Gray, W.G. (1985) *Stability and Accuracy of Spatial Approximations for Wave Equation Tidal Motions*. Journal of Computational Physics, Vol. 60 , No. 3 , 447-466 .

[3] Werner ,F.E. and Lynch , D.R. , (1986) *Field Studies with the Wave Equation Formulation* . in Proceeding of the 6 th International Conference on Finite Elements Water Resources, (eds. A. Sa de Costa et al.) , Springer-Verlag, 547-560.

[4] Werner , F.E. and Lynch , D.R. (1987) *Field Verification of Wave Equation Tidal Dynamics in the English Channel and Southern North Sea*. Advances in Water Resources, Volume 10, 115-130.

[5] Laible, J.P. (1984) *A Finite Element-Finite Difference Wave Model for Depth Varying Nearly Horizontal Flow*. Advances in Water Resources, Volume 7.

[6] Laible, J.P. (1984) *Recent Developments In the Use of the Wave Equation for Finite Element Modeling of Three Dimensional Flow*. in Proceedings of the 4 th International Conference on Applied Numerical Modeling , (eds. Hsia ,H.M. et al.), National Cheng Kung University, Republic of China , 300-306.

[7] Laible, J.P. (1984) *A Modified Wave Equation Model for 3D Flow in Shallow Bodies of Water*. in Proceedings of the 4 th International Conference on Finite Elements in Water Resources, (eds. Laible ,J.P. et al.), Springer-Verlag , 609-620.

[8] Gray , W.G. , and Kinnmark I.P.E. (1986) *Evolution of Two Dimensional FE Wave Equation Models* . in Proceeding of the 6 th International Conference on Finite Elements in Water Resources, (eds. A. Sa de Costa et al.) , Springer-Verlag, 29-47.

A Simple Staggered Finite Element Scheme for Simulation of Shallow Water Free Surface Flows

S. Sigurdsson
Science Institute, University of Iceland, Reykjavik, Iceland
S.P. Kjaran and G.G. Tómasson
Vatnaskil Consulting Engineers, Reykjavik, Iceland

INTRODUCTION

Numerical schemes for simulation of free surface flows that make use of the idea of replacing the first order continuity equation with the second order wave equation have received considerable attention in recent years. We present an analagous approach where we first derive a finite element approximation in space to the continuity equation and then differentiate this approximate equation in time and combine it with the momentum equations. Although the resulting equation may indeed be viewed as a space approximation to the wave equation the proposed approach simplefies the derivation with respect to eg. boundary conditions and offers added insight into the treatment of non-linear terms.

On the basis of this approach we go on to derive a finite element scheme for the shallow water equations that uses linear approximations over triangular elements for surface elevation but piecewise constant approximations for the velocity components within the elements. The resulting scheme is similar to finite difference schemes based on so called staggered grid approach, so we refer to it as a staggered finite element scheme.

The computational attraction of the staggered finite element scheme is that in the absence of convective terms and for a given approximation to the elevation values the momentum equations can be integrated independently within the elements. By treating the discontinuites in the approximate velocity values across element boundaries appropriately it is none the less found that the scheme is sufficiently accurate for many simulation purposes.

BASIC EQUATIONS

The following equations, derived from the primary equations of conservation of mass and momentum, often serve as the basis for shallow water free surface flow simulation:

$$\frac{\partial \eta}{\partial t} = - \nabla \cdot (H\bar{V}) \tag{1}$$

$$\frac{\partial \bar{V}}{\partial t} = - g \nabla \eta - \beta \bar{V} - f\bar{k} \times \bar{V} + \frac{1}{H} \bar{W} - (\bar{V} \circ \nabla)\bar{V} \tag{2}$$

along with boundary conditions of either prescribed head or no normal flow. Here η is surface elevation above mean sea level, D is depth corresponding to mean sea level, $H = D+\eta$ is total depth of flow, $\bar{V}$ is the vertically averaged flow velocity, g is gravitational acceleration, β is friction coefficient, f is Coriolis parameter, $\bar{W}$ is windstress, and $\bar{k}$ is the unit vector in the vertical direction.

Rather than integrating Eqs. (1) and (2) as they stand, an alternative approach is to differentiate Eq. (1) with respect to time, substituting for $\partial \bar{V}/\partial t$ using Eq. (2) and in turn for $\nabla \cdot (H\bar{V})$ using Eq. (1), thus obtaining the so-called wave equation:

$$\frac{\partial^2 \eta}{\partial t^2} + \beta \frac{\partial \eta}{\partial t} + \nabla \cdot \left(\frac{\partial \eta}{\partial t} \bar{V}\right) = g \nabla(H \nabla \eta) + H \nabla \beta \cdot \bar{V}$$
$$- f(\nabla \times H\bar{V})_z - \nabla \cdot \bar{W} + \nabla \cdot ((H\bar{V} \cdot \nabla)\bar{V}) \tag{3}$$

where $(\nabla \times (a_x, a_y))_z$ denotes $\partial/\partial x\, a_y - \partial/\partial y\, a_x$. We then integrate Eq. (3) along with Eq. (2) as advocated by eg. Gray and Kinnmark[1] or even a differentiated form of eq. (2) (Gray and Kinnmark[2]). Analysis presented by them indicates the improved stability characteristics of such a procedure. We further note that as an equation in η, Eq. (3) is less coupled with Eq. (2) than Eq. (1) which is important if we advance a timestep by first integrating one and then the other.

DERIVATION OF FINITE ELEMENT SCHEME

We proceed to derive a finite element scheme that is based on the ideas above in a slightly modified form. Contrary to most such schemes we only use piecewise linear approximations for the surface elevation and piecewise constant approximations for the velocity. (See, however, Urban and Zielke[3]). Our first step is, in fact, to derive a discrete space approximation to Eq. (1). Using a weighted residual approach on an irregular triangular nesh the basis functions φ_i for the η approximation will be patch functions that take the value 1 at a nodal point, i, and fall linearly to zero on the surrounding triangular elements, i.e. their support will correspond to the total region shown in figure 1. The coefficients of these basis functions will thus correspond to the approximate surface values, η_i, at the nodal points between triangles. The velocities are on the other hand approximated by constants within the triangular elements. These constants may be viewed as giving the best approximation at the centroids of the triangles. In this sense our scheme has a close resemblance with the staggered grid approach which has been used for a considerable time in the context of finite difference approximations. We therefore choose to refer to our scheme as a staggered finite element scheme. Finally, we consider two types of test functions, ψ_i, in the weighted residual approach. (i) The test functions are the same as the basis trial functions, φ_i. This is in accordance with the classical Galerkin method and we refer to it as the Galerkin variation of the scheme. (ii) The test functions are discontinuous, ψ_i taking the value 1 within the shaded region in figure 1 that extends to the midpoints of the edges and the centroids of the triangles and the value 0 outside. This corresponds closely to so-called box methods or control volume methods as used eg. in the code SIMPLE(R) of Patankar and Spalding (Patankar[4]). We refer to this case as the box variation of our scheme. A third variation of some interest that we, however, do not consider further here is an upstream variation where the test functions are in a sense shifted in a direction opposite to that of the fluid velocity.

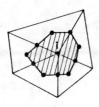

FIGURE 1

In order to describe the scheme in greater detail we may consider Eqs. (1) and (2) applied to a single triangular element Δ as shown in figure 2 with an arbitrary normal outflow function, q, specified along the boundary $\partial \Delta$. The approximate constant velocity within the element, denoted by $\bar{V} = (u,v)$, does not have to agree with q on $\partial \Delta$. There are three linear basis functions φ_0, φ_1, and φ_2 taking the value 1 at the corresponding corner points and 0 at the opposite edges and the approximate surface elevation may be expressed as:
$\eta(x,y,t) = \eta_0(t)\varphi_0(x,y) + \eta_1(t)\varphi_1(x,y) + \eta_2(t)\varphi_2(x,y)$
and the approximate total depth as
$H(x,y,t) = \eta(x,y,t) + D_0\varphi_0(x,y) + D_1\varphi_1(x,y) + D_2\varphi_2(x,y)$.

For the Galerkin variation with $\psi_i = \varphi_i$, $i = 0,1,2$, the conditions to be satisfied are:

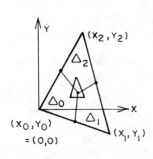

FIGURE 2

$$\iint_\Delta \frac{\partial \eta}{\partial t} \varphi_i \, dxdy = -\iint_\Delta \nabla \cdot (H\bar{V})\varphi_i \, dxdy =$$

$$= -\iint_\Delta \nabla H \cdot \bar{V} \varphi_i \, dxdy - \iint_{\partial \Delta} (q - H\bar{V} \cdot \bar{n}) \varphi_i \, ds$$

resulting in the system:

$$M \frac{d}{dt} \bar{\eta} = -(uL_x - vL_y) \bar{H} - \bar{q}. \tag{4}$$

where $\bar{\eta} = [\eta_i]$, $\bar{H} = [\eta_i + D_i]$, $\bar{q} = [\int_{\partial \Delta} q \varphi_i \, ds]$

$$M = \left[\iint_\Delta \varphi_j \varphi_i \, dxdy \right] = \frac{|\Delta|}{12} \begin{vmatrix} 2 & 1 & 1 \\ 1 & 2 & 1 \\ 1 & 1 & 2 \end{vmatrix}$$

$$L_x = \left[\iint_\Delta \frac{\partial \varphi_j}{\partial x} \varphi_i \, dxdy - \int_{\partial \Delta} \varphi_j \varphi_i n_x \, ds \right] = \frac{1}{6} \begin{vmatrix} (y_2-y_1) & (y_2-y_1) & (y_2-y_1) \\ -y_2 & -y_2 & -y_2 \\ y_1 & y_1 & y_1 \end{vmatrix}$$

and L_y is the same as L_x except that y and x are to be exchanged. Here $\bar{n} = (n_x, n_y)$ denotes the unit normal to $\partial \Delta$ and $|\Delta| = 1/2(x_1 y_2 - x_2 y_1)$ is the area of Δ.

In the box variation the conditions to be satisfied are

$$\iint_{\Delta_i} \frac{\partial \eta}{\partial t} \, dxdy = -\iint_{\Delta_i} \nabla \cdot (H\bar{V}) \, dxdy = -\int_{\partial_I \Delta_i} H\bar{V} \cdot \bar{n} \, ds - \int_{\partial_O \Delta_i} q \, ds$$

where Δ_i are the subregions shown on figure 1, $\partial_I \Delta_i$ is the part of the boundary of Δ_i that is within the triangular element and $\partial_O \Delta_i$ is the part of the boundary that coincides with the boundary of the triangular element. The resulting system will again be Eq. (4) except that now:

$$M = \left[\iint_{\Delta_i} \varphi_j \, dxdy \right] = \frac{|\Delta|}{108} \begin{vmatrix} 22 & 7 & 7 \\ 7 & 22 & 7 \\ 7 & 7 & 22 \end{vmatrix}$$

$$L_x = \left[\int_{\partial_I \Delta_i} \varphi_j n_x \, ds \right] = \frac{1}{24} \begin{vmatrix} 5(y_2-y_1) & (4y_2-3y_1) & (3y_2-4y_1) \\ (-4y_2+y_1) & -5y_2 & -(y_1+3y_2) \\ (4y_1-y_2) & (y_2+3y_1) & 5y_1 \end{vmatrix}$$

and L_y is the same as L_x except that x and y are to be exchanged. The close similarity between the two variations should be observed since it does not always seem to be fully appreciated.

In the case of a region composed of many triangular elements the element contributions of Eq. (4) are assembled in the usual way into a global system. In particular, the arbitrary q values will not enter into this global system since they cancel out on common boundaries between elements due to the requirement of continous flow, reduce to zero on boundaries with no normal flow, and do not enter into the element contribution on boundaries with specified head.

Along with the global counterpart of Eq. (4) approximating Eq. (1) we have Eq. (2). We now restrict our attention to the simplified case when the final convective term in Eq. (2) may be neglected. The importance of this restriction within the framework of our scheme with the velocity $\bar{V} = (u,v)$ being approximated by a constant inside each element is that the velocity may be integrated <u>independently</u> for each element for a given approximation to the water level η, i.e. we integrate for each element:

$$\frac{du}{dt} = -g \frac{\partial \eta}{\partial x} - \beta u + fv + \frac{W_x}{H_m}$$

$$\frac{dv}{dt} = -g \frac{\partial \eta}{\partial y} - \beta v - fu + \frac{W_y}{H_m} \qquad (5)$$

where $H_m = \frac{1}{3}(H_0 + H_1 + H_2)$

$$\frac{\partial \eta}{\partial x} = -\frac{(y_2-y_1)\eta_0 - y_2\eta_1 + y_1\eta_2}{2|\Delta|}$$

$$\frac{\partial \eta}{\partial y} = \frac{(x_2-x_1)\eta_0 - x_2\eta_1 + x_1\eta_2}{2|\Delta|}$$

using the fact that η is approximated by a linear function within the element.

However, in analogy with the wave equation formulation, we do not integrate the global counterpart of Eq. (4) as it stands. Rather, differentiating through Eq. (4) with respect to time we obtain:

$$M \frac{d^2}{dt^2} \bar{\eta} = -(uL_x - vL_y)\frac{d}{dt}\bar{\eta} - \left(\frac{du}{dt}L_x - \frac{dv}{dt}L_y\right)\bar{H} - \frac{d\bar{q}}{dt}$$

and substituting for du/dt and dv/dt using Eq. (5) and then for $(uL_x-vL_y)\bar{H}$ using Eq. (4) we get:

$$M \frac{d^2}{dt^2} \bar{\eta} + [\beta M + (uL_x-vL_y)]\frac{d}{dt} \eta = g\left(\frac{\partial \eta}{\partial x} L_x - \frac{\partial \eta}{\partial y} L_y\right)\bar{H}$$

$$- \beta\bar{q} - f(vL_x + uL_y)\bar{H} - \frac{1}{H_m}(w_x L_x - w_y L_y)\bar{H} - \frac{d\bar{q}}{dt} \qquad (6)$$

When assembled into a global equation the term $d\bar{q}/dt$ does not enter by a similar argument that applies to the term $\bar{q}$ in Eq. (4). If the β value varies between two adjacent triangles Δ and Δ' there will, on the other hand, be a contribution from the term $\beta\bar{q}$ ammounting to:

$$\frac{\ell}{2}((\beta - \beta')(H_m \bar{V} + H_m' \bar{V}') \cdot \bar{n}$$

where ℓ is the length of the common boundary and $\bar{n}$ the unit normal on that boundary from Δ into Δ'.

The assembled counterpart of Eq. (6) along with Eq.(5) for each element form the basis of the staggered finite element scheme. It should be noted that Eq. (6) may indeed be derived directly from the wave equation (Eq. (3)) provided due care is taken in dealing with the discontinuities in the approximated velocity and the boundary conditions. It should further be noted that schemes that supposedly integrate Eqs. (1) and (2) as they stand, in fact often make indirect use of the same basic ideas in their numerical implementation eg. the fractional step scheme used in the simulation system CYTHERE-ESI (Benqué et al.[5]).

IMPLEMENTATION

When implementing the staggered finite element scheme we have in addition to omitting the final convective term in Eq. (2) usually made the additional simplifications of considering βM to be negligible in comparison with $uL_x - vL_y$ in Eq. (6) as well as approximating $\bar{H}$ with $[D_m\ D_m\ D_m]^T$ where $D_m = 1/3[D_0+D_1+D_2]$. Although not essential they simplify considerably the integration of Eq. (6). The omission of the convective term is more critical, and although we have experimented with various ideas in order to include this term, it is as yet an unresolved problem how to do this in the most satisfactory way within the framework of our simple scheme.

When integrating Eqs. (5) and (6) we have in most applications found it sufficient first to advance one timestep with the global system corresponding to Eq. (6) using velocity approximations from the previous timestep and subsequently advancing one timestep with Eq. (5) for each element. The integration scheme used has usually been a single step backward difference method along with its two step counterpart for integrating a second order system. With the simplifications referred to above the matrix to be inverted when advancing one timestep with the global system will be positive definite and independent of time so that we only have to carry out a Cholesky factorization at the beginning and when the timestep is changed. The adequacy of this approach has been demonstrated in a number of test runs, some of which are presented in a separate paper (Kjaran et al.[6]). In the tests carried out so far we have usually applied the box variation of the method but the difference between it and the Galerkin variation seems, in fact, to be negligible.

ACKNOWLEDGEMENTS

This research has been supported by Science Institute, Vatnaskil Consulting Engineers, and through grants from the National Science Foundation in Iceland.

REFERENCES

1. Gray W. G. and Kinnmark I.P.E. (1986), Evolution of Two-Dimensional FE Wave Equation Models, Proc. of VI Intl. Conf. on Finite Elements in Water Resources, Springer Verlag, Berlin, pp. 29-47.

2. Gray W.G. and Kinnmark I.P.E. (1987), Finite Element Surface Flow Simulation using Wave Equation, Proc. of XIIth Congress IAHR, Technical Session B (eds. Cunge, J.A. and Ackers, P.), ÉPFL, Lausanne, pp. 113-119.

3. Urban, C. and Zielke, W. (1986), A Finite Element Method with Linear Trial Functions for Steady-State Two-Dimensional Flows, Proc. of VI Intl. Conf. on Finite Elements in Water Resources, Springer Verlag, Berlin, pp. 791-797.

4. Patankar S.V. (1980). Numerical Heat Transfer and Fluid Flow, McGraw-Hill, New York.

5. Benqué J.P., Cunge J.A., Feuillet, J., Hauguel, A. and Holly, F.M. (1982), New Method for Tidal Current Computation, J. of The Waterway, Port, Coastal and Ocean Division, Proc. of ASCE, Vol. 108, No. WW3, pp.396-417.

6. Kjaran, S.P., Hólm, S.L. and Sigurdsson, S. (1988), Testing of Finite Element Schemes for Linear Shallow Water Equations, Proc. of VII Intl. Conf. on Computational Methods in Water Resources (these proceedings).

Improved Stability of the "CAFE" Circulation Model
E.A. Zeris and G.C. Christodoulou
Department of Civil Engineering, National Technical University of Athens, Greece

ABSTRACT

The CAFE finite element scheme for the Shallow Water Equations is shown to be non-linearly unstable due to the type of discretization applied to the advective terms. An improved time-integration scheme is proposed which stabilizes advection, as verified by simple numerical experiments.

BACKGROUND

Two-dimensional nearly horizontal flow can be described by the Shallow Water Equations (SWE). One of the earliest and best known finite element solution schemes for the SWE was developed by Wang and Connor[6] and given the acronym CAFE (Pagenkopf et al[3]). In this scheme, the equation of continuity and the linear part of the equations of motion (i.e. the terms modeling wave propagation) are discretized by the leapfrog method, with the advection, resistance, diffusion and Coriolis terms being computed at the lower time level of the integration step. The scheme is time-centered for the linear case, but accuracy is reduced to $O(\Delta t)$ in the complete non-linear form. Linear finite elements are used for discretization in space. Solutions for specific discharges and water levels are obtained at alternating time levels, as follows:

$$\underline{M}_1 (\underline{\zeta}^{n+\frac{1}{2}} - \underline{\zeta}^{n-\frac{1}{2}}) + \Delta t \underline{G}_1 \underline{Q}^n = \Delta t \underline{P}_1(\underline{\zeta}^{n-\frac{1}{2}}, \underline{Q}^n) \tag{1}$$

$$\underline{M}_2 (\underline{Q}^{n+1} - \underline{Q}^n) + \Delta t \underline{G}_2 \underline{\zeta}^{n+\frac{1}{2}} = \Delta t \underline{P}_2(\underline{\zeta}^{n+\frac{1}{2}}, \underline{Q}^n) - \Delta t (\underline{E} + \underline{C}) \underline{Q}^n \tag{2}$$

where $\underline{\zeta}$ is the vector of surface elevations at the nodes
$\underline{Q} = [q_x, q_y]$ is the vector of the specific discharges
$\underline{M}_1, \underline{M}_2$ relate to the geometry of the grid
$\underline{G}_1, \underline{G}_2$ contain spatial gradient operators
$\underline{E}, \underline{C}$ contain the eddy viscosity and Coriolis terms
$\underline{P}_1, \underline{P}_2$ are the forcing terms, the latter including the bottom friction and the advective terms.

The stability properties of the scheme have yet to be fully delineated. Wang and Connor[6] originally suggested that the scheme is unconditionally stable for the linear case. Subsequently, Wang[7] proposed an approximate stability constraint on the Courant number ($Cr = \Delta t \sqrt{2gh}/\Delta s$) between 0.5 and 1.5[2], a constraint deduced from operating experience. Gray and Lynch[2] examined a linearized x-t reduction of CAFE and suggested a Courant number limitation close to unity.

A study of the effect of individual terms on the overall stability of CAFE was undertaken by the authors. It was found that the scheme is non-linearly unstable due to the discretization of the advective terms by the forward explicit method. In the following, this is demonstrated by means of a typical numerical experiment. Analogy is drawn with the behaviour of similarly structured finite difference schemes. Corrective action is also suggested and applied successfully to the CAFE scheme.

ADVECTION-INDUCED INSTABILITY IN CAFE

The advection—induced instability inherent in the CAFE scheme can be demonstrated by the following numerical experiment. The scheme was applied to the typical free oscillation test shown in Figure 1. This test requires computation of the flow generated after a harmonic wave profile is released from rest, within a closed rectangular basin of uniform depth. The scheme was applied in its complete, non-linear form but with the resistance, Coriolis and diffusion terms suppressed. In other words, the equations solved contain only the terms modeling wave propagation and advection. The scheme was also applied in linear form, (wave propagation only), for comparison. Stability is assessed by monitoring the variation of the energy (kinetic plus potential) of the computed solutions with time. The initial condition applied identifies with the fundamental mode of oscillation for this basin shape: $\zeta = A\cos(\pi x/L_x)\cos(\pi y/L_y)$, having directional wave numbers $N = \Lambda/L_x = \Lambda/L_y = 8$ in both directions (Λ being the wavelength -see Fig. 1) The test parameters are as follows: uniform nodal spacing $\Delta s = 150$ m, depth $h = 10.19$ m, wave amplitude $A = 1.0$ m. Solutions were attempted at Courant numbers $Cr = \Delta t \sqrt{2gh}/\Delta s = [0.2, 0.4, 0.6, 0.8, 1.0]$. The results of those stability tests are presented in Table 1 and Fig. 2.

The destabilizing role of the advective terms in CAFE is clearly demonstrated. When the linearized equations are used, stability is preserved for Courant numbers up to unity. When the advective terms are included, the energy grows ie. instability sets in the computation. It is to be noted, however, that the strength of the instability (i.e. the rate of energy growth) appears to be related to the size of the time step (cf. Fig.2). Consequently, it might be possible that for small enough Δt, (or Cr), this instability is controlled and masked by the dissipating action of the resistance and diffusion terms. In such circumstances, however, model calibration is unreliable and the credibility of its predictive capacity is eroded. Improvement is clearly required on the discretization of the advective terms in the CAFE scheme.

ANALOGIES WITH FINITE DIFFERENCE SCHEMES

The behaviour of CAFE, as far as non-linear instability is concerned, is markedly similar to that of a class of similarly structured finite difference schemes for the SWE. Consider, as an example, the following simplified form of one of Dronkers[1], tidal schemes, developed on a space-time staggered grid in x-t space, in which only the terms modeling wave propagation and advection are retained (see Figure 3):

$$\frac{\zeta_i^{n+\frac{1}{2}} - \zeta_i^{n-\frac{1}{2}}}{\Delta t} + \frac{q_{i+1}^n - q_{i-1}^n}{\Delta x} = 0 \qquad (3)$$

$$\frac{q_{i+1}^{n+1} - q_{i+1}^n}{\Delta t} + gd_{i+1}^{n+\frac{1}{2}} \frac{\zeta_{i+2}^{n+\frac{1}{2}} - \zeta_i^{n+\frac{1}{2}}}{\Delta x} + u_{i+1}^n \frac{q_{i+3}^n - q_{i-1}^n}{2\Delta x} = 0 \qquad (4)$$

in which : $d_{i+1} = h_{i+1} + \frac{1}{2}(\zeta_{i+2} + \zeta_i)$ is the total depth and

$u_{i+1} = (q/d)_{i+1}$ is the local velocity

This scheme employs $O(\Delta t^2)$ leapfrog differencing for the linear part of the equations, and $O(\Delta t)$ forward explicit differencing for the advective term. It is identical to CAFE, inasmuch as discretization in time is concerned. A von Neuman stability analysis of the quasi-linearized form of eqs. (3) and (4) has been performed by Stelling[5]. It was shown that this scheme, being linearly stable for Courant numbers up to unity, becomes unconditionally unstable due to the forward explicit differencing of the advective term. The analogy with the instability exhibited by CAFE is clear.

One of the options available for stabilizing advection in eq. (4) is to perform the discretization by the following two-step formula:

$$\frac{\overline{q_{i+1}^{n+1}} - q_{i+1}^n}{\Delta t} + gd_{i+1}^{n+\frac{1}{2}} \frac{\zeta_{i+2}^{n+\frac{1}{2}} - \zeta_i^{n+\frac{1}{2}}}{\Delta x} + u_{i+1}^n \frac{q_{i+3}^n - q_{i-1}^n}{2\Delta x} = 0 \qquad (5)$$

$$\frac{q_{i+1}^{n+1} - q_{i+1}^n}{\Delta t} + gd_{i+1}^{n+\frac{1}{2}} \frac{\zeta_{i+2}^{n+\frac{1}{2}} - \zeta_i^{n+\frac{1}{2}}}{\Delta x} + u_{i+1}^{\overline{n+1}} \frac{\overline{q_{i+3}^{n+1}} - \overline{q_{i-1}^{n+1}}}{2\Delta x} = 0 \qquad (6)$$

$\overline{q^{n+1}}$ being regarded as a provisional solution at $t = (n+1)\Delta t$. This formulation follows from the two-step scheme of Matsuno, originally used for numerical solutions of the advection equation (see Roache[4]). Very similar iterative techniques have also been used to stabilize advection in Alternating Direction Implicit schemes for the SWE (see Zeris[8]). The two-step formula defined by eqs. (5) and (6) was applied to stabilize the CAFE scheme.

AN IMPROVED TIME-INTEGRATION SCHEME

Implementation of the two-step formula defined above to the equations of motion as discretized in the CAFE scheme, cf.eq(2), is straightforward. The resulting scheme was applied to the free oscillation test described previously, with the model equations containing only the wave propagation and the advective terms. The scheme remained stable at all Courant numbers tested, except Cr=1.0 (cf. Table 1), and exhibited dissipative behaviour (cf. Fig. 2). Two further sets of stability tests were performed on the same test problem, but with the free oscillation now initiated from the following initial conditions: a) $\zeta = A\cos(2\pi x/L_x)\cos(2\pi y/L_y)$, this corresponding to the second mode of oscillation, and b) $\zeta = A\cos(4\pi x/L_x)\cos(4\pi y/L_y)$, this being comprised of the shortest waves resolvable by the grid (wavelength $\Lambda = 2\Delta s$). In both cases, stability was preserved for Courant numbers ($Cr = \Delta t \sqrt{2gh}/\Delta s$) up to $Cr=0.8$.

On the basis of these preliminary tests, it appears that the new time-integration scheme described above can lead to a nonlinearly stable exponent of the CAFE circulation model.

CONCLUSIONS

The CAFE finite element scheme for the Shallow Water Equations was shown to be non-linearly unstable due to the discretization of the advective terms by the forward explicit method. A two-step time-integration formula was proposed which stabilizes advection. Solutions at Courant numbers close to unity were obtained for the cases tested. It is to be noted that this improvement can be implemented with only little modification of the original computer program.

REFERENCES

1. Dronkers, J.J.(1969), Tidal Computations for Rivers, Coastal Areas and Seas, J. of Hydraulics Div.,ASCE, Vol.95, No.HY 1, pp. 29-77.

2. Gray,W. G.,and Lynch, D.R.(1977), Time Stepping Schemes for Finite Element Tidal Model Computations, Adv.Water Resources, Vol.1, Dec. 77.

3. Pagenkopf,J.R., Christodoulou, G.C., Pearce, B.R., and Connor, J.J.(1976), A User'sManual for CAFE-1, a Two-Dimensional Finite Element Circulation Model, R.M.Parsons Laboratory, Report No 217, M.I.T., Cambridge,Massachusetts.

4. Roache, P.J. (1972), Computational Fluid Dynamics, Hermosa Publ., Albuquerque, N. Mexico.

5. Stelling, G. S. (1980), Improved Stabilityof Dronkers'Tidal Schemes, J. of the Hydraulics Div.,ASCE, Vol.106, No.HY8, pp.1365-1379.

6. Wang, J.D.,and Connor, J.J.(1975), Mathematical Modeling of Near Coastal Circulation, R.M.Parsons Lab. Report No.200, M.I.T., Cambridge, Massachusetts.

7. Wang J.D.(1978), Real-time Flow in Unstratified Shallow Water, J. of the Waterway, Port, Coastal and Ocean Div,ASCE,Vol. 104, No. WW1, pp.53-68.

8. Zeris, E.A. (1986), Investigation of Certain Implicit Finite Difference Schemes for Integration of the Long Wave Equations, Thesis presented at the University of Strathclyde in fulfillment of the requirements for the degree of Doctor of Philosophy.

$\Delta t(s)$	Cr	Linear CAFE	Non-linear CAFE	Two-step Scheme
2.0	0.2	S	U	S
4.0	0.4	S	U	S
6.0	0.6	S	U-1195	S
8.0	0.8	S	U-525	S
10.0	1.0	S	U-226	U-1966

Table 1. Results of stability tests with the CAFE scheme and its two-step exponent (S = stable; U = unstable; U-i = termination of computation at i-th step due to 'overflow condition')

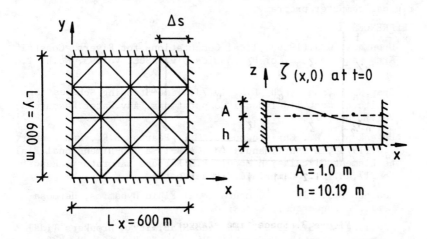

Figure 1. Description of the test problem

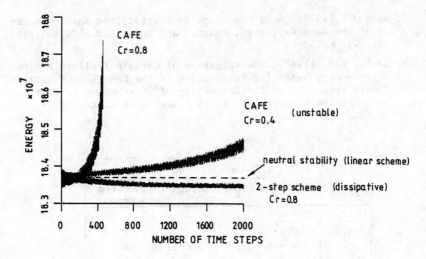

Figure 2. Results of stability tests with CAFE and the two-step scheme (propagation+advection)

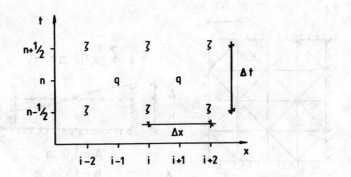

Figure 3. Space-time staggered grid

SECTION 3C - OPEN CHANNEL FLOW AND SEDIMENTATION

An Implicit Factored Scheme for the Simulation of One-Dimensional Free Surface Flow

A.A. Aldama, J. Aparicio and C. Espinosa
Mexican Institute of Water Technology, Insurgentes 4, Jiutepec, Mor. 62550, Mexico

ABSTRACT

A second order-accurate, noniterative implicit scheme for the numerical solution of the governing equations for one-dimensional free surface flow is presented and its use is illustrated through an example.

INTRODUCTION

One of the main difficulties associated with the numerical simulation of one-dimensional, free surface flow is the presence of nonlinearities in the governing equations. These nonlinearities make it necessary to perform iterations whenever a conventional implicit time integration scheme is used.

In this paper, a second order-accurate, noniterative implicit scheme for the solution of the above mentioned equations is presented and its use is illustrated through a numerical example.

GOVERNING EQUATIONS

The mass and momentum conservation equations governing the one-dimensional, free surface flow in a rectangular channel are respectively (Cunge et al.[1]):

$$\frac{\partial h}{\partial t} + u \frac{\partial h}{\partial x} + h \frac{\partial u}{\partial x} = 0 \qquad (1)$$

$$\frac{\partial u}{\partial t} + u\frac{\partial u}{\partial x} + g\frac{\partial h}{\partial x} - g\left(S_o - S_f\right) = 0 \tag{2}$$

where h = flow depth, u = mean flow velocity, g = gravitational acceleration, S_o = bottom slope, S_f = friction slope, x = spatial coordinate and t = time.

The friction slope is parametrized in this paper by means of the Manning formula, i.e.,

$$S_f = \frac{n^2 \, u|u|}{R^{4/3}} \tag{3}$$

in which n is the Manning's roughness coefficient and R is the hydraulic radius.

SPACE DISCRETIZATION

A Galerkin finite-element formulation with quadratic expansions (Pinder & Gray[2]) was used for space discretization. Due to the lack of space, the corresponding formulation will not be described here, allowing a detailed description of the time integration scheme.

IMPLICIT FACTORED SCHEME

This scheme was implemented basically as is described in Beam & Warming[3,4] and Aldama[5]. The corresponding semidiscrete time approximation of the continuity equation (1) is

$$\frac{h^{n+1}-h^n}{\Delta t} + \frac{1}{2}\left(u^n\frac{\partial h^{n+1}}{\partial x} + u^{n+1}\frac{\partial h^n}{\partial x}\right) + \frac{1}{2}\left(h^n\frac{\partial u^{n+1}}{\partial x} + h^{n+1}\frac{\partial u^n}{\partial x}\right) = 0 \tag{4}$$

where the superindex n denotes time level (not to be confused with Manning's roughness coefficient). Note that this scheme is linear in the unknowns h^{n+1} and u^{n+1} and, as may be readily shown, is $O(\Delta t^2)$-accurate.

For the momentum equation (2), it becomes necessary to find a way to discretize linearly the nonlinear term S_f (see equation 3).

Let
$$\frac{u|u|}{R^{4/3}} = N(u,h)u \tag{5}$$

where $N(u,h)$ is the nonlinear function in u and h

$$N = \frac{|u|}{R^{4/3}} \tag{6}$$

Now, let the product Nu be approximated at $t=(n+1/2)\Delta t$ by the $O(\Delta t^2)$-accurate trapezoidal rule:

$$Nu = \frac{1}{2}\left[N^{n+1} u^{n+1} + N^n u^n\right] \tag{7}$$

Expanding the nonlinear product $N^{n+1}u^{n+1}$ in a Taylor series about the time level $n+1$, the following is obtained:

$$N^{n+1}u^{n+1} = N^n u^n + \Delta t\, u^n \frac{\partial N^n}{\partial t} + \Delta t\, N^n \frac{\partial u^n}{\partial t} \tag{8}$$

where terms of $O(\Delta t^2)$ have been neglected. But

$$\frac{\partial N^n}{\partial t} = \left[\frac{\partial N}{\partial u}\right]^n_x \frac{\partial u^n}{\partial t} + \left[\frac{\partial N}{\partial h}\right]^n_x \frac{\partial h^n}{\partial t} \tag{9}$$

and, from (6),

$$\frac{\partial N}{\partial u} = \frac{|u|}{u\, R^{4/3}} = \frac{N}{u} \tag{10}$$

$$\frac{\partial N}{\partial h} = -\frac{4}{3} \frac{|u|}{R^{7/3}} \frac{\partial R}{\partial h} \tag{11}$$

where, for a rectangular channel,

$$\frac{\partial R}{\partial h} = \left(\frac{b}{b+2h}\right)^2 \tag{12}$$

The time derivatives in equations (8) and (9) may be discretized in the following form, without affecting the overall $O(\Delta t^2)$ accuracy:

$$\frac{\partial u^n}{\partial t} = \frac{u^{n+1} - u^n}{\Delta t} \; ; \; \frac{\partial h^n}{\partial t} = \frac{h^{n+1} - h^n}{\Delta t} \qquad (13)$$

Using equations (8) through (13) in (7), the following result is obtained:

$$Sf = n^2 \, Nu = n^2 \left[\frac{|u^n|}{(R^n)^{4/3}} u^{n+1} - \frac{2}{3} \frac{|u^n|u^n}{(R^n)^{7/3}} \left(\frac{b}{b+2h^n} \right)^2 \cdot \left[h^{n+1} - h^n \right] \right] \qquad (14)$$

thus, the semidiscrete momentum equation becomes

$$(15)$$

$$\frac{u^{n+1} - u^n}{\Delta t} + \frac{1}{2}\left\{ u^n \frac{\partial u^{n+1}}{\partial x} + u^{n+1} \frac{\partial u^n}{\partial x} \right\} + \frac{g}{2}\left\{ \frac{\partial h^{n+1}}{\partial x} + \frac{\partial h^n}{\partial x} \right\} =$$

$$= gS_o - gn^2 \left[\frac{|u^n|}{(R^n)^{4/3}} u^{n+1} - \frac{2}{3} \frac{|u^n|u^n}{(R^n)^{7/3}} \left(\frac{b}{b+2h^n} \right)^2 \left[h^{n+1} - h^n \right] \right]$$

which, as may be observed, is linear in h^{n+1} and u^{n+1}, a fact that, added to the linearity of the semidiscrete continuity equation (4) in the same variables, avoids the need of iterations for their solution.

NUMERICAL EXAMPLE

In order to illustrate the use of the previously described implicit factored scheme, a rectangular channel 1 Km long and 2 m wide with a bottom slope $S_o = 0.0002$ was considered. The channel receives a hydrograph at the upstream boundary while a constant depth is maintained at the downstream boundary.

The upper boundary condition was taken into account in the implicit scheme by substituting the continuity equation corresponding to the first node by the following:

$$Q_1^{n+1/2} = \frac{b}{2}\left\{ u_1^{n+1} h_1^n + u_1^n h_1^{n+1} \right\} \qquad (16)$$

where $Q_1^{n+1/2}$ is the known discharge at the upstream end of the channel in time $n+1/2$. Equation (16) may be shown to have $O(\Delta t^2)$ accuracy. The lower boundary condition was imposed by making

$$h_m^{n+1} = h_c \qquad (17)$$

where h_c is the constant, known depth at the downstream end of the channel. The triangular input hydrograph shown in figure 1 was routed through the channel.

Values of $\Delta x = 10$ m and $\Delta t = 60, 120, 240$ and 360 s were used. The resulting downstream hydrograph is also shown in figure 1 for $\Delta t = 360$ s. The results obtained for the other values of Δt were practically the same.

CONCLUSIONS

The implicit-factored scheme is a very suitable method for one-dimensional free-surface flow simulation, because it eliminates the need for iterations, thus saving considerable CPU time. More numerical experiments in one and two dimensions will be carried out and the results will be published later.

REFERENCES

1. Cunge, J.A., Holly, F.M. Jr., Verwey, A. (1980). Practical Aspects of Computational River Hydraulics, Pitman, London

2. Pinder, G.F. Gray, G.W. (1977). Finite Element Simulation in Surface and Subsurface Hydrology, Academic Press, New York

3. Beam, R.M., Warming, R.F. (1976). An Implicit Finite-Difference Algorithm for Hyperbolic Systems in Conservation-Law Form, J. Comp. Phys, 22, 87-110

4. Beam, R.M., Warming, R.F. (1978). An Implicit Factored Scheme for the Compressible Navier-Stokes Equations, AIAA Journal 16, 393-402

5. Aldama, A. (1985). Theory and Application of Two- and Three-scale Filtering Approaches for Turbulent Flow Simulation, Ph.D. Thesis, M.I.T.

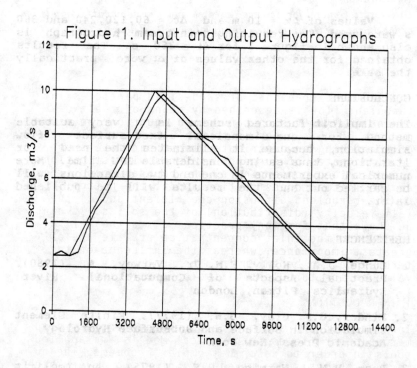

Figure 1. Input and Output Hydrographs

Practical Aspects for the Application of the Diffusion-Convection Theory for Sediment Transport in Turbulent Flows

W. Bechteler
Institute for Hydromechanics, University of the Armed Forces Munich, D-8014 Neubiberg, Germany

W. Schrimpf
Directorate General for Science Research and Development, Commission of the European Communities, B-1049 Brussels, Belgium

ABSTRACT

The sediment transport of sand-sized particles in turbulent flow is described by means of the two-dimensional diffusion-convection equation. The main purpose of this paper is to evaluate the influence of scattering or distributed input parameters on settling rates of suspended material. Investigations are described concerning a) the boundary conditions at the initial cross section and at the bottom, b) the turbulent diffusion coefficient and c) the fall velocity (distribution) of the solids. It can be shown that for many practical cases the fall velocity and the inlet concentration profile are of greatest importance, whereas the influence of the turbulence can be taken into account by an integral value of the turbulent diffusion coefficient.

1 INTRODUCTION

In hydraulic and sedimentation engineering as in numerous other fields mathematical models are applied tools in order to describe complex problems. The numerical analysis is driven very often to a very high level whereas for practical applications of the models the input parameters and the boundary conditions cannot be descibed with a comparable quality. Thus extreme differences in results are possible for the same problem according to the assumptions which have to be made.

In practice the sensitivity of a model has to be known in regard to the different input variables. A sensitivity study can clarify which of these parameters "dominate" the process and should be determined as thoroughly as possible. For practical applications it is often more appropriate to have relatively simple models which are easily to "control". These more basic models can give in many cases very appropriate results and assist the engineer in finding the weak points in the lay-out of a plant or a process.

2 THEORETICAL REFERENCE

The diffusion-convection theory is widely applied for the prediction and calculation of transport phenomena of suspended solids in turbulent flowing water. Solutions of the diffusion-convection equation provide e.g. the evolution of concentration profiles or the prediction of sedimentation rates. The two-dimensional (x,y) distribution of the solid concentration c can be obtained for steady state conditions by solutions of the following set of equations:

$$-u\frac{\delta c}{\delta x} + v_s \frac{\delta c}{\delta y} + \frac{\delta}{\delta y}(\varepsilon_{sy}\frac{\delta c}{\delta y}) + \frac{\delta}{\delta x}(\varepsilon_{sx}\frac{\delta c}{\delta x}) = 0 \quad (1)$$

$$\varepsilon_{sy}\frac{\delta c}{\delta y} + \alpha v_s c + S_E = 0 \qquad y=0 \quad (2)$$

$$\varepsilon_{sy}\frac{\delta c}{\delta y} + v_s c = 0 \qquad y=h \quad (3)$$

$$c = c_0(y) \qquad x=0 \quad (4)$$

The governing input parameters are:

- The flow velocity profile $u = f_1(...)$
- the fall velocity of the solids $v_s = f_2(...)$
- the turbulent diffusion coefficients $\varepsilon_{sx}, \varepsilon_{sy} = f_3(..)$
- the resuspension coefficient $\alpha = f_4(...)$
- the sediment entrainment $S_E = f_5(...)$

An explicit finite-difference scheme is applied for the solution of this partial differential equation. The algorithm is based on a version which is described by Bechteler and Schrimpf (1984).

3 SENSITIVITY ANALYSIS

The sensitivity analysis is done separately for each of the parameters mentioned above in order to find out their particular influence. In the design stage of a settling basin for example a complete sensitivity analysis is useful (e.g. Schrimpf (1986/2)).

3.1 The Influence of the Initial Concentration Profile

As an example in Fig. 1 the plots for the removal ratios are shown for three different idealized initial concentration profiles.

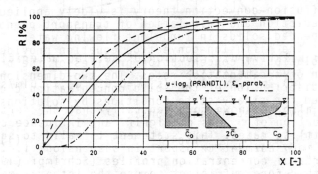

Fig. 1: Removal ratio R=f(X) for different initial concentration profiles,
$\bar{u}$=0.2m/s, v_s=0.5cm/s, λ=0.02, u_{o*}=0.01m/s.

In this example for a removal ratio of R=90% the maximum difference in the necessary length is about 50%. In general the inlet zones or inlet channels of settling basins need a particular investigation in order to have the relevant conditions at hand. This necessity is often underestimated.

3.2 The Influence of the Turbulent Diffusion Coefficient

Numerous relations are derived either from theoretical or from experimental investigations. Parabolic profiles can be derived by applying the flow velocity profiles of Prandtl and v. Kármán, respectively. In Fig. 2 the relative distribution of settled material along the dimensionless flow axis is shown for one numerical example applying two parabolic and one constant profile for ε_s=f(Y). The initial solid concentration profile has its maximum at the water surface because it can be assumed that for this case the effect of the different distributions for ε_s would be most significant.

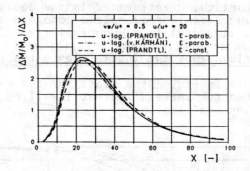

Fig. 2: Relative distribution of settled material along the X-axis;
$\bar{u}$=0.2m/s, v_s=0.5cm/s, λ=0.02, u_{o*}=0.01m/s.

It can also be shown more generally that the vertical distribution of ε_s has no significant influence on the settling rates. This statement is valid to some extent for sediment transport in suspension, i.e. for the vertical concentration profiles (Schrimpf (1986/1) It is therefore sufficient to use the integral mean value of the turbulent diffusion coefficient for the calculation of many practical sediment transport problems.

3.3 The Influence of Resuspension

If open channel is considered the flow conditions at the critical stage for keeping particles in suspension is characterized with $u_{o*}=u_{o*1}$. Furthermore the critical shear stress τ_{o1} is taken to be constant for each individual grain size. In a first approximation the bottom shear stress is assumed to be normally distributed. A grain of specific grain size is kept in suspension when the local shear stress exceeds the the critical value. In this case the theoretical resuspension rate is 0.5 (50%). In Fig. 3 the removal ratio is plotted for resuspension rates of 0% and 50% for a grain diameter of 0.2mm. The relation of Bagnold(1966) is taken for the suspension criterion. The results are nearly identical. That means particles are not kept in suspension over a significantly longer distance in case of critical flow conditions. However, it has to be pointed out that for this example a time-dependent variation of the bed shape and a reduction of the fall velocity due to an increasing sediment concentration near the bed are not taken into account. For demonstration the settling rates are also calculated for increasing shear stresses $\tau_o > \tau_{o1}$ ($\alpha > 50\%$). In order to get quantitative results the standard deviation of τ has to be known. Gessler (1973) reports the bottom shear stress to be normally distributed with a coefficient of variation of 0.57. This value is taken for calculations with $\tau_o > \tau_{o1}$.

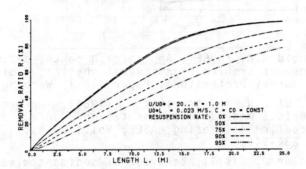

Fig. 3: Removal ratio for different resuspension rates; inlet concentration $c = \bar{c}_o = $ const.

4 THE METHOD OF TWO-POINT ESTIMATES

Another possibility to evaluate the influence of input parameter scatter is to apply the method of two-point estimates in probalities as proposed by Rosenblueth (1974). This method can be applied with many problems in practical engineering. As far as sediment transport problems are concerned it should be worthwhile to evaluate the influence of e.g. the fall velocity distribution on concentration profiles, settling lengths etc.. For demonstration the vertical concentration profiles are calculated with the common Rouse equation for a grain size distribution $0.125 < d \leq 0.25$ mm. The fall velocity distribution is got by means of a computerized sedimentation tube. The mean value is measured to $\bar{v}_s = 1.408$ cm/s and the standard deviation is calculated to $\sigma = 0.462$. In Fig. 4 the concentration profiles are shown for a calculation with $\bar{v}_s$ and the method of two point estimates. The third curve is obtained for a calculated fall velocity $v_s(d_{max} + d_{min}/2)$ and a particle Shape Factor SF=0.7. These values are often used when the fall velocity cannot be obtained be experiments.

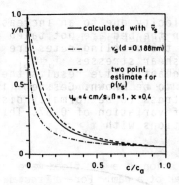

Fig. 4: Vertical concentration profiles for sand grains $0.125 < d \leq 0.25$ mm, $T = 24.4°C$

1. Bagnold R.A. (1966), An approach to the sediment transport problem from general physics, Geological Survey Professional Paper 422-I, Washington.

2. Bechteler W. and Schrimpf W. (1984), Improved numerical model for sedimentation, Journal of Hydraulic Engineering, ASCE, Vol.110, No.3 pp. 234-246.

3. Gessler J. (1973), Behavior of sediment mixtures in rivers, pp. A35.1-A35.11 Vol. I, in Proceedings of the IAHR Symposium on River Mechanics, Bangkok Thailand.

4. Rosenblueth E. (1974), Aproximaciones de segundus momentos in probalidades, Boletin del Instituto Mexicano de Planeación y Operación de Sistemas, 26,1.

5. Schrimpf W. (1986), The Influence of the Mixing Length Assumption on Calculating Suspended Sediment Transport, in River Sedimentation Vol. III (Ed. S.Y. Wang, H.W. Shen, L.Z. Ding) pp.757-765, Proceedings of the Third International Symposium on River Sedimentation, Jackson, Mississippi, USA, 1986.

6. Schrimpf W. (1986), Ein Beitrag zur Berechnung der Sedimentation von Feststoffen in horizontal durchströmten Sandfängen, Dissertation, Universität der Bundeswehr München, Institut für Wasserwesen.

Computing 2-D Unsteady Open-Channel Flow by Finite-Volume Method

C.V. Bellos, J.V. Soulis and J.G. Sakkas
Department of Civil Engineering, Faculty of Engineering, Democrition University of Thrace, Xanthi, 67100, Greece

ABSTRACT

This paper deals with the computation of a 2-D flood wave resulting from the instantaneous break of a dam. For a rectangular computational grid the flow equations require a specific finite difference formulation in the flow area close to the solid boundaries. This leads to numerical approximations which can not accurately represent the flow characteristics in the area near the boundaries. In particular, in two-dimensional free-surface flow computations, the difficulties and inaccuracies associated with the determination of flow characteristics near these boundaries are well known. An attempt has been made here to overcome the aforementioned difficulties. The general technique used is a combination of the finite element and the finite difference method. First a transformation is introduced through which quadrilaterals in the physical domain are mapped into squares in the computational domain. The governing system of equations are thus transformed into an equivalent system applied over a square-grid network. In the second phase, the McCormack explicit finite difference numerical scheme is used for the solution of the transformed system of equations. This two-step (predictor-corrector) scheme has been proved to be well suited for flow computations where discontinuities may be present. The approach described above has been applied for the computation of the propagation of a flood wave resulting from the instantaneous break of a dam. Comparison between computed and experimental data shows a satisfactory agreement. The most important prospect of the method lies in its capability to facilitate computation of unsteady flows in natural (irregular) channels using a relatively simple method of finite differences.

INTRODUCTION

In this work the movement on a dry bed of a 2-dimensional flood wave resulting from the break of a dam is numerically examined. The associated flow problem has been the subject of research work for more than a century. The main efforts have been concentrated on the one dimensional flood wave problem and only recently 2-dimensional approaches have been reported, amongst others, by Katopodes[1], Schmitz[2], Di Monaco[3] and Garcia[4]. Most of the 2-dimensional approaches utilize the numerical methods of finite differences or finite elements. From the analysis of the dam break flood-wave movement two main problems arise: a) the proper mathematical formulation of the wave front region, where the St. Venant equations are not any more applicable due to the abrupt changes of physical quantities of the flow (water depth, velocities) and b) the proper satisfaction of the boundary conditions of a geometrically complex channel.

The first problem was attacked here using a two-step numerical scheme with proven ability in computing hydraulic jumps (Fennema[5]). This scheme was developed by McCormack[6,7] in compressible fluid flow analysis. The second problem was solved using a coordinate transformation system so as the physical flow domain could be mapped into an orthogonal flow domain without altering the form of the partial differential equations governing the flow.

FLOW EQUATIONS AND NUMERICAL FORMULATION.

The equations which govern the 2-dimensional, unsteady flow are written down in matrix form as

$$W_t + F_x + G_y = D \qquad (1)$$

where:

$$W = \begin{bmatrix} h \\ hu \\ hv \end{bmatrix} \quad F = \begin{bmatrix} hu \\ hu^2 + gh^2/2 \\ huv \end{bmatrix} \quad G = \begin{bmatrix} hv \\ huv \\ hv^2 + gh^2/2 \end{bmatrix} \quad D = \begin{bmatrix} 0 \\ gh(S_{ox} - S_{fx}) \\ gh(S_{oy} - S_{fy}) \end{bmatrix}$$

The system of Eqs.(1) is of the conservative form capable in dealing also with hydraulic jumps. The essence of the present numerical scheme is that quadrilaterals in the physical domain will be separately mapped into squares in the computational domain by independent transformations from Cartesian x,y to local ξ,η coordinates as illustrated in Fig.1.

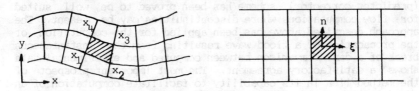

Figure 1. Flow domain transformation.

If x_1, y_1 are the Cartesian coordinates of the corners of a cell then the coordinates of any point of this cell can be expressed as

$$x = \sum_1^4 N_1 x_1 \qquad y = \sum_1^4 N_1 y_1 \qquad (2)$$

where N_1 are the shape functions associated with the finite volume nodes. The shape functions are defined in terms of a local non-orthogonal coordinate system ξ,η, as

$$\begin{aligned} N_1 &= (1-\xi)(1-\eta)/4 \\ N_2 &= (1+\xi)(1-\eta)/4 \\ N_3 &= (1+\xi)(1+\eta)/4 \\ N_4 &= (1-\xi)(1+\eta)/4 \end{aligned} \qquad (3)$$

Let J^{-1} be the transformation matrix from the physical system to the computational local coordinate system,

$$J^{-1} = \begin{bmatrix} x_\xi & x_\eta \\ y_\xi & y_\eta \end{bmatrix} \quad (4)$$

The following relations hold (Steger[8]),

$$x_\xi = J^{-1} \eta_y, \quad x_\eta = -J^{-1} \xi_y, \quad y_\xi = -J^{-1} \eta_x, \quad y_\eta = J^{-1} \xi_x \quad (5)$$

Under the aforementioned transformation of Eqs.(1) into the local coordinate system ξ, η they assume the form

$$W'_t + F'_\xi + G'_\eta = D' \quad (6)$$

where

$$W' = J^{-1}\begin{bmatrix} h \\ hu \\ hv \end{bmatrix} \quad F' = J^{-1}\begin{bmatrix} hU \\ hUu + \xi_x gh^2/2 \\ hUv + \xi_y gh^2/2 \end{bmatrix} \quad G' = J^{-1}\begin{bmatrix} hV \\ hVu + \eta_x gh^2/2 \\ hVv + \eta_y gh^2/2 \end{bmatrix} \quad D' = J^{-1}\begin{bmatrix} 0 \\ gh(S_{ox}-S_{fx}) \\ gh(S_{oy}-S_{fy}) \end{bmatrix}$$

and U,V are the velocity components in the computational domain while the velocity components u,v in the physical domain are

$$\begin{bmatrix} u \\ v \end{bmatrix} = J^{-1}\begin{bmatrix} U \\ V \end{bmatrix} \quad (7)$$

For the numerical solution of Eqs.(6) the arithmetic scheme developed by McCormack[7] was used. This is a two-step explicit scheme with second order accuracy. Application of a complete two-step cycle advances the solution by one time increment Δt. In the first step, approximate values of the unknown variables are determined. These approximations are then used in the second step to determine the final values of the variables at the end of the time increment considered. The algorithm, as was used in this work, is as follows:
predictor algorithm

$$\tilde{W}_{i,j} = W_{i,j} - \lambda_x(F_{i+1,j} - F_{i,j}) - \lambda_y(G_{i,j+1} - G_{i,j}) + D\Delta t \quad (8)$$

corrector algorithm

$$W_{i,j} = 0.5[W_{i,j} + \tilde{W}_{i,j} - \lambda_x(\tilde{F}_{i,j} - \tilde{F}_{i-1,j}) - \lambda_y(\tilde{G}_{i,j} - \tilde{G}_{i,j-1}) + \tilde{D}\Delta t] \quad (9)$$

where $\lambda_x = \Delta t/\Delta x$ and $\lambda_y = \Delta t/\Delta y$
Combinations of backward and forward spatial finite differences lead to four equivalent variants of the numerical scheme and their cyclic alterations correct its inherent assymetry.

INITIAL AND BOUNDARY CONDITIONS

The following relations have been taken as initial conditions for the dam-break problem:
a) Upstream of the dam
$h(x,y,t_0) = h_0(x,y,t_0), \quad u(x,y,t_0) = 0, \quad v(x,y,t_0) = 0$
The values of $h_0(x,y,t_0)$ depend on the upstream bottom slope.
b) At the dam site, denoted by subscript d

$$h(x_d,y_d,t_o)=4/9h_o(x_d,y_d,t_o), \quad u(x_d,y_d,t_o)=\sqrt{gh(x_d,y_d,t_o)},$$
$$v(x_d,y_d,t_o)=0$$

The above equations represent the Ritter solution for the one-dimensional frictionless flow.

c) Downstream of the dam (dry bed condition)
$$h(x,y,t_o)=0, \quad u(x,y,t_o)=0, \quad v(x,y,t_o)=0$$

Flow characteristcs at time $t+\Delta t$ at a solid boundary are determined by solving the normal flow equations using a node beyond the boundary. Flow conditions at time t at this node are determined by a quadratic extrapolation (Sparis[9]). It is assumed that the flux through the solid boundary is zero. The upstream boundary of the flow field is realized by a vertical wall (no flow). Regarding the downstream flow boundary it is the nature of the problem under consideration that the depth of flow reduces to zero there. Under such an assumption no specific boundary conditions and equations expressing them are necessary: the normal flow equations are applied sweepingly throughout the computational field including not only the actual flow field but also the space downstream of it where no actual flow exists at the moment. Computational difficulties as well as diffusion or smearing of the wave front arise due to very small or occasionally negative depths at the boundaries of the flow field as a result of wave front displacement when grid points with zero flow depth are being considered. To avoid these unfavorable consequences computed depths at a newly covered grid point are being accepted when they become larger than a minimum value of the order of $0.01h_o$, otherwise they are set equal to zero. The same procedure is applied, inversely, at the trailing edge of the wave in the case the flow depth at the upstream boundary reduces to zero.

The stability of the numerical scheme is ensured by the application of the Courant-Friedrichs-Lewy (CFL) criterion
$$\Delta t < \min[\Delta x/u+c, \Delta y/v+c] \tag{10}$$
where c is the celerity of a small flow disturbance.

APPLICATIONS

For the purpose of comparing results with computational and experimental data obtained by others, the technique developed in the present work was applied to certain simple dam-break situations.

The first case concerns to a series of experiments for the dam-break problem conducted at the Waterways Experiment Station (WES), U.S.Corps of Engineers[10]. For these experiments a rectangular, wooden flume, lined with plastic-coated plywood, 122m long and 1.22m wide and having a bottom slope $S_o=0.005$ was used. The model dam was placed at midsection, impounding water to a depth of 0.305m. Average value of Manning roughness coefficient n=0.050 was used. In Fig.2 stage hydrographs at certain locations along the flume are given. Except for the first few moments of flow, especially at the dam site, agreement between computed and experimental results is very satisfactory.

The second case refers to a converging-diverging open channel flume configuration with an hypothetical dam located right at the throat of the flume, Fig.3a. The experiments were made in the Hydraulics Laboratory, Democrition University of Thrace. Comparison between the measurements and finite volume predicted results is shown in Fig.3b. The agreement is satisfactory.

The third case refers to a curved open channel flume configuration, shown in Fig.4a. Computed results for the water depth along the convex and the concave sides of the flume are shown in Fig.4b.

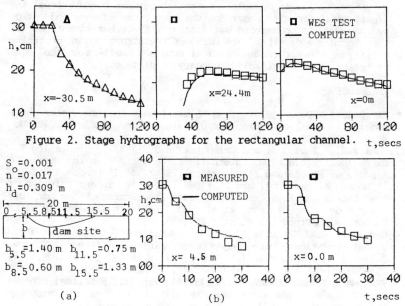

Figure 2. Stage hydrographs for the rectangular channel.

(a) (b)

Figure 3. Converging-diverging channel.
a) Geometry. b) Stage hydrographs.

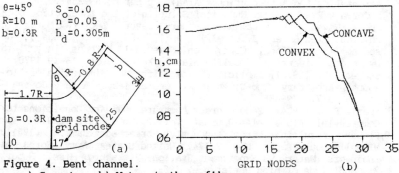

Figure 4. Bent channel.
a) Geometry. b) Water depth profiles.

CONCLUSIONS

The McCormack two step explicit scheme with second order accuracy was employed for the solution of the 2-dimensional unsteady flow equations written in a conservation-law form. The technique was applied for determining the propagation and the profile of a flood wave resulting from the instantaneous and complete collapse of a storage dam. A body fitted non orthogonal, local coordinate system was used to avoid the problems associated with a conventional grid system. Two experimental cases and a computational one of dam break problem were considered for the purpose of comparing results. Agreement detween computational and experimental results regarding the wave front advance and stage hydrographs is considered satisfactory. The great advantages of the technique developed are based on the strong shock-capturing ability of the McCormack numerical scheme as well as on the ease of treatment of exact solid boundary conditions.

REFERENCES.

1. Katopodes N.D. (1979), Two Dimensional Shallow Water Wave Models, Journal of Hydraulics Division, proc. ASCE, vol.104 No. HY 9, 1269-1288.
2. Shmitz G., Seus G.J. and Czirwitzky H.J. (1983), Simulating Two Dimensional Fluid Flow, International Conference on the Hydraulic Aspects of Floods and Flood Control, Sept. 13-15 London, 195-206.
3. Di Monaco A. and Molinaro P. (1982), Finite Element Solution of the Lagrangian Equations of Unsteady Free-Surface Flows on Dry River Beds, In Holz,k.p.,et al.(eds.), Finite Elements in Water Resources, Springer, Berlin, 4.25-4.35.
4. Garcia R. and Kahawita R. (1986), Numerical Solution of the St. Venant Equations with the McCormack Finite Difference Scheme, International Journal for Numerical Methods in Fluids, 6, 259-274.
5. Fennema R.J. and Chaudhry M.H. (1986), Explicit Numerical Schemes for Unsteady Free Surfaces Flows with Shocks, Water Resources Research, 22(13), 1923-1930.
6. McCormack R.W. (1969), The Effect of Viscosity in Hypervelocity Impact Cratering, Journal of the American Institute of Aeronautics and Astronautics, (AIAA), paper No 69-354+.
7. McCormack R.W. (1970). Numerical Solution of the Interaction of a Shock Wave with a Laminar Boundary Layer, Lecture Notes in Physics, (ed.Ehlers J.,at al.), pp 151-163, Proceedings of the second International Conf. on Numerical Methods in Fluid Dynamics, University of California, Berkeley, Sept. 15-19,1970.
8. Steger J.L. (1978), Implicit Finite-Difference Simulation of Flow about Arbitrary Two-Dimensional Geometries, AIAA Journal, Vol.16,No 7, 679-686.
9. Sparis P.D. (1984), Second-order Accurate Boundary Conditions for Compressible Flows, AIAA Journal, Vol.22, No 9, 1222-1228.
10. Floods resulting from suddenly breached dams. (1961), Miscellaneous Paper No. 2-374, United States Army Corps of Engineers, Waterways Experiment Station, Vicksburg, Miss., Report 2: Conditions of high Resistance, 1961.

Eulerian-Lagrangian Linked Algorithm for Simulating Discontinuous Open Channel Flows

S.M.A. Moin
Inland Waters/Lands Directorate, Environment Canada, Burlington, Ontario, Canada

D.C.L. Lam
National Water Research Institute, Environment Canada, Burlington, Ontario, Canada and Department of Civil Engineering and Engineering Mechanics, McMaster University, Hamilton, Ontario, Canada

A.A. Smith
Department of Civil Engineering and Engineering Mechanics, McMaster University, Hamilton, Ontario, Canada

ABSTRACT

Standard finite difference and finite element methods result in poor and unsatisfactory solutions for near discontinuous open channel flows. The presence of oscillatory waves in the proximity of the discontinuous region is commonly addressed by adding external or internal dissipating mechanisms. A space-time finite element method based on the Galerkin formulation of the divergent form of the open channel flow equations is presented. The proposed Eulerian-Lagrangian linked algorithm demonstrates remarkable shock capturing properties while maintaining a high order of accuracy. No extraneous parameters are required to dissipate the parasitic oscillations. Results are presented for a number of flow conditions that lead to the formation of discontinuity and shocks. The improvement over other techniques is demonstrated by comparing the solution with finite element and finite difference methods.

INTRODUCTION

Discontinuities in open channel hydraulics develop as a result of the sudden release of water following the breach of a dam or rejection of the load at a turbine. The literature is abundant in both theoretical development of procedures for dam-breach/floodwave routing and their practical applications.

The problems in solving rapidly varying flows reside in the oscillatory behaviour of many discretized difference forms of the open channel flow equations. In fluid mechanics, similar problems of shock analysis required the development of numerous numerical techniques. When analyzed critically, all the techniques boil down to devising novel dissipative interfaces and numerical damping devices. These devices are biased toward the short wavelengths and selectively neutralize them.

This paper describes the development of a novel finite element method for solving near discontinuous rapidly varying open channel flows. The divergent form of the St.Venant Equations describes the conservation of volume (strictly speaking mass) and momentum of the open channel flow. These equations are solved by a moving space-time finite element method.

The technique is first applied to two numerical experiments with known analytical solution. Comparison with the finite element and finite difference methods demonstrates the superiority of the scheme.

MODEL DEVELOPMENT

The divergent form of the open channel flow equations in a frictionless horizontal channel is given as follows:

Mass Conservation

$$\frac{\partial Q}{\partial x} + \frac{\partial A}{\partial t} = 0 \tag{1}$$

Momentum Conservation

$$\frac{\partial Q}{\partial t} + \frac{\partial}{\partial x}(Qu + gAh) = 0 \tag{2}$$

where x = distance, t = time, y = depth of flow, Q = discharge rate, A = cross-sectional area (a function of y), u = mean velocity, h = depth of centroid of cross-section, g = acceleration due to gravity.

A new technique is proposed for solving Eqs. (1) and (2) involving a space-time finite element. The mathematical approach was first presented by Bannerot and Jamet (1974) for convection dominated flows. This approach was furthered by Voroglu and Finn(1980).

Application of the Galerkin's method to Eqs. (1) and (2) results in the vanishing of the weighted residual of the two equations with respect to a continuous basis function $\phi(x,t)$ which can be presented as

$$\int_{t^n}^{t^{n+1}} \int_{x_L}^{x_R} \phi(x,t) \left[\frac{\partial \bar{Q}}{\partial x} + \frac{\partial \bar{A}}{\partial t} \right] dx\, dt = 0 \tag{3}$$

$$\int_{t^n}^{t^{n+1}} \int_{x_L}^{x_R} \phi(x,t) \left[\frac{\partial \bar{Q}}{\partial t} + \frac{\partial}{\partial x}(\overline{Qu} + g\overline{Ah}) \right] dx\, dt = 0 \tag{4}$$

where the variables with a bar are approximations of dependent and derived variables.

The solution domain is shown in Fig. 1. The area bounded by $x_L \leq x \leq x_R$ and $t^n \leq t \leq t^{n+1}$ is discretized by the trapezoidal space-time elements $K_1^n, K_2^n, K_3^n, \ldots K_{N-1}^n$. After performing the numerical integration, the continuity and momentum equations are written as (5) and (6). For brevity, only contributions from interior nodes are presented here.

Continuity Equation

$$\frac{k}{4}(Q_{i+1}^n + Q_{i+1}^{n+1} - Q_{i-1}^n - Q_{i-1}^{n+1}) - \frac{1}{4}\left[(A_{i+1}^n + A_{i+1}^{n+1})\delta x_{i+1}\right.$$

$$\left. - (A_{i-1}^n + A_{i-1}^{n+1})\delta x_{i-1}\right] + \frac{1}{2}(A_i^{n+1}\Delta x^{n+1} - A_i^n \Delta x^n) = 0$$

(5)

Momentum Equation

$$-\frac{1}{4}\left[(Q_{i+1}^n + Q_{i+1}^{n+1})\delta x_{i+1} - (Q_{i-1}^n - Q_{i-1}^{n+1})\delta x_{i-1}\right]$$

$$+ \frac{k}{4}\left[(Qu + gAh)_{i+1}^n + (Qu + gAh)_{i+1}^{n+1} - (Qu + gAh)_{i-1}^n\right.$$

(6)

$$\left. - (Qu + gAh)_{i-1}^{n+1}\right] + \frac{1}{2}(Q_i^{n+1}\Delta x^{n+1} - Q_i^n \Delta x^n) = 0$$

where

$$k = t^{n+1} - t^n$$

$$\Delta x^{n+1} = x_{i+1}^{n+1} - x_{i-1}^{n+1}$$

$$\delta x_{i+1} = x_{i+1}^{n+1} - x_{i+1}^n$$

$$x_{i+1}^{n+1} = x_{i+1}^n + u_{i+1}^n k$$

The resulting equations lead to a bi-tridiagonal matrix which is solved by the double sweep technique adopted by Cooley and Moin (1976).

Eulerian-Lagrangian Solution
In this research, the strength of a superior, Lagrangian mode solution was captured at any given time step by discretizing the equations along ABQP, BCRQ, etc. as shown in Fig. 2. Before advancing the solution to the next level, however, the dependent variables Q, and y, were obtained by interpolation at the Eulerian points A B C .... The Lagrangian step is again implemented by allowing rays to emit at the Eulerian grid points. Because the elements are allowed to move this scheme is also called the Moving (finite) Element Method.

NUMERICAL TESTING

In the absence of rigorous field or experimental data, it was decided to test the technique against a known analytical solution. This provides an excellent bench mark. The Moving Element (ME) model is first compared against the analytical solution in terms of shape, size, location and speed of the surge. The superiority of the moving element model is next demonstrated by a comparison with the finite difference (FD) and finite element (FE) methods.

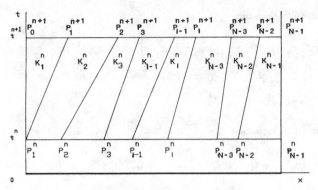

Figure 1 Lagrangian Solution Domain

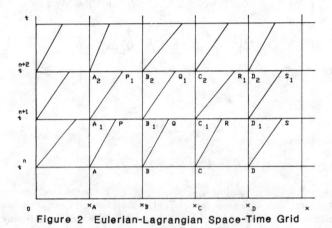

Figure 2 Eulerian-Lagrangian Space-Time Grid

Supercritical Floodwave in a Frictionless Channel - Test 'A'
This experiment was studied by Vasiliev et al.(1965) and Terzidis and Strelkoff (1970). A very sharp supercritical front is propagated through a horizontal frictionless channel of rectangular cross-section 200 m wide. Initially there is no flow at a depth of 2.0 m. Almost instantaneously (in 0.5 sec) the flow is increased to 28000 m^3/s and the depth of flow set at 10.1 m at the upstream boundary. The depth at the downstream end is maintained at 2.0 m. The channel is 1500 m. long.

The channel was divided into 150 elements each 10 m in length with a 0.5 sec time step. The program was run for 120 time steps. Figure 3 depicts the analytical and numerical results at 60 seconds. Apart from the spike at the front, the quality of numerical results indicate excellent wave capturing properties in terms of the shape, size, location and speed of the surge.

Subcritical Surge in a Frictionless Channel - Test 'B'
This test has not been reported by others. Although this does not provide a comparison with other available techniques, the results when compared with FE and FD computations are of interest.

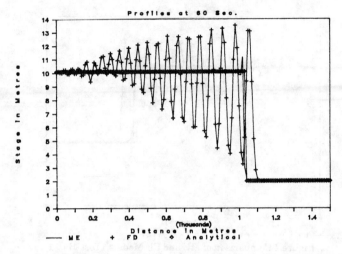

Figure 3 Comparison of ME and FD Models (Test 'A')

The test involves a horizontal, frictionless channel of rectangular section 100 m wide and 1000 m long subjected to a positive surge. The initial flow is zero at 2.0 m depth. At the upstream boundary, Q is increased to 2000 m³/s in 10 seconds. The zero flow constraint is imposed on the downstream boundary.

The channel was divided into 100 elements each 10 m. long with 1.0 second as the time step. The program was operated for 90 time steps. The quality of wave capturing properties is again evident from Figure 4.

Comparison with Existing Methods The proposed ME model was compared against existing finite difference and finite element schemes. The FD scheme is a second order central difference model with Crank-Nicholson's temporal weighting of 0.5. Similarly, the finite element model developed by Cooley and Moin (1976) was modified and adopted. This temporal centering provided consistency with the ME model.

The FE and FD models were implementd for the two tests reported for the ME model. For Test 'B' the rate of increase of upstream hydrograph was slowed 10 fold to permit the finite element model to operate. The results from the ME model are compared with the FD scheme for Test 'A' and with the FE scheme for Test 'B' in Figures 3 and 4 respectively.

The quality of improvement in results is immediately evident. Both FE and FD results are made meaningless by the presence of parasitic oscillations.

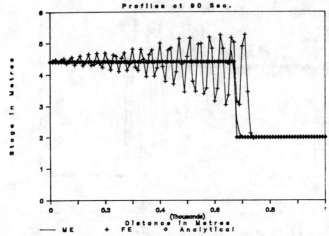

Figure 4 Comparison of ME and FE Models (Test 'B')

CONCLUSIONS

Based on the foregoing, the following important conclusions are deduced.

(1) A powerful new technique in solving the unsteady flow equations especially for the sharply rising flood wave has been developed.

(2) The Eulerian-Lagrangian based method is superior to any other technique in the purest form i.e. without adding diffusion terms or by advancing the Crank-Nicholson centered weighting.

(3) The Lagrangian scheme in itself does not provide for a non-dissipative interface. Dissipation is provided by the Eulerian regridding.

REFERENCES

(1) Bonnerot R. and Jamet P. (1974), A Second Order Finite Element Method for the One-Dimensional Stefan Problem, Int. J. Num. Meth. Eng. V8, pp 811-820.

(2) Cooley R.L. and Moin S.A. (1976), Finite Element Solution of the St. Venant Equations. J. Hydraulics Division, ASCE V102, HY6, pp 759-775.

(3) Terzidis G. and Strelkoff T. (1970), Computation of open Channel Surges and Shocks, J. Hydraulics Division, ASCE V96 HY1, pp 223-252.

(4) Vasiliev O.F., Gladyshev M.T., Privites N.A. and Sudobicher V.G. (1965), Methods for the Calculation of Shock Waves in Open Channels, Proceedings, 11the Congress IAHR, Leningrad, Paper 3.44, 14 p.

(5) Voroglu E. and Finn W.D.L. (1980), Finite Elements Incorporating Characteristics for One-Dimensional Diffusion-Convection Equation. J. Computational Physics 34 pp 371-389.

SECTION 4 - SPECIAL SESSION ON REMOTE SENSING AND SIGNAL PROCESSING FOR HYDROLOGICAL MODELING

On Thin Ice: Radar Identification of thin and not so thin Layers in Hydrological Media

K. O'Neill

U.S. Army Cold Regions Research and Engineering Laboratory, Hanover, NH 03755, USA

ABSTRACT

Many remote sensing systems have been tried or proposed for identifying interfaces in surface and subsurface media.[3,4] As an example we consider here the use of short pulse radar applied to ice layers. A computational system is required which will clearly mark reflections from interfaces while suppressing other unwanted responses as much as possible, so that automated layer detection is possible. This is achieved using a straight forward, band limited inverse filter, modified by a time dependent penalty weighting procedure. Contrasting model dependent techniques must be applied to reflections from thin ice, in which returns from top and bottom surfaces and multiples are not separated. The same techniques which succeed here should be applicable to other sufficiently non-dissipative media featuring layers with contrasting dielectric constants.

INTRODUCTION

Over the past few decades, considerable attention has been devoted to the detection of layering in subsurface media and to the remote sensing problems posed by it. Fundamental signal processing progress in this area has been stimulated by the petroleum industry, for which the primary task has often been the clarification of seismic records and the elimination in them of "ghost" echoes (reverberation "multiples") between layer interfaces. Radar techniques were chosen over seismic methods for the investigation reported here, in part because in the former one can often characterize the radiated signal rather well. This helps substantially in interpreting recorded returns. The returns will feature the same sort of reverberation sequences as appear in seismic records. In the treatment here of relatively thin layers, we will include all multiples in the considerations necessary to determine layer thickness.

The response of a single layer to an incident radar signal is illustrated in Figure 1. The incident radiation produces a primary reflection (R_o) from the top of the layer. Because some of the incident energy also penetrates the upper surface of the layer, a portion of the signal propagates downward, reflecting off the layer bottom (R_1). Because returns from the bottom reflect off the top of the layer on their way up, a sequence of multiples is generated (R_2, R_3, R_4,....), each smaller in magnitude than the previous one. If one could input a perfect impulse (and if media properties are such that the signal remains undistorted), then an observer above the layer would record a sequence like that shown at the top of the figure. Given the electromagnetic wave velocity in the layer, one can measure the separation in time between, say, R_o and R_1, and then calculate the thickness of the layer.

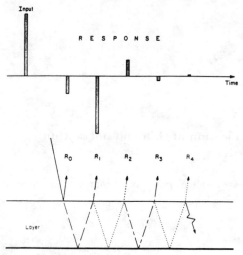

Figure 1. Response to incident radar by ice layer over perfect reflector.

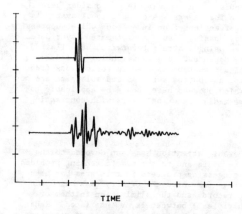

Figure 2. Top: Incident wavelet. Bottom: Field record of response by river ice layer.

Radar antennas commonly used in surface and subsurface exploration cannot produce a perfect impulse. Figure 2 (bottom) shows the recorded response to an incident wavelet (top) by a layer of ice on a river. A reference incident wavelet such as this may be extracted from records like that in the figure, may be estimated from calibration shots, or may be known a priori. Our first task is to devise an algorithm which can recognize this wavelet in a general, wiggley, possibly noisy signal and then transform it into some clear and compact shape, while suppressing other uninteresting content nearby. This shape can then serve as an unambiguous time marker, which ideally can be discerned by automatic equipment. Our second task derives from the fact that the incident pulse has a finite duration which may be equal to or greater than the time separation between responses in the echo sequence. This means that R_0, R_1, R_2, etc. may overlap. This has been the case for "thin" (30 - 50cm) ice layers of interest in this study. Hence a method was required for inferring the thickness of a layer when it produces an amalgam of superposed responses.

WELL SEPARATED REFLECTIONS

The by now classical approach to wavelet identification in digital records is the least squares Wiener filter and its descendents. This has been discussed extensively in the literature[4], and will only be reviewed briefly here. Basically, one begins with the desired shape d_t into which the known input wavelet s_t is to be converted, where the subscript t is a discrete time index. This conversion will be done via a filtering operation of the form

$$y_t = \sum_{k=-m}^{m} c_k s_{t-k} . \qquad (1)$$

The filtered signal y_t is the result at any time point t from applying the (initially unknown) filter coefficients c_k about that point in the signal record. To determine the filter coefficients, we construct an error measure

$$E = \sum_t (d_t - y_t)^2 \qquad (2)$$

and minimize it by setting its partial derivatives with respect to each c_i equal to zero. This leads to 2m+1 equations for an equal number of unknown coefficients

$$\sum_{k=-m}^{m} c_k \sum_t s_{t-k} s_{t-i} = \sum_t d_t s_{t-i} \qquad (3)$$

$$i = -m, -m+1, \ldots, m-1, m$$

or in matrix notation

$$[S] \{c\} = \{D\} . \qquad (4)$$

We may solve eq (4) for the vector of filter coefficients {c}.

In principle one may derive a filter for virtually any desired wave shape d_t. However here we gain the most from a "spiking filter," which will convert the wavelet into an easily distinguishable impulse. The left side of Figure 4 shows the results of applying such a Wiener filter with m=25 to a typical wavelet of length 38.

One gains insight into the action of this sort of inverse filter by considering its operations in the frequency domain. By the Fourier integral convolution theorem, we know that convolution in the time domain is equivalent to multiplication in the frequency domain. Thus the equivalent of eq (1) is

$$Y(f) = C(f) * S(f) \qquad (5)$$

where the variables are the Fourier transforms of their respective lower case counterparts in eq (1), and f denotes frequency. The Fourier transform of a perfect spike (delta function) is uniformly equal to a constant over all frequencies; It is real and has zero phase, assuming that the spike is at t=0. Thus $C(f)$ is the inverse of $S(f)$ and its action is to produce a flattened spectrum in $Y(f)$ when it operates on the wavelet. The amplitude spectra of the wavelet and of the filter used in Figure 4 are shown in Figure 3. The corresponding phase spectra are such that they combine to produce approximately zero.

As formulated above, spiking Wiener filters have a number of limitations. The autocorrelation in equation (3) which produces the governing matrix [S] wipes out information on wavelet phase content. For "mixed delay" type signals such as those considered here, one must search about for an acceptable spike location over the duration of the wavelet, in order to compensate for the particular distribution of energy within the wavelet. Also, one wants the filter to respond to the wavelet and, ideally, only to the wavelet. However other signals with substantial content within the main frequency band of the wavelet may still produce significant unwanted responses. And worse, noise outside of the main band may produce a large response, because the amplitude of the filter is quite high there (Figure 3). If one adds even very faint high frequency noise to the wavelet, then an otherwise good filter can produce terrible

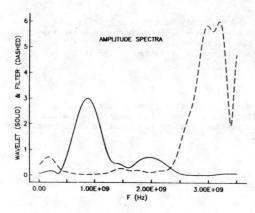

Figure 3. Amplitudes of wavelet and its Wiener spiking filter vs frequency.

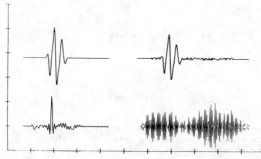

Figure 4. Left: Wavelet before (top) and after (bottom) processing by spiking Wiener filter. Right: Same in the presence of noise.

results (Figure 4). Under realistic conditions, additional special measures must be taken and the Wiener filter strategy must be changed somewhat to deal with inevitable noise[1].

In this work an acceptable inverse filtering system was formulated by operating in the frequency domain, simply using the inverse of the wavelet spectrum directly for $C(f)$, and restricting operations to the main frequency band. One proceeds by marching through the signal, excising for treatment at each time t a segment x_τ of the same length as the wavelet, where τ equals zero at the beginning of the segment. To test for presence of the wavelet one divides the segment transform $X(f)$ by $S(f)$, and inverts the result only for $\tau = 0$.

$$U(f) = X(f)/S(f)$$

(6)

$$y_t = u|_{\tau=0} = \sum_{f=f_1}^{f_2} U(f)$$

where f_1 and f_2 are chosen to restrict calculation to the range of significant amplitude in $S(f)$ (see Figure 3). If x_τ consists of a wavelet, then the result should be a spike at u_o, and hence at y_t. While hypersensitive, the Wiener filter is optimal in the least squares sense. To compensate for loss of optimality in our band limited procedure and to increase suppression of unwanted responses, we intensify the spike when agreement between $X(f)$ and $S(f)$ is found and we penalize it when it is not. This is accomplished by multiplying y_t by a time dependent weight w_t, such as

$$w_t = \text{avg} \left\{ \frac{\exp[-2|\phi_x - \phi_s|/2\pi]}{1 + |1 - A_x/A_s|} \right\}^3 \tag{7}$$

where A_x and ϕ_x are respectively the amplitude and phase of $X(f)$ and similarly for $S(f)$, and avg indicates an average over the frequency band. Other such weights may be used successfuly[2], the only requirement being that w_t equal 1 when $X(f) = S(f)$ and that it decline sharply otherwise. Without this factor one obtains unwanted side lumps about the spike in y_t as he windows through the signal, because portions of the wavelet may have spectra similar to $S(f)$.

When noise is added to the wavelet by a random number generator, this modified inverse system is still successful when the peak signal energy to average noise power ratio (SNR) is 100 or even 10 (Figure 5), whereas the SNR on the right in Figure 4 is 1000. When the modified inverse system is applied to the central portion of the signal at the bottom of Figure 2, two distinct reflections emerge with a time separation equivalent to about 48 cm of ice thickness (Figure 6). This is within about 10% of values measured in the field.

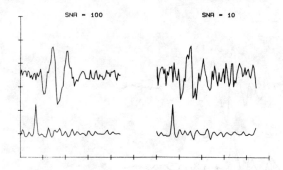

Figure 5. Top: Noisy wavelets. Bottom: Modified inverse filter results.

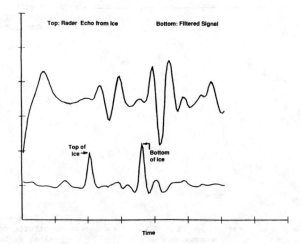

Figure 6. Portion of recorded (top) and processed signal (bottom) using modified inverse filter.

THIN ICE AND OVERLAPPING REFLECTIONS

An altogether new approach is required for thinner ice. With equipment used for the case in Figure 6, ice much thinner than about 50 cm will produce an overlapping not only of R_o and R_1, but also of higher multiples. No clear spikes will result from the filtering operations. For this case we begin by using a reference wavelet to construct a set of synthetic returns, each corresponding to some possible ice thickness. This can be done by summing the sequence R_i for the wavelet, or by applying a unified transfer function which accounts for all multiples[2]. Given the set of synthetic returns, one seeks the best match to the actual record.

Figure 7a shows a real and a synthetic reflection from a sheet of ice about 17 cm thick on top of foil. Clearly they match quite well; each is more distended than the original wavelet, with R_o and R_2 effects visible on either side of the larger, central R_1. To analyze a thin layer response, one runs through the set of comparable synthetic reflections, calculating the correlation func-

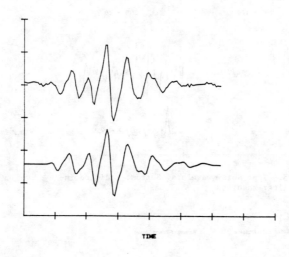

a. Actual (top) and synthetic reflections (bottom) from 17 cm ice layer over foil.

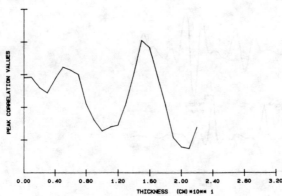

b. Peak actual – synthetic correlation values vs thickness.

Figure 7.

tion of each with the actual reflection. Some shift in each correlation calculation will produce a peak value for that possibility. The highest such peak over all correlations will presumably identify an autocorrelation; that is, it will point to the correct thickness. Figure 7b shows that, while the content of reflections from a range of thin layers is close to that from 17 cm, the highest correlation is indeed in the right locale. Similar tests at other thicknesses have been equally successful, even for layers less than 5 cm.

CONCLUSION

For sufficiently thick, non-dissipative hydrological layers, well separated impulse radar reflections may be obtained corresponding to returns from interfaces and their multiples. By suitably modifying a straight forward, band limited inverse filter, one can distinguish these returns in field records and successfully estimate the thickness of ice and presumably other layers. For thin layers which produce overlapping interface returns, one obtains successful thickness estimations by referring to a set of synthetic total reflections including all multiples. The synthetic signals correspond to various possible thicknesses, and the best correlation between measured and synthetic reflections points to the correct thickness.

REFERENCES

1. Cooper, R.C. and McGillem, C.D., (1986). Modern Communications and Spread Spectrum, McGraw-Hill, New York.

2. O'Neill, K. (1988). Signal Processing of Radar Reflections from Fresh Water Ice, CRREL Report, in preparation.

3. Rossiter, J.R. and Bazeley, D.P. (Eds) (1980). Proc. Int'l. Wkshp. Remote Estim. Sea Ice Thickness, St. John's, C-Core Publ. No. 80-5.

4. Webster, G.M. (Ed) (1981). Deconvolution, SEG Geophysics Reprint Series No. 1, Tulsa, Oklahoma.

Satellite Observations of Oceans and Ice

K.C. Jezek and W.D. Hibler

U.S.A. Cold Regions Research and Engineering Laboratory and Thayer School of Engineering, Dartmouth College, Hanover, NH 03755, USA

Introduction

   Data collected by satellite-borne instruments are being employed increasingly in studies of oceans and ice. Indeed, optical and infrared observations are used routinely for projects such as estimating the edge and condition of the arctic ice pack, but at high latitudes, optical instruments are limited frequently by cloud cover and the long polar night. Consequently, instruments operating in the microwave region have been receiving considerable attention in high latitude studies. By relying on either active probing or on natural emission, these instruments function day or night and by appropriate choice of band are not strongly affected by cloud cover.

   The significance of microwave remote sensing instruments lies in the promise of regular and routine observations of the polar regions that is on a scale suitable for understanding the role of the polar regions in global processes. For example, whereas surface campaigns to map sea ice type may be done once a year during a portion of the year optimal for ship based operations, satellite data can be acquired daily over periods as short as several days and as long as years with coverage extending over the entire ice covered regions of the earth. This is not to say satellite observations will replace more focused and detailed surface or limited aircraft observations; rather remote sensing data are a new element offering for the first time a contextual frame work into which the more concentrated information can be incorporated or compared.

   Several satellites presently in orbit are instrumented to collect data relevant to studies of oceans and ice. Recently deployed examples include the U.S. Navy's Geosat for ocean topography, the Defense Meteorological Satellite Programs Special Sensor Microwave Imager (SSM/I) used for ocean-surface wind speed estimates, land-surface snow-cover, and sea ice concentration, and the Japanese MOS-1 satellite which also carries on board a passive microwave imager. A series of new satellites is planned by several nations for deployment in the early and mid 1990s. These platforms will include such onboard instrumentation as Synthetic Aperture Radar (SAR) for sea ice motion and ocean wave observations, radar altimeters for ocean topography, scatterometers for wind speed and color sensors for biomass.

   So, on the one hand, we can anticipate an unprecedented variety of environmental data repeatedly acquired over the worlds oceans. On the other hand, there is a growing body of numerical models that predict large scale ocean behavior that ultimately require such data for initialization and validation. For example, a dynamic/thermodynamic model of arctic sea ice (Hibler[1]) is used by the science community to study circulation in the arctic basin and by the U.S. Navy to support its operational activities with ice motion and thickness predictions. However, the data base presently utilized by these models is limited and the challenge now faced is developing efficient means for assimilating the large satellite data base into global scale models of ocean and ice dynamics. The objective of this paper is to review the expected satellite capabilities, to outline the types and potential of the data to be collected and to discuss issues and progress associated with casting these data into geophysical parameters that can be assimilated by the models.

Overview of Selected Satellite Systems

Passive microwave imagers have been successfully deployed in space since the early 1970's. (A detailed compilation of data from the Electronically Scanning Microwave Imager can be found in Zwally, et. al. [2]; Parkinson, et. al.[3].) These instruments sense the radiant flux from the earth's surface in a narrow frequency band. The radiance (or in microwave terminology the brightness temperature) is related both to the actual surface temperature and to the emissivity of the surface. At present the Special Sensor Microwave Imager (SMM/I) collects 7 channels of data (19, 37, and 85 GHz at orthogonal polarizations; 22 GHz at vertical polarization) with surface footprints ranging from 15 km at 85 Ghz to 70 by 40 km at 19 Ghz. The swath width is about 1400 km. On-board tape recorders store data for subsequent retransmission to a ground receiving station. Passive microwave data are well suited to global scale observations, but their large footprint precludes their application to many more detailed observations. High resolution (about 10 m) can be achieved with a class of active microwave sensors. These devices, called Synthetic Aperture Radars, are sensitive either to geometric or dielectric roughness of the surface. High range resolution is achieved by transmitting a long duration, swept frequency pulse. Range compression is affected by matched filter techniques. Along track resolution is achieved by recording the doppler history of the received signals and then employing a matched filter constructed using either the measured or computed doppler characteristics of a point target.

The first satellite-borne SAR developed for geophysical observations of the earth was carried on-board the SEASAT satellite launched in 1978. Data from that mission demonstrated the capability to use SAR for, among other objectives, observations of ocean waves, lake and river ice, glacier mapping and sea ice. SAR misssions are planned by several nations through the 1990's including the the First European Remote Sensing Satellite with a C-band SAR, the Japanese Earth Resources Satellite with an L-band SAR, the three frequency, dual polarization Shuttle Imaging Radar - C, and the Canadian Radarsat, a C-band SAR with a maximum planned swath width of 500 km.

A central problem inherent in both classes of image data is sheer data volume. The polar regions alone are imaged twice daily by SSM/I and given that future DMSP platforms are planned to carry SSM/I instruments, the eventual retrospective archive will be immense (Weaver, et. al. [4]). SAR raw data volumes are on the order of 200 Mbits /second and a single ground receiving station being implemented in Alaska is planned to receive 40 minutes of data per day (Carsey, et. al. [5]). Definition and development of analysis and archive facilities for these data sets by several, narrowly focused scientific disciplines are now underway. As regards oceans and ice, much progress has been made in implementing algorithms that routinely extract ice concentration, ice type and ice edge from passive microwave data (e.g. Swift and Cavalieri [6]). Work is in progress to develop an algorithm for ice motion derived from repeat SAR scences. This work, based on both hierarchical correlation (Filly and Rothrock [7]) or edge detection/correlation (Vesecky et al. [8]) schemes will be implemented as part of the Alaska SAR Facility Project. The next crucial step will be to insure that the derived geophysical products can be assimilated by numeric and analytic models of geophysical processes.

Numerical Models

Large scale data bases as provided by satellite observations interact with numerical and analytic models on three levels. First, the data can be used to calibrate the model. For example, areas known to be 100% ice free and areas known to be 100% ice covered are used to calibrate radiation transport models used for extracting sea ice concentration. Second, the data can be used to validate models by providing crucial tests of model predictions. For example, models that input data on ocean surface wind speed and geostrophic currents make predictions about ice motion that may be testable against SAR data. Finally, satellite data can be used to initialize models by providing start up conditions such as the location of the ice edge (Prellor [9]). Alternatively, the results of models can be used iteratively as a test of the original interpretation of the remote sensing imagery. Recently, Zwally and Walsh[10] have shown a consistency between passive microwave derived ice concentrations and dynamic/thermodynamic model results for regions of the central arctic ice pack which were characterized with the microwave data to exhibit diverging flow. Results like those are significant because they enable an independent assessment of the validity of a remote sensing conclusion without recourse to a surface validation campaign.

Our work is proceeding towards using remote sensing data to intialize numerical models as well as iteratively applying numerical models to remote sensing observations to reinforce our interpretation of the remote sensing data. In the former case, key issues of polar oceanography addressable with passive microwave data include the physics of ice edge advance and retreat, ice rheology, and ice ocean dynamics. For example, a localized modeling study of Weddell Sea ice pack motion by Hibler and Ackley[11] identified the critical role played by ice dynamics in the decay of the ice pack both by advecting ice into warmer waters and increasing the fraction of open water available for radiation absorption. Ice edge and ice concentration data derived from SSM/I form a natural validation product for this study but details regarding the temporal and spatial sampling of derived geophysical products, the density of grid points and grid configurations require futher investigation. As regards the latter point, we are exploring statistical methods for characterizing SAR imagery as those statistics relate to different sea ice types. The motivation is that one important ice type, frazil ice, forms under turbulent ocean conditions whereas another important ice type, congelation ice, grows thermodynamically under quiescent ocean conditions. We seek to characterize with SAR, large (100's of km) regions of the polar ice packs in terms of these ice types for use as proxy indicators of ocean conditions.

References

1. Hibler, W.D., (1979). A dynamic thermodynamic sea ice model, J. Geophys. Res, Vol. 88, p815-846.

2. Zwally, H.J., J.C. Comiso, C.L. Parkinson, W.J. Campbell, F.D. Carsey, and P. Gloersen (1983). Antarctic Sea Ice, 1973-76: Satellite passive microwave observations, NASA Spec. Publ. SP459, 206 p.

3. Parkinson, C.L., J.C. Comiso, H.J. Zwally, D.J. Cavalieri, P. Gloersen, and W.J. Campbell (1988). Arctic Sea Ice, 1973-76: Satellite passive microwave observations, NASA Spec. Publ.

4. Weaver, R., C. Morris, R.G. Barry (1987). Passive microwave data for snow and ice research: Planned products for the DMSP SSM/I System, EOS, Vol. 68, no. 39, p 769,776-777.

5. Carsey, F.D., K.C. Jezek, J. Miller, W.Weeks, and G.Weller (1987). The Alaska Synthetic Aperture Radar (SAR) Facility Project, EOS, Vol. 68, no. 25, p. 593-596.

6. Swift, C.T. and D.J. Cavalieri (1985). Passive microwave remote sensing for sea ice research, EOS, Vol. 66, no. 49, p.1210-1212.

7. Filly,M. and D.A. Rothrock (1986). Extracting sea ice data from satellite SAR imagery, IEEE Trans. Geosci. Remote Sensing, Vol. GE-24, p849-854.

8. Vesecky, J.F., R. Samadani, M.P. Smith, J.M. Daida, R.N. Bracewell (1988). Observations of sea ice dynamics using Synthetic Aperture Radar images: Automated analysis, IEEE Trans. Geosci. Remote Sensing, Vol. 26, no. 1, p.38-48.

9. Prellor, R.H. (1985). The NORDA/FNOC polar ice prediction system (PIPS) - Arctic: A technical description, NORDA Report 108, 60 p.

10. Zwally, H.J. and J.E. Walsh (1987). Comparison of observed and modeled ice motion in the Arctic Ocean, Annals of Glaciology, VOl. 9, p. 136-144.

11. Hibler, W.D., and S.F. Ackley (1983). Numerical simulation of the Weddell Sea pack ice, J. Geoph. Res., Vol. 88, p2873-2877.

Applications of Remote Sensing in Hydrology

T. Schmugge and R.J. Gurney
USDA Hydrology Laboratory, Beltsville, MD 20705, USA
R.J. Gurney
Hydrological Sciences Branch, Laboratory for Terrestrial Physics, NASA Goddard Space Flight Center, Greenbelt, MD 20771, USA

ABSTRACT

Remote sensing has the capability of observing several variables of hydrological interest over large areas on a repetitive basis. These variables include surface soil moisture, surface temperature, albedo/land cover, snow water equivalent and snow cover area. With the possible exception of the last item there has not been extensive use of remotely sensed data in hydrological models. Nevertheless, remotely sensed data are essential for global and continental applications for which useful estimates of hydrologic parameters can be made.

INTRODUCTION

The physical basis of remote sensing depends on the inference of land surface characteristics from measurements of the emitted or reflected electromagnetic energy from the earth. The intensity of this energy at different wavelengths can provide information about hydrologically related parameters. These include: surface temperature using thermal infrared wavelengths; soil moisture using microwave wavelengths; snow using both microwave and visible wavelengths; vegetation using visible, infrared and microwave wavelengths; and components of the energy balance using visible and thermal wavelengths. In considering the application of remote sensing techniques to hydrologic problems it is important to keep in mind that these techniques can not measure quantities as well as in-situ techniques but that they can make repetitive measurements over large and often inaccessible areas. The measurements are few in type and often require numerical interpretation to provide hydrological variables. However, the ability of the measurements to allow some account to be taken of spatially variable quantities, many of which are hard to measure conventionally, is very attractive. A particular example of this is soil moisture where remote sensing techniques can measure only the surface layer approximately 5 cm thick, but the sensors can make repetitive measurements over very large areas. The task remains of how best to use this less detailed but larger area measurement.

In this short paper it is hard to give examples of all the ways remotely sensed data can be used in hydrological models. It is clear that the conventional hydrological models which are optimized on frequent but spatially limited conventional data are rather different from those required for use with remotely sensed data, which use spatially dense but temporally infrequent data. It is clear that many of the most successful initial uses of remotely sensed data will be in models that are based on physical principles even if statistical fitting is involved. This is illustrated using one type of model where remotely sensed have been applied, for estimation of evapotranspiration (ET), after an illustration of how one type of hydrologically relevant data can be extracted, for soil moisture.

MICROWAVE SOIL MOISTURE SENSING

The remote sensing of soil moisture through the use of microwave sensors (wavelengths between 3 and 30 cm) results from the strong sensitivity of a soil's dielectric properties on its moisture content. The dielectric constant of soil changes from 3 - 4 in the dry state to greater than 20 in the wet state producing changes in the surface emissivity from 0.9 to 0.6 respectively, with corresponding changes in the radar backscatter.

Microwave radiometers measure the thermal emission coming up from the soil which is proportional to the physical temperature of the soil, with the constant of proportionality being the emissivity of the soil which depends very strongly on its moisture content. Other factors which affect the microwave emission from the soil surface include soil texture, surface roughness and vegetation cover; their magnitude has been estimated reasonably well in the past several years. Vegetation is the most significant because it can completely shield the soil surface from observation. It has been found that a mature corn crop (2 meters high) is about the limiting situation for radiometric observations of the surface at a wavelength of 21 cm (Jackson et al.[1]). As a result of many studies it has been concluded that a 21-cm radiometer is best for soil moisture observations (Jackson and Schmugge[2]). The longer wavelengths give better penetration into the soil and through an overlying vegetation canopy. However there are limitations imposed by extensive manmade sources of stray radiation at the longer wavelengths, e.g. TV's UHF band, and by the larger antennae required for adequate spatial resolution. For example, a 10 m antenna would be required to obtain 10 km resolution from a 500 km orbit.

A radar system emits electromagnetic radiation and then measure that portion of the emitted pulse which is reflected directly back, i.e. the backscattered energy. The intensity of the backscatter depends on the incidence angle, surface roughness and dielectric properties, and vegetation cover. Compared to passive approaches, angle and roughness factors are more important while vegetation is less so. As with the passive approach, radars operating at the longer wavelengths are better suited for soil moisture sensing from vegetation and sampling depth considerations. However, the longer wavelengths are more sensitive to surface roughness and look angle and as a result the best compromise is a 5 or 6 cm wavelength at an incidence angle of 10 to 20° (Dobson and Ulaby[3]). A major advantage of the active microwave approach is the use of synthetic aperture radar (SAR) techniques to obtain high resolution (10 m) even from space. Results from test flights of radars on the space shuttle have demonstrated sensitivity to soil moisture variations (Dobson and Ulaby[4] and Wang et al.[5]), while airborne sensors have shown the same for passive measurements.

For both active and passive sensing, measurements are thus limited in some way. For example, the sampling depth is linked to wavelength: longer wavelengths allow the estimation of the soil moisture in the top 5 cm of the soil. If time series of data are used it is possible to relate this near surface soil moisture to that in the whole profile if ET is also estimated. Recent work has thus coupled soil moisture observations with other remote sensing and conventional observations, as will be discussed in the next section.

EVAPOTRANSPIRATION

Three conditions are necessary for ET to occur: 1) the energy for the change of phase of water; 2) a source of water, i.e. adequate soil moisture in the surface layer or in the root zone of the plant and 3) a sink for the water, i.e. a moisture deficit in the air above the surface. Remote sensing can contribute to all of these factors by enabling us to make estimates of a number of surface and boundary layer parameters. These include the incoming solar radiation (Diak and Gautier[6]), surface albedo, vegetation cover (Tucker et al.[7]; Jackson[8]), surface temperature (Price[9]; Carlson[10]), atmospheric temperature and water vapor, and surface soil moisture. Most of the approaches using surface temperature observations rely on the solution to the energy balance equation. In these approaches the components of the energy flux at the surface are expressed in terms of the surface temperature and a set of routinely available meteorological variables. The energy balance equation is then solved iteratively until the calculated surface temperature agrees with the observed. There have been several recent reports of the application of the techniques using satellite data, primarily the AVHRR data from the NOAA series of satellites (Carlson[10]; and Taconet et al.[11]). Most of these models rely heavily on ground based meteorological observations and further research is required to evaluate how locality independent synoptic meteorological data can be used in combination with the remotely sensed data. A recent and promising new set of measurements that could aid this are atmospheric moisture and temperature which could be obtained using atmospheric lidar techniques. Recently there have been two approaches which make use of repetitive surface soil moisture observations (Prevot et al.[12]) or simultaneous observations of both surface temperature and soil moisture (Camillo et al. [13]) to study both the energy and moisture balances at the surface. The former uses microwave determinations of surface soil moisture as the upper boundary condition in a water balance model to obtain estimates of the surface flux. The latter adds a radiative transfer model to calculate the expected microwave brightness and iteratively solves for the fluxes.

The next step will be to combine the estimates of incoming solar and longwave radiation and albedo from visible and near infrared observations; surface temperature estimates from thermal infrared observations; and surface soil moisture from microwave observations to develop a method for estimating evapotranspiration from remotely sensed data.

CONCLUSIONS

This brief summary of the derivation of soil moisture from remotely sensed data shows that even though the use of these data in hydrologically useful ways is not straightforward, the potential benefits are great. The use of these data may be expected to increase because of their global nature, which is essential for a whole class of hydrological problems related to climatic and other changes.

REFERENCES

1. Jackson T.J. Schmugge T.J. and Wang J.R. (1982), Passive Microwave Sensing of Soil Moisture Under Vegetative Canopies, Water Resour. Res., 18, pp 1137-1142.

2. Jackson T.J. and Schmugge T.J. (1986), Passive Microwave Remote Sensing of Soil Moisture, in Advances in Hydroscience, 14, pp 123-159, Academic Press Inc.

3. Dobson M.C. and Ulaby F.T. (1986), Active microwave soil moisture research, IEEE Trans. Geoscience Remote Sensing, GE-24, 23-36.

4. Dobson M.C. and Ulaby F.T. (1986), Preliminary evaluation of the SIR-B response to soil moisture, surface roughness, and crop canopy cover, IEEE Trans. Geoscience and Remote Sensing, GE-24, 517-526.

5. Wang J.R. Engman E.T. Shiue J.C. Rusek M. and Steinmeier C. (1986), The Sir-B Observations of Microwave Backscatter Dependence on Soil Moisture, Surface Roughness and Vegetation

Covers, IEEE Trans. Geoscience and Remote Sensing, GE-24, 510-516.

6. Diak G.R. and Gautier C. (1983), Improvements to a simple Physical Model for Estimating Insolation from GOES Data, J. Clim. Appl. Meteorl., 22, pp 505-508.

7. Tucker C.J. Townshend J.R.G. and Goff T.E. (1985), African Landcover Classification Using Satellite Data, Science, Vol. 227, pp 369-375.

8. Jackson R.D. (1985), Evaluating Evapotranspiration at Local and Regional Scales, Proc. IEEE, 73, pp 1086-1096.

9. Price J.C. (1983), Estimation of surface temperature from satellite thermal infrared data, Rem. Sens. Environm., 13, 353-361,.

10. Carlson T.N. (1986), Regional Scale Estimates of Surface Moisture Availability and Thermal Inertia Using Remote Thermal Measurements, Remote Sensing Reviews, 1, pp 197-247.

11. Taconet O. Bernard R. and Vidal-Madjar D. (1986), Evapotranspiration over an Agricultural Region Using a Surface Flux/Temperature Model Based on NOAA AVHRR Data, J. Clim. Appl. Meteorol., 25, 284-307.

12. Prevot L. Bernard R. Taconet O. Vidal-Madjar D. and Thony J.L. (1984), Evaporation from a Bare Soil Field Using a Soil Water Transfer Model and Remotely Sensed Surface Soil Moisture Data, Water Resour. Res., 20, pp 311-316.

13. Camillo, P.J. O'Neill P.E. and Gurney R.J. (1986), Estimating Soil Hydraulic Parameters Using Passive Microwave Data, IEEE Trans. Geosci. Remote Sensing, GE-24, pp 930-936.